T0073106

Mathematical 数 Olympiad in China (2021–2022)

Problems and Solutions

Mathematical Olympiad Series

ISSN: 1793-8570

Published

Vol. 22 *Mathematical Olympiad in China (2021–2022): Problems and Solutions*
editor-in-chief Bin Xiong (East China Normal University, China)

Vol. 21 *Problem Solving Methods and Strategies in High School Mathematical Competitions*
edited by Bin Xiong (East China Normal University, China) & Yijie He (East China Normal University, China)

Vol. 20 *Hungarian Mathematical Olympiad (1964–1997): Problems and Solutions*
by Fusheng Leng (Academia Sinica, China),
Xin Li (Babeltime Inc., USA) &
Huawei Zhu (Shenzhen Middle School, China)

Vol. 19 *Mathematical Olympiad in China (2019–2020): Problems and Solutions*
edited by Bin Xiong (East China Normal University, China)

Vol. 18 *Mathematical Olympiad in China (2017–2018): Problems and Solutions*
edited by Bin Xiong (East China Normal University, China)

Vol. 17 *Mathematical Olympiad in China (2015–2016): Problems and Solutions*
edited by Bin Xiong (East China Normal University, China)

Vol. 16 *Sequences and Mathematical Induction: In Mathematical Olympiad and Competitions Second Edition*
by Zhigang Feng (Shanghai Senior High School, China)
translated by: Feng Ma & Youren Wang

The complete list of the published volumes in the series can be found at
http://www.worldscientific.com/series/mos

Vol. 22 | Mathematical Olympiad Series

Mathematical Olympiad in China (2021–2022)

Problems and Solutions

Editor-in-Chief

Xiong Bin

East China Normal University, China

English Translators

Chen Haoran

Xi'an Jiaotong-Liverpool University, China

Zhao Wei

East China Normal University, China

Copy Editors

Ni Ming

Kong Lingzhi

Wan Yuanlin

East China Normal University Press, China

Published by

East China Normal University Press
3663 North Zhongshan Road
Shanghai 200062
China

and

World Scientific Publishing Co. Pte. Ltd.
5 Toh Tuck Link, Singapore 596224
USA office: 27 Warren Street, Suite 401-402, Hackensack, NJ 07601
UK office: 57 Shelton Street, Covent Garden, London WC2H 9HE

Library of Congress Control Number: 2024008422

British Library Cataloguing-in-Publication Data
A catalogue record for this book is available from the British Library.

Mathematical Olympiad Series — Vol. 22
MATHEMATICAL OLYMPIAD IN CHINA (2021–2022)
Problems and Solutions

ISBN 978-981-12-8400-7 (hardcover)
ISBN 978-981-12-8454-0 (paperback)
ISBN 978-981-12-9219-4 (ebook for institutions)
ISBN 978-981-12-9220-0 (ebook for individuals)

For any available supplementary material, please visit
https://www.worldscientific.com/worldscibooks/10.1142/13614#t=suppl

Typeset by Stallion Press
Email: enquiries@stallionpress.com

Preface

The first time China participate in IMO (International Mathematical Olympiad) was in 1985, when two students were sent to the 26th IMO. Since 1986, China has a team of 6 students at every IMO except in 1998. So far, up to 2022, China has achieved the number one ranking in team effort 23 times. A great majority of students received gold medals. The fact that China obtained such encouraging result is due to, on one hand, Chinese students' hard work and perseverance, and on the other hand, the effort of the teachers in schools and the training offered by national coaches. We believe this is also a result of the education system in China, in particular, the emphasis on training of the basic skills in science education.

The materials of this book come from two volumes (Vol. 2021 and Vol. 2022) of a book series in Chinese "走向 IMO: 数学奥林匹克试题集锦" (*Forward to IMO*: *A Collection of Mathematical Olympiad Problems*). It is a collection of problems and solutions of the major mathematical competitions in China. It provides a glimpse of how the China national team is selected and formed. First, there is the China Mathematical Competition, a national event. It is held on the second Sunday of September every year. Through the competition, about 550 students are selected to join the China Mathematical Olympiad (commonly known as the winter camp), or in short CMO, in November. CMO lasts for five days. Both the type and the difficulty of the problems match those of IMO. Similarly, students are given three problems to solve in 4.5 hours each day. From CMO, 60 students are selected to form a national training team. The training takes place for two weeks in the month of March. After four to six tests plus two qualifying examinations, six students are finally selected to form the national team, taking part in IMO in July of that year.

In view of the differences in education, culture and economy of the western part of China with the coastal part in eastern China, mathematical competitions in West China did not develop as fast as the rest of the country. In order to promote the activity of mathematical competition, and to enhance the level of mathematical competition, starting from 2001, China Mathematical Olympiad Committee has been organizing the China Western Mathematical Olympiad.

Since 2012, the China Western Mathematical Olympiad has been renamed the China Western Mathematical Invitation. The competition dates have been changed from the first half of October to the middle of August since 2013.

The development of this competition reignited the enthusiasm of Western students for mathematics. Nowadays, the figure of Western students often appeared in the national team.

Since 1995, there had been no female students in the Chinese national team. In order to encourage more female students participating in the mathematical competition, starting from 2002, China Mathematical Olympiad Committee has been conducting the China Girls' Mathematical Olympiad. Again, the top 15 winners will be admitted directly into the CMO.

The authors of this book are coaches of the China national team. They are Xiong Bin, Xiao Liang, Yu Hongbing, Yao Yijun, Qu Zhenhua, Li Ting, Ai Yinhua, Wang Bin, Fu Yunhao, He Yijie, Zhang Sihui, Wang Xinmao, Lin Tianqi, Xu Disheng, *et al.* Those who took part in the translation work are Chen Haoran and Zhao Wei. We are grateful to Qiu Zonghu, Wang Jie, Zhou Qin and Pan Chengbiao for their guidance and assistance to the authors. We are grateful to Ni Ming, Kong Linzhi of East China Normal University Press. Their effort has helped make our job easier. We are also grateful to Tan Rok Ting and Liu Nijia of World Scientific Publishing for their hard work leading to the final publication of the book.

Authors
October 2023

Introduction

Early Days

The International Mathematical Olympiad (IMO), founded in 1959, is one of the most competitive and highly intellectual activities in the world for high school students.

Even before IMO, there were already many countries which had mathematical competitions. They were mainly the countries in Eastern Europe and in Asia. In addition to the popularization of mathematics and the convergence in educational systems among different countries, the success of mathematical competitions at the national level provided a foundation for the setting-up of IMO. The countries that asserted great influence are Hungary, the former Soviet Union, and the United States. Here is a brief history of the IMO and mathematical competitions in China.

In 1894, the Department of Education in Hungary passed a motion and decided to conduct a mathematical competition for secondary schools. The well-known scientist, *J. von Etövös*, was the Minister of Education at that time. His support for the event had made it a success and thus it was well publicized. In addition, the success of his son, *R. von Etövös*, who was also a physicist, in proving the principle of equivalence of the general theory of relativity by *A. Einstein* through experiment, had brought Hungary to the world stage in science. Thereafter, the prize for mathematical competition in Hungary was named "*Etövös* prize". This was the first formally organized mathematical competition in the world. In what follows, Hungary had indeed produced a lot of well-known scientists including *L. Fejér*, *G. Szegö*, *T. Radó*, *A. Haar* and *M. Riesz* (in real analysis), *D. König* (in combinatorics), *T. von Kármán* (in aerodynamics), and *J. C. Harsanyi* (in game theory), who had also won the Nobel Prize for Economics in 1994. They all were the winners of Hungary mathematical competition.

The top scientific genius of Hungary, *J. von Neumann*, was one of the leading mathematicians in the 20th century. *Neumann* was overseas while the competition took place. Later he did the competition himself and it took him half an hour to complete. Another mathematician worth mentioning is the highly productive number theorist *P. Erdös*. He was a pupil of *Fejér* and a winner of the Wolf Prize. *Erdös* was very passionate about mathematical competitions and setting competition questions. His contribution to discrete mathematics was unique and of great significance. The rapid progress and development of discrete mathematics over the subsequent decades had indirectly influenced the types of questions set in IMO. An internationally recognized prize was named after *Erdös* to honor those who had contributed to the education of mathematical competition. Professor *Qiu Zonghu* from China had won the prize in 1993.

In 1934, a famous mathematician *B. Delone* conducted a mathematical competition for high school students in Leningrad (now St. Petersburg). In 1935, Moscow also started organizing such events. Other than being interrupted during World War II, these events had been carried on until today. As for the Russian Mathematical Competition (later renamed as the Soviet Mathematical Competition), it was not started until 1961. Thus, the former Soviet Union and Russia became the leading powers of Mathematical Olympiad. A lot of grandmasters in mathematics, including the great *A. N. Kolmogorov*, were all very enthusiastic about the mathematical competition. They would personally involve themselves in setting the questions for the competition. The former Soviet Union even called it the Mathematical Olympiad, believing that mathematics is the "gymnastics of thinking". These points of view had a great impact on the educational community. The winner of the Fields Medal in 1998, *M. Kontsevich*, was once the first runner-up of the Russian Mathematical Competition. *G. Kasparov*, the international chess grandmaster, was once the second runner-up. *Grigori Perelman*, the winner of the Fields Medal in 2006 (which he declined), who solved the Poincaré's Conjecture, was a gold medalist of IMO in 1982.

In the United States of America, due to the active promotion by the renowned mathematician *G. D. Birkhoff* and his son, together with *G. Pólya*, the Putnam mathematics competition was organized in 1938 for junior undergraduates. Many of the questions were within the scope of high school curriculum. The top five contestants of the Putnam mathematical competition would be entitled to the membership of Putnam. Many of these eventually became outstanding mathematicians. There were the famous

R. Feynman (winner of the Nobel Prize for Physics, 1965), *K. Wilson* (winner of the Nobel Prize for Physics, 1982), *J. Milnor* (winner of the Fields Medal, 1962), *D. Mumford* (winner of the Fields Medal, 1974), and *D. Quillen* (winner of the Fields Medal, 1978).

In 1972, in order to prepare for the IMO, the United States of America Mathematical Olympiad (USAMO) was established. The standard of questions posed was very high, parallel to that of the Winter Camp in China. Prior to this, the United States had organized American High School Mathematics Examination (AHSME) for the high school students since 1950. This was at the junior level and yet the most popular mathematics competition in America. Originally, it was intended to select about 100 contestants from AHSME to participate in USAMO. However, due to the discrepancy in the level of difficulty between the two competitions and other restrictions, from 1983 onwards, an intermediate level of competition, namely, American Invitational Mathematics Examination (AIME), was introduced. Henceforth both AHSME and AIME became internationally well-known. Since 2000, AHSME was replaced by AMC 12 and AMC 10. Students who perform well on the AMC 12 and AMC 10 are invited to participate in AIME. The combined scores of the AMC 12 and the AIME are used to determine approximately 270 individuals that will be invited back to take the USAMO, while the combined scores of the AMC 10 and the AIME are used to determine approximately 230 individuals that will be invited to take the USAJMO (United States of America Junior Mathematical Olympiad), which started in 2010 and follows the same format as the USAMO. A few cities in China had participated in the competition and the results have been encouraging.

Similar to the case of the former Soviet Union, the Mathematical Olympiad education was widely recognized in America. The book "How to Solve it" written by *George Polya* along with many other titles had been translated into many different languages. *George Polya* provided a whole series of general heuristics for solving problems of all kinds. His influence in the educational community in China should not be underestimated.

International Mathematical Olympiad

In 1956, the East European countries and the Soviet Union took the initiative to organize the IMO formally. The first International Mathematical Olympiad (IMO) was held in Brasov, Romania, in 1959. At that time, there were only seven participating countries, namely, Romania, Bulgaria,

Poland, Hungary, Czechoslovakia, East Germany and the Soviet Union. Subsequently, the United States of America, United Kingdom, France, Germany, and also other countries including those from Asia joined. Today, the IMO has managed to reach almost all the developed and developing countries. Except in the year 1980 due to financial difficulties faced by the host country, Mongolia, there have been 59 Olympiads held yearly and with 107 countries and regions participating nowadays.

The mathematical topics in the IMO include algebra, combinatorics, geometry, number theory. These areas have provided guidance for setting questions for the competitions. Other than the first few Olympiads, each IMO is normally held in mid-July every year, and the test paper consists of 6 questions in total. The actual competition lasts for 2 days for a total of 9 hours, where participants are required to complete 3 questions each day. Each question is 7 points, which totals up to 42 points. The full score for a team is 252 marks. About half of the participants will be awarded a medal, where 1/12 will be awarded a gold medal. The numbers of gold, silver and bronze medals awarded are in the ratio of 1:2:3, approximately. In the case when a participant provides a better solution than the official answer, a special award is given.

Each participating country and region will take turns to host the IMO. The cost is borne by the host country. China had successfully hosted the 31st IMO in Beijing. The event had made a great impact on the mathematical community in China. According to the rules and regulations of the IMO, all participating countries are required to send a delegation consisting of a leader, a deputy leader and 6 contestants. The problems are contributed by the participating countries and are later selected carefully by the host country for submission to the international jury set up by the host country. Eventually, only 6 problems will be accepted for use in the competition. The host country does not provide any questions. The shortlisted problems are subsequently translated, if necessary, in English, French, German, Spanish, Russian, and other working languages. After that, the team leaders will translate the problems into their own languages.

The answer scripts of each participating team will be marked by the team leader and the deputy leader. The team leader will later present the scripts of their contestants to the coordinators for assessment. If there is any dispute, the matter will be settled by the jury. The jury is formed by the various team leaders and an appointed chairman by the host country. The jury is responsible for deciding the final 6 problems for the competition. Their duties also include finalizing the grading standard, ensuring the accuracy of the translation of the problems, standardizing replies to written queries raised by participants during the competition, synchronizing differences in grading between the team leaders and the coordinators, and also deciding on the cut-off points for the medals depending on the contestants' results as the difficulties of problems each year are different.

China had participated informally in the 26th IMO in 1985. Only two students were sent. Starting from 1986, except in 1998 when the IMO was held in Taiwan, China had always sent 6 official contestants to the IMO. Today, the Chinese contestants not only performed outstandingly in the IMO, but also in the International Physics, Chemistry, Informatics, and Biology Olympiads. This can be regarded as an indication that China pays great attention to the training of basic skills in mathematics and science education.

Winners of the IMO

Among all the IMO medalists, there were many of them who eventually became great mathematicians. They were also awarded the Fields Medal, Wolf Prize and Nevanlinna Prize (a prominent mathematics prize for computing and informatics). In what follows, we name some of the winners.

G. Margulis, a silver medalist of IMO in 1959, was awarded the Fields Medal in 1978. *L. Lovasz*, who won the Wolf Prize in 1999, was awarded the Special Award in IMO consecutively in 1965 and 1966. *V. Drinfeld*, a gold medalist of IMO in 1969, was awarded the Fields Medal in 1990. *J.-C. Yoccoz* and *T. Gowers*, who were both awarded the Fields Medal in 1998, were gold medalists in IMO in 1974 and 1981, respectively. A silver medalist of IMO in 1985, *L. Lafforgue*, won the Fields Medal in 2002. A gold medalist of IMO in 1982, *Grigori Perelman* from Russia, was awarded the Fields Medal in 2006 for solving the final step of the Poincaré conjecture. In 1986, 1987, and 1988, *Terence Tao* won a bronze, silver, and gold medal respectively. He was the youngest participant to date in the IMO, first competing at the age of ten. He was also awarded the Fields Medal in 2006. Gold medalist of IMO 1988 and 1989, *Ngo Bau Chao*, won the Fields Medal in 2010, together with the bronze medalist of IMO 1988, *E.Lindenstrauss*. Gold medalist of IMO 1994 and 1995, *Maryam Mirzakhani* won the Fields Medal in 2014. A gold medalist of IMO in 1995, Artur Avila, won the Fields Medal in 2014. Gold medalist of IMO 2005, 2006 and 2007, Peter Scholze, won the Fields Medal in 2018. A Bronze medalist of IMO in 1994, Akshay Venkatesh, won the Fields Medal in 2018.

A silver medalist of IMO in 1977, *P. Shor*, was awarded the Nevanlinna Prize. A gold medalist of IMO in 1979, *A. Razborov*, was awarded the Nevanlinna Prize. Another gold medalist of IMO in 1986, *S. Smirnov*, was awarded the Clay Research Award. *V. Lafforgue*, a gold medalist of IMO in 1990, was awarded the European Mathematical Society prize. He is *L. Lafforgue*'s younger brother.

Also, a famous mathematician in number theory, *N. Elkies*, who is also a professor at Harvard University, was awarded a gold medal of IMO in 1982. Other winners include *P. Kronheimer*, awarded a silver medal in 1981, and *R. Taylor* a contestant of IMO in 1980.

Mathematical competition in China

Due to various reasons, mathematical competition in China started relatively late but is progressing vigorously.

"We are going to have our own mathematical competition too!" said *Hua Luogeng*. *Hua* is a household name in China. The first mathematical competition was held concurrently in Beijing, Tianjin, Shanghai, and Wuhan in 1956. Due to the political situation at the time, this event was interrupted a few times. It was not until 1962, when the political environment started to improve, that Beijing and other cities started organizing the competition, though not regularly. In the era of Cultural Revolution, the whole educational system in China was in chaos. The mathematical competition came to a complete halt. In contrast, the mathematical competition in the former Soviet Union was still on-going during the war and at a time of difficult political situation. The competitions in Moscow were interrupted only 3 times between 1942 and 1944. It was indeed commendable.

In 1978, it was the spring of science. *Hua Luogeng* conducted the Middle School Mathematical Competition for 8 provinces in China. The mathematical competition in China was then making a fresh start and embarked on a road of rapid development. *Hua* passed away in 1985. To commemorate him, a competition named *Hua Luogeng* Gold Cup was set up in 1986 for students in Grades 6 and 7, and it has had a great impact.

The mathematical competitions in China before 1980 can be considered as the initial period. The problem sets were within the scope of middle school textbooks. After 1980, the competitions gradually moved towards the senior middle school level. In 1981, the Chinese Mathematical Society decided to conduct the China Mathematical Competition, a national event for high schools.

In 1981, the United States of America, the host country of IMO, issued an invitation to China to participate in the event. Only in 1985, China sent two contestants to participate informally in the IMO. The results were not encouraging. In view of this, another activity called the Winter Camp was conducted after the China Mathematical Competition. The Winter Camp was later renamed as the China Mathematical Olympiad or CMO. The winning team would be awarded the *Chern Shiing-Shen* Cup. Based on the outcome at the Winter Camp, a selection would be made to form the 6-member national team for IMO. From 1986 onwards, other than the year when IMO was organized in Taiwan, China has been sending a 6-member team to IMO. Up to 2018, China had been awarded the overall team champion for 19 times.

In 1990, China successfully hosted the 31st IMO. It showed that the standard of mathematical competition in China has leveled that of other leading countries. Firstly, the fact that China achieves the highest marks at the 31st IMO for the team is evidence for the effectiveness of the pyramid approach in selecting the contestants in China. Secondly, the Chinese mathematicians had simplified and modified over 100 problems and submitted them to the team leaders of the 35 countries for their perusal. Eventually, 28 problems were recommended. At the end, 5 problems were chosen (IMO requires 6 problems). This is also evidence to show that China has achieved the highest quality in setting problems. Thirdly, the answer scripts of the participants were marked by the various team leaders and assessed by the coordinators who were nominated by the host countries. China had formed a group 50 mathematicians to serve as coordinators who would ensure the high accuracy and fairness in marking. The marking process was completed half a day earlier than it was scheduled. Fourthly, that was the first ever IMO organized in Asia. The outstanding performance by China had encouraged the other developing countries, especially those in Asia. The organizing and coordinating work of the IMO by the host country was also reasonably good.

In China, the outstanding performance in mathematical competition is a result of many contributions from all quarters of the mathematical community. There are the older generation of mathematicians, middle-aged mathematicians, and also the middle and elementary school teachers. There is one person who deserves a special mention, and he is *Hua Luogeng*. He initiated and promoted mathematical competitions. He is also the author of the following books: Beyond *Yang hui*'s Triangle, Beyond the *pi* of *Zu Chongzhi*, Beyond the Magic Computation of *Sun-zi*,

Mathematical Induction, and Mathematical Problems of Bee Hive. These were his books derived from mathematical competitions. When China resumed mathematical competitions in 1978, he participated in setting problems and giving critique to solutions of the problems. Other outstanding books derived from the Chinese mathematical competitions are: Symmetry by *Duan Xuefu*, Lattice and Area by *Min Sihe*, One Stroke Drawing and Postman Problem by *Jiang Boju*.

After 1980, the younger mathematicians in China had taken over from the older generation of mathematicians in running the mathematical competitions. They worked and strived hard to bring the level of mathematical competition in China to a new height. *Qiu Zonghu* is one such outstanding representative. From the training of contestants and leading the team 3 times to IMO, to the organizing of the 31th IMO in China, he had contributed prominently and was awarded the *P. Erdös* prize.

Preparation for IMO

Currently, the selection process of participants for IMO in China is as follows.

First, the China Mathematical Competition, a national competition for high schools, is organized on the second Sunday in September every year. The objectives are to increase the interest of students in learning mathematics, to promote the development of co-curricular activities in mathematics, to help improve the teaching of mathematics in high schools, to discover and cultivate the talents, and also to prepare for the IMO. This has been happening since 1981. Currently, there are about 500,000 participants taking part in it.

Through the China Mathematical Competition, around 550 students are selected to take part in the China Mathematical Olympiad or CMO, that is, the Winter Camp. The CMO lasts for 5 days and is held in November every year. The types and difficulties of the problems in CMO are very much similar to the IMO. There are also 3 problems to be completed within 4.5 hours each day. However, the score for each problem is 21 marks which adds up to 126 marks in total. Starting from 1990, the Winter Camp instituted the *Chern Shiing-Shen* Cup for team championship. In 1991, the Winter Camp was officially renamed as the China Mathematical Olympiad (CMO). It is similar to the highest national mathematical competition in the former Soviet Union and the United States.

The CMO awards the first, second and third prizes. Among the participants of CMO, about 60 students are selected to participate in the training for IMO. The training takes place in March every year. After 6 to 8 tests and another 2 rounds of qualifying examinations, only 6 contestants are short-listed to form the China IMO national team to take part in the IMO in July.

Besides the China Mathematical Competition (for high schools), the Junior Middle School Mathematical Competition is also developing well. Starting from 1984, the competition is organized in April every year by the Popularization Committee of the Chinese Mathematical Society. The various provinces, cities and autonomous regions would rotate to host the event. Another mathematical competition for the junior middle schools is also conducted in April every year by the Middle School Mathematics Education Society of the Chinese Educational Society since 1998 till 2014.

The *Hua Luogeng* Gold Cup, a competition by invitation, has also been successfully conducted since 1986. The participating students comprise elementary-six and junior-middle-one students. The format of the competition consists of a preliminary round, semi-finals in various provinces, cities and autonomous regions, then the finals.

Mathematical competitions in China provide a platform for students to showcase their talents in mathematics. It encourages learning of mathematics among students. It helps identify talented students and to provide them with differentiated learning opportunities. It develops co-curricular activities in mathematics. Finally, it brings about changes in the teaching of mathematics.

Contents

China Mathematical Competition (First Round)

2020

While the scope of the test questions in the first round of the 2020 China Mathematical Competition does not exceed the teaching requirements and content specified in the "General High School Mathematics Curriculum Standards (2017)" promulgated by the Ministry of Education of China in 2017, the methods of proposing the questions have been improved. The emphasis placed on to test the students' basic knowledge and skills, and their abilities to integrate and use flexibly of them. Each test paper includes eight fill-in-the-blank questions and three answer questions. The answer time is 80 minutes, and the full score is 120 marks.

The scope of the test questions in the Second Round (Complementary Test) is in line with the International Mathematical Olympiad, with some expanded knowledge, plus a few contents of the Mathematical Competition Syllabus. Each test paper consists of four answer questions, including a plane geometry one, and the answering time is 170 minutes. The full score is 180 marks.

Test Paper A
(8:00 – 9:20; September 13, 2020)

Part I Short-Answer Questions (Questions 1–8, eight marks each)

1 Given geometric sequence $\{a_n\}$, $a_9 = 13$, $a_{13} = 1$, then the value of $\log_{a_1} 13$ is ________.

Solution By the properties of geometric sequence, we have $\dfrac{a_1}{a_9} = \left(\dfrac{a_9}{a_{13}}\right)^2$, and thus $a_1 = \dfrac{a_9^3}{a_{13}^2} = 13^3$.

Consequently, $\log_{a_1} 13 = \dfrac{1}{3}$. □

2 In ellipse Γ, A is an endpoint of the major axis, B is an endpoint of the minor axis, and F_1, F_2 are the foci. If $\overline{AF}_1 \cdot \overline{AF}_2 + \overline{BF}_1 \cdot \overline{BF}_2 = 0$, then the value of $\dfrac{|AB|}{|F_1F_2|}$ is ________.

Solution Without loss of generality, suppose the equation of Γ is $\dfrac{x^2}{a^2} + \dfrac{y^2}{b^2} = 1$ $(a > b > 0)$, and $A(a, 0)$, $B(0, b)$, $F_1(-c, 0)$, $F_2(c, 0)$. By the given conditions, we get

$$\begin{aligned}\overrightarrow{AF_1} \cdot \overrightarrow{AF_2} + \overrightarrow{BF_1} \cdot \overrightarrow{BF_2} &= (-c - a)(c - a) + (-c^2 + b^2)\\ &= a^2 + b^2 - 2c^2 = 0.\end{aligned}$$

Therefore, $\dfrac{|AB|}{|F_1F_2|} = \dfrac{\sqrt{a^2 + b^2}}{2c} = \dfrac{\sqrt{2c^2}}{2c} = \dfrac{\sqrt{2}}{2}$. □

3 Suppose that $a > 0$ and the minima of function $f(x) = x + \dfrac{100}{x}$ on intervals $(0, a]$ and $[a, +\infty)$ are m_1, m_2, respectively. If $m_1 m_2 = 2020$, then the value of a is ________.

Solution Note that $f(x)$ is monotonically decreasing on $(0, 10]$ and monotonically increasing on $[10, +\infty)$. When $a \in (0, 10]$, $m_1 = f(a)$, $m_2 = f(10)$; when $a \in [10, +\infty)$, $m_1 = f(10)$, $m_2 = f(a)$. Therefore, there is always

$$f(a)f(10) = m_1 m_2 = 2020,$$

namely, $a + \dfrac{100}{a} = \dfrac{2020}{20} = 101$. The solution is $a = 1$ or $a = 100$. □

4 Let z be a complex number. If $\dfrac{z-2}{z-\mathrm{i}}$ is a real number (i is the imaginary unit), then the minimum of $|z+3|$ is ________.

Solution 1 Suppose $z = a + b\mathrm{i}\,(a, b \in \mathbb{R})$. By the given condition we can find

$$\begin{aligned}\mathrm{Im}\left(\frac{z-2}{z-\mathrm{i}}\right) &= \mathrm{Im}\left(\frac{(a-2)+b\mathrm{i}}{a+(b-1)\mathrm{i}}\right)\\ &= \frac{-(a-2)(b-1)+ab}{a^2+(b-1)^2}\\ &= \frac{a+2b-2}{a^2+(b-1)^2} = 0,\end{aligned}$$

and thus $a + 2b = 2$. Therefore,

$$\sqrt{5}|z+3| = \sqrt{(1^2+2^2)((a+3)^2+b^2)} \geq |(a+3)+2b| = 5,$$

namely, $|z+3| \geq \sqrt{5}$. When $a = -2, b = 2$, $|z+3|$ takes the minimum $\sqrt{5}$.

Solution 2 From $\dfrac{z-2}{z-\mathrm{i}} \in \mathbb{R}$ and the geometric meaning of complex division, it is known that the point corresponding to z on the complex plane lies on the line connecting the points corresponding to 2 and i (excluding the point corresponding to i), so the minimum of $|z+3|$ is the distance from point $(-3, 0)$ to line $x + 2y - 2 = 0$ in plane rectangular coordinate system xOy, i.e., $\dfrac{|-3-2|}{\sqrt{1^2+2^2}} = \sqrt{5}$. □

5 In $\triangle ABC$, $AB = 6$, $BC = 4$, the median to side AC is $\sqrt{10}$. Then the value of $\sin^6 \dfrac{A}{2} + \cos^6 \dfrac{A}{2}$ is ________.

Solution Let M be the midpoint of AC. By the median formula we have

$$4BM^2 + AC^2 = 2(AB^2 + BC^2),$$

and thus

$$AC = \sqrt{2(6^2 + 4^2) - 4 \cdot 10} = 8.$$

By the law of cosines, we obtain $\cos A = \dfrac{CA^2 + AB^2 - BC^2}{2CA \cdot AB} = \dfrac{8^2 + 6^2 - 4^2}{2 \cdot 8 \cdot 6} = \dfrac{7}{8}$. Therefore,

$$\begin{aligned}
\sin^6 \frac{A}{2} + \cos^6 \frac{A}{2} &= \left(\sin^2 \frac{A}{2} + \cos^2 \frac{A}{2}\right)\left(\sin^4 \frac{A}{2} - \sin^2 \frac{A}{2} \cos^2 \frac{A}{2} + \cos^4 \frac{A}{2}\right) \\
&= \left(\sin^2 \frac{A}{2} + \cos^2 \frac{A}{2}\right)^2 - 3\sin^2 \frac{A}{2} \cos^2 \frac{A}{2} \\
&= 1 - \frac{3}{4}\sin^2 A \\
&= \frac{1}{4} + \frac{3}{4}\cos^2 A \\
&= \frac{211}{256}.
\end{aligned}$$

□

6 Suppose all the edges of regular triangular pyramid $P - ABC$ have length 1 and L, M, N are the midpoints of edges PA, PB and PC, respectively. The area of the cross section of the circumscribed sphere of this regular triangular pyramid intercepted by plane LMN is ________.

Solution The given conditions show that plane LMN is parallel to plane ABC and the ratio of the distances from point P to planes LMN and ABC is $1 : 2$. Let H be the centroid of face ABC in regular triangular pyramid $P - ABC$ and PH intersects plane LMN at point K. Then $PH \perp ABC$ and $PK \perp LMN$, and thus $PK = \frac{1}{2}PH$.

The regular triangular pyramid $P - ABC$ can be regarded as a regular tetrahedron. Let O be the centre of its circumscribed sphere). Then O is on PH and by the properties of regular tetrahedron we know that $OH = \frac{1}{4}PH$. By combining $PK = \frac{1}{2}PH$ we know that $OK = OH$, that is, point O is equally distant to planes LMN and ABC. This shows that the section circle of the circumscribed sphere of this regular triangular pyramid intercepted by planes LMN and ABC is equal in size.

As a result, the area of the required cross section is equal to the area of the circumcircle of $\triangle ABC$, namely, $\pi \cdot \left(\frac{AB}{\sqrt{3}}\right)^2 = \frac{\pi}{3}$. □

7 Suppose $a, b > 0$. The equation $\sqrt{|x|} + \sqrt{|x+a|} = b$ for x has exactly three different real solutions, namely x_1, x_2, x_3, and $x_1 < x_2 < x_3 = b$. Then the value of $a + b$ is ________.

Solution Let $t = x + \frac{a}{2}$. Then the equation $\sqrt{\left|t - \frac{a}{2}\right|} + \sqrt{\left|t + \frac{a}{2}\right|} = b$ for t has exactly three different real solutions $t_i = x_i + \frac{a}{2}$ $(i = 1, 2, 3)$.

Since $f(t) = \sqrt{\left|t - \frac{a}{2}\right|} + \sqrt{\left|t + \frac{a}{2}\right|}$ is an even function, the three real solutions of equation $f(t) = b$ are symmetrically distributed about the origin of the number axis, so that there must be $b = f(0) = \sqrt{2a}$. In the following, we will find the real solutions of equation $f(t) = \sqrt{2a}$.

When $|t| \leq \frac{a}{2}$, $f(t) = \sqrt{\frac{a}{2} - t} + \sqrt{\frac{a}{2} + t} = \sqrt{a + \sqrt{a^2 - 4t^2}} \leq \sqrt{2a}$ and the equal sign holds if and only if $t = 0$; when $|t| > \frac{a}{2}$, $f(t)$ is monotonically increasing, and when $t = \frac{5a}{8}$, $f(t) = \sqrt{2a}$; when $t < -\frac{a}{2}$, $f(t)$ is monotonically decreasing, and when $t = -\frac{5a}{8}$, $f(t) = \sqrt{2a}$.

Thus, equation $f(t) = \sqrt{2a}$ has exactly three real solutions $t_1 = -\frac{5}{8}a$, $t_2 = 0$, $t_3 = \frac{5}{8}a$.

By the given conditions, we can find $b = x_3 = t_3 - \frac{a}{2} = \frac{a}{8}$. Combining $b = \sqrt{2a}$, we get $a = 128$.

Consequently, $a + b = \frac{9a}{8} = 144$. □

8 There are 10 cards, each of which has two numbers, numbered 1, 2, 3, 4, 5, written on it, and the numbers on any two cards are not exactly identical. The 10 cards are placed in five boxes labelled 1, 2, 3, 4, 5, and card with i and j written on them can only be placed in box i or j. One placement is called "good" if there are more cards in box 1 than in each of the other boxes. Then the total number of the "good" placements is ________.

Solution Denote the card with i, j written on it as $\{i, j\}$. It is easy to know that these 10 cards are exactly $\{i, j\}(1 \leq i < j \leq 5)$.

Consider the "good" placements of the cards. There are 10 cards in the five boxes, so there are at least 3 cards in box 1. The only cards that will fit in box 1 are $\{1, 2\}$, $\{1, 3\}$, $\{1, 4\}$ and $\{1, 5\}$.

Case 1: The 4 cards are all placed in box 1, and at this point it is no longer possible to have 4 cards in each of the remaining boxes. Therefore, no matter how the remaining 6 cards are placed, they will fit the requirements and there are $2^6 = 64$ "good" placements.

Case 2: There are exactly 3 of the 4 cards in box 1 and the rest of each box contains at most 2 cards.

Consider the number N of placements of $\{1, 2\}$, $\{1, 3\}$, $\{1, 4\}$ in box 1 and $\{1, 5\}$ in box 5.

There are 8 possible ways to place the cards $\{2, 3\}$, $\{2, 4\}$, $\{3, 4\}$, with 6 of them being two cards placed in one of the boxes $2, 3, 4$, and the remaining 2 being one card placed in each box of $2, 3, 4$.

If there are two cards of $\{2, 3\}$, $\{2, 4\}$ and $\{3, 4\}$ in a box, suppose that $\{2, 3\}$ and $\{2, 4\}$ are in box 2, and then $\{2, 5\}$ can only be in box 5. Therefore, box 5 already has $\{1, 5\}$ and $\{2, 5\}$, so $\{3, 5\}$ and $\{4, 5\}$ are in boxes 3 and 4 respectively, namely, the placement of $\{2, 5\}$, $\{3, 5\}$ and $\{4, 5\}$ is unique.

If one card of $\{2, 3\}$, $\{2, 4\}$ and $\{3, 4\}$ is placed in each box of 2, 3, 4, then there are at most 2 cards in each box of 2, 3, 4. It is only necessary to make sure that there are no more than 2 cards in box 5, namely, there are 0 or 1 card of $\{2, 5\}$, $\{3, 5\}$ and $\{4, 5\}$ in box 5, and the number of the corresponding placements is $C_3^0 + C_3^1 = 4$.

As a result, $N = 6 \times 1 + 2 \times 4 = 14$. By symmetry, there are $4N = 56$ "good" placements in *case 2*.

To sum up, there is a total of $64 + 56 = 120$ "good" placements.

Part II Word Problems (16 marks for Question 9, 20 marks for Question 10 and 11, and then 56 marks in total)

9 (16 marks) In $\triangle ABC$, $\sin A = \dfrac{\sqrt{2}}{2}$. Find the range of $\cos B + \sqrt{2} \cos C$.

Solution Denote $f = \cos B + \sqrt{2} \cos C$.

By the given conditions, we know that $A = \frac{\pi}{4}$ or $A = \frac{3\pi}{4}$.

When $A = \frac{\pi}{4}$, it follows that $B = \frac{3\pi}{4} - C$, where $0 < C < \frac{3\pi}{4}$. And there are

$$\begin{aligned} f &= \cos\left(\frac{3\pi}{4} - C\right) + \sqrt{2}\cos C \\ &= \frac{\sqrt{2}}{2}\sin C + \frac{\sqrt{2}}{2}\cos C \\ &= \sin\left(C + \frac{\pi}{4}\right) \in (0, 1]. \end{aligned}$$

When $A = \frac{3\pi}{4}$, we have $B = \frac{3\pi}{4} - C$, where $0 < C < \frac{\pi}{4}$. Therefore,

$$\begin{aligned} f &= \cos\left(\frac{\pi}{4} - C\right) + \sqrt{2}\cos C \\ &= \frac{\sqrt{2}}{2}\sin C + \frac{3\sqrt{2}}{2}\cos C \\ &= \sqrt{5}\sin(C + \varphi), \end{aligned}$$

where $\varphi = \arctan 3$.

Note that $\varphi \in \left(\frac{\pi}{4}, \frac{\pi}{2}\right)$. Function $g(x) = \sqrt{5}\sin(x + \varphi)$ is monotonically increasing on $\left[0, \frac{\pi}{2} - \varphi\right]$ and monotonically decreasing on $\left[\frac{\pi}{2} - \varphi, \frac{\pi}{4}\right]$. And since $g(0) = \frac{3\sqrt{2}}{2} > 2 = g\left(\frac{\pi}{4}\right)$, $g\left(\frac{\pi}{2} - \varphi\right) = \sqrt{5}$, it follows that $f \in (2, \sqrt{5}]$.

In conclusion, the range of $\cos B + \sqrt{2}\cos C$ is $(0, 1] \cup (2, \sqrt{5}]$. □

10 (20 marks) For positive real number n and real number $x (0 \leq x < n)$, we define

$$f(n, x) = (1 - \{x\}) \cdot \mathrm{C}_n^{[x]} + \{x\} \cdot \mathrm{C}_n^{[x]+1},$$

where $[x]$ denotes the largest integer that does not exceed real number x and $\{x\} = x - [x]$. If integers $m, n \geq 2$ satisfy

$$f\left(m, \frac{1}{n}\right) + f\left(m, \frac{2}{n}\right) + \cdots + f\left(m, \frac{mn-1}{n}\right) = 123,$$

find the value of $f\left(n, \frac{1}{m}\right) + f\left(n, \frac{2}{m}\right) + \cdots + f\left(n, \frac{mn-1}{m}\right)$.

Solution For $k = 0, 1, \ldots, m-1$, there is

$$\sum_{i=1}^{n-1} f\left(m, k+\frac{i}{n}\right) = \mathrm{C}_m^k \cdot \sum_{i=1}^{n-1}\left(1-\frac{i}{n}\right) + \mathrm{C}_m^{k+1} \cdot \sum_{i=1}^{n-1} \frac{i}{n}$$
$$= \frac{n-1}{2} \cdot (\mathrm{C}_m^k + \mathrm{C}_m^{k+1}).$$

Therefore,

$$f\left(m, \frac{1}{n}\right) + f\left(m, \frac{2}{n}\right) + \cdots + f\left(m, \frac{mn-1}{n}\right)$$
$$= \sum_{j=1}^{m-1} \mathrm{C}_m^j + \sum_{k=0}^{m-1}\sum_{i=1}^{n-1} f\left(m, k+\frac{i}{n}\right)$$
$$= 2^m - 2 + \frac{n-1}{2} \cdot \left(\sum_{k=0}^{m-1} \mathrm{C}_m^k + \sum_{k=0}^{m-1} \mathrm{C}_m^{k+1}\right)$$
$$= 2^m - 2 + \frac{n-1}{2} \cdot (2^m - 1 + 2^m - 1)$$
$$= (2^m - 1)n - 1.$$

Similarly, we can obtain

$$f\left(n, \frac{1}{m}\right) + f\left(n, \frac{2}{m}\right) + \cdots + f\left(n, \frac{mn-1}{m}\right) = (2^n - 1)m - 1.$$

By the given conditions, we have $(2^m - 1)n - 1 = 123$, namely, $(2^m - 1)n = 124$. Thus, $(2^m - 1) \mid 124$. And since $m \geq 2$, there is $2^m - 1 \in \{3, 7, 15, 31, 63, 127, \ldots\}$. $2^m - 1 = 31$ is a factor of 124 only when $m = 5$, and thus we have $n = \dfrac{124}{31} = 4$.

Therefore,

$$f\left(n, \frac{1}{m}\right) + f\left(n, \frac{2}{m}\right) + \cdots + f\left(n, \frac{mn-1}{m}\right) = (2^4 - 1) \cdot 5 - 1 = 74.$$

□

11 (20 marks) In a plane rectangular coordinate system xOy, points A, B and C are on hyperbola $xy = 1$ satisfying $\triangle ABC$ is an isosceles right triangle. Find the minimum of the area of $\triangle ABC$.

Solution As shown in Fig. 11.1, assume that vertices A, B, C of the isosceles right triangle $\triangle ABC$ is arranged in anticlockwise direction and A is the right angle vertex.

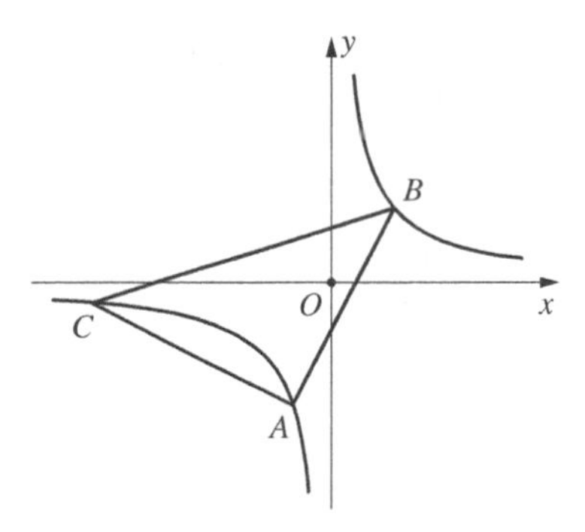

Fig. 11.1

Let $\overrightarrow{AB} = (s, t)$. Then $\overrightarrow{AC} = (-t, s)$ and the area of $\triangle ABC$ is

$$S_{\triangle ABC} = \frac{1}{2}|\overrightarrow{AB}|^2 = \frac{s^2 + t^2}{2}.$$

Note that A is on hyperbola $xy = 1$. Let $A\left(a, \frac{1}{a}\right)$, and then

$$B\left(a + s, \frac{1}{a} + t\right), \quad C\left(a - t, \frac{1}{a} + s\right).$$

Since B and C are on hyperbola $xy = 1$, we know that

$$(a + s)\left(\frac{1}{a} + t\right) = (a - t)\left(\frac{1}{a} + s\right) = 1,$$

and this is equivalent to

$$\frac{s}{a} + at = -st, \qquad ①$$

$$-\frac{t}{a} + as = st. \qquad ②$$

The sum of ① and ② gives $\frac{s - t}{a} + a(t + s) = 0$, namely,

$$a^2 = \frac{t - s}{t + s}. \qquad ③$$

By multiplying ① and ②, and making use of ③, we get

$$\begin{aligned}-s^2t^2 &= \left(\frac{s}{a} + at\right)\left(-\frac{t}{a} + as\right)\\ &= \left(a^2 - \frac{1}{a^2}\right)st + s^2 - t^2\end{aligned}$$

$$= \left(\frac{t-s}{t+s} - \frac{t+s}{t-s}\right) st + s^2 - t^2$$
$$= \frac{4st}{s^2-t^2} \cdot st + s^2 - t^2$$
$$= \frac{(s^2+t^2)^2}{s^2-t^2}.$$

By the inequality of arithmetic and geometric means, it follows that

$$\begin{aligned}(s^2+t^2)^4 &= (-s^2t^2(s^2-t^2))^2 \\ &= \frac{1}{4} \cdot 2s^2t^2 \cdot 2s^2t^2 \cdot (s^2-t^2)^2 \\ &\le \frac{1}{4} \cdot \left(\frac{2s^2t^2 + 2s^2t^2 + (s^2-t^2)^2}{3}\right)^3 \\ &= \frac{(s^2+t^2)^6}{108}, \qquad \text{④}\end{aligned}$$

and thus $s^2 + t^2 \ge \sqrt{108} = 6\sqrt{3}$.

In the following we will take a set of real numbers (s, t, a) satisfying the conditions such that $s^2 + t^2 = 6\sqrt{3}$ (and hence by s, t, a we can determine a $\triangle ABC$ satisfying the conditions such that $S_{\triangle ABC} = \frac{s^2+t^2}{2} = 3\sqrt{3}$).

Considering the condition that ④ takes the equal sign, we have $2s^2t^2 = (s^2-t^2)^2$, i.e., $\frac{s^2}{t^2} = 2 \pm \sqrt{3}$.

Suppose $0 < s < t$. Combining $s^2 + t^2 = 6\sqrt{3}$, we get

$$s = \sqrt{3(\sqrt{3}-1)}, \quad t = \sqrt{3(\sqrt{3}+1)}.$$

From ① we know that $a < 0$, so by ③ we get $a = -\sqrt{\frac{t-s}{t+s}}$, where $t = \sqrt{\frac{\sqrt{3}+1}{\sqrt{3}-1}} s = \frac{\sqrt{3}+1}{\sqrt{2}} s$. Therefore,

$$a = -\sqrt{\frac{\sqrt{3}+1-\sqrt{2}}{\sqrt{3}+1+\sqrt{2}}}.$$

In conclusion, the minimum of the area of $\triangle ABC$ is $3\sqrt{3}$.

Test Paper B
(8:00 – 9:20; September 13, 2020)

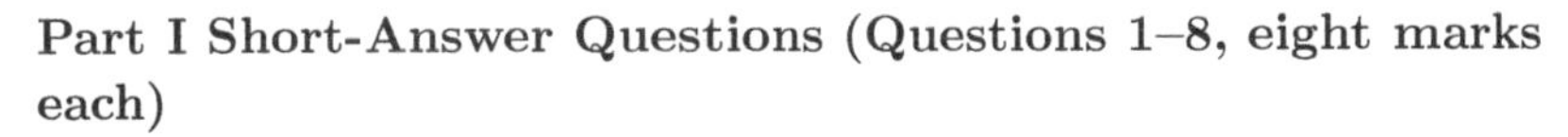

Part I Short-Answer Questions (Questions 1–8, eight marks each)

1 If real number x satisfies $\log_2 x = \log_4(2x) + \log_8(4x)$, then the value of x is ________.

Solution By the given condition, we have

$$\log_2 x = \log_4 2 + \log_4 x + \log_8 4 + \log_8 x = \frac{1}{2} + \frac{1}{2}\log_2 x + \frac{2}{3} + \frac{1}{3}\log_2 x,$$

and its solution is $\log_2 x = 7$. Therefore, $x = 128$. □

2 In a plane rectangular coordinate system xOy, circle Ω passes through points $(0,0)$, $(2,4)$, $(3,3)$. Then the maximum of the distance from a point on circle Ω to the origin is ________.

Solution Denote $A(2,4)$, $B(3,3)$. Then circle Ω passes through points O, A and B. Note that $\angle OBA = 90^\circ$ (the slopes of lines OB and AB are 1 and -1, respectively), so OA is a diameter of circle Ω. Consequently, the maximum of the distance from a point on circle Ω to the origin O is $|OA| = 2\sqrt{5}$. □

3 Suppose set $X = \{1, 2, \ldots, 20\}$. A is a subset of X. The number of the elements of A is at least 2 and all the elements of A can be arranged as consecutive positive integers. Then the number of such set A is ________.

Solution Each set A satisfying the above conditions can be uniquely determined by its minimum element a and maximum element b, where $a, b \in X$ and $a < b$. The total number of such ways of taking (a, b) is $\mathrm{C}_{20}^2 = 190$, so the number of such sets A is 190. □

4 In triangle ABC, $BC = 4$, $CA = 5$ and $AB = 6$. Then the value of $\sin^6 \frac{A}{2} + \cos^6 \frac{A}{2}$ is ________.

Solution By the law of cosines, we get $\cos A = \dfrac{CA^2 + AB^2 - BC^2}{2CA \cdot AB} = \dfrac{5^2 + 6^2 - 4^2}{2 \times 5 \times 6} = \dfrac{3}{4}$. Therefore,

$$\begin{aligned}\sin^6 \frac{A}{2} + \cos^6 \frac{A}{2} &= \left(\sin^2 \frac{A}{2} + \cos^2 \frac{A}{2}\right)\left(\sin^4 \frac{A}{2} - \sin^2 \frac{A}{2} \cos^2 \frac{A}{2} + \cos^4 \frac{A}{2}\right) \\ &= \left(\sin^2 \frac{A}{2} + \cos^2 \frac{A}{2}\right)^2 - 3 \sin^2 \frac{A}{2} \cos^2 \frac{A}{2} \\ &= 1 - \frac{3}{4} \sin^2 A \\ &= \frac{1}{4} + \frac{3}{4} \cos^2 A = \frac{43}{64}.\end{aligned}$$

□

5 Let the 9-element set $A = \{a + b\mathrm{i} \mid a, b \in \{1, 2, 3\}\}$, with i being the imaginary unit. $\alpha = (z_1, z_2, \ldots, z_9)$ is a permutation of all the elements in A, satisfying $|z_1| \leq |z_2| \leq \cdots \leq |z_9|$. The number of such permutations α is ________.

Solution Since

$$\begin{aligned}|1 + \mathrm{i}| < |2 + \mathrm{i}| = |1 + 2\mathrm{i}| < |2 + 2\mathrm{i}| < |3 + \mathrm{i}| \\ = |1 + 3\mathrm{i}| < |3 + 2\mathrm{i}| = |2 + 3\mathrm{i}| < |3 + 3\mathrm{i}|,\end{aligned}$$

it follows that

$$\begin{gathered}z_1 = 1 + \mathrm{i}, \{z_2, z_3\} = \{2 + \mathrm{i}, 1 + 2\mathrm{i}\}, \\ z_4 = 2 + 2\mathrm{i}, \{z_5, z_6\} = \{3 + \mathrm{i}, 1 + 3\mathrm{i}\}, \\ \{z_7, z_8\} = \{3 + 2\mathrm{i}, 2 + 3\mathrm{i}\}, z_9 = 3 + 3\mathrm{i}.\end{gathered}$$

By the multiplication principle, the number of permutations α satisfying the condition is $2^3 = 8$. □

6 Given a regular triangular prism, the length of each edge is 3. Then the volume of its circumscribed sphere is ________.

Solution As shown in Fig. 6.1, let the centroids of faces ABC and $A_1B_1C_1$ be O and O_1, respectively. Denote the midpoint of segment OO_1 as P. By symmetry, we know that P is the centre of the circumscribed sphere of the regular triangular prism and PA is its radius.

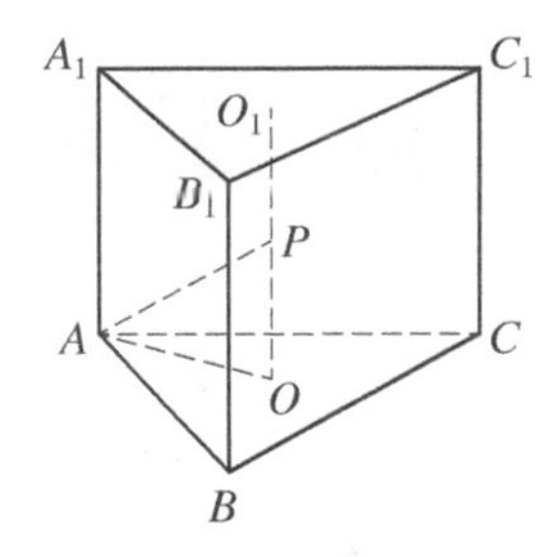

Fig. 6.1

It is easy to find that $PO \perp AO$, and thus

$$PA = \sqrt{PO^2 + AO^2} = \sqrt{\left(\frac{3}{2}\right)^2 + (\sqrt{3})^2} = \frac{\sqrt{21}}{2}.$$

Therefore, the volume of the circumscribed sphere is $\frac{4}{3}\pi\left(\frac{\sqrt{21}}{2}\right)^3 = \frac{7\sqrt{21}}{2}\pi$. □

7 In convex quadrilateral $ABCD$, $\overrightarrow{BC} = 2\overrightarrow{AD}$. Point P is on the plane of quadrilateral $ABCD$, satisfying $\overrightarrow{PA} + 2020\overrightarrow{PB} + \overrightarrow{PC} + 2020\overrightarrow{PD} = \overrightarrow{0}$. Let s and t be the areas of quadrilateral $ABCD$ and $\triangle PAB$, respectively. Then the value of $\frac{t}{s}$ is ________.

Solution We may assume that $AD = 2$ and $BC = 4$. As shown in Fig. 7.1, denote M, N, X, Y as the midpoints of AB, CD, BD, AC, respectively. Then M, X, Y, N are collinear in order and $MX = XY = YN = 1$.

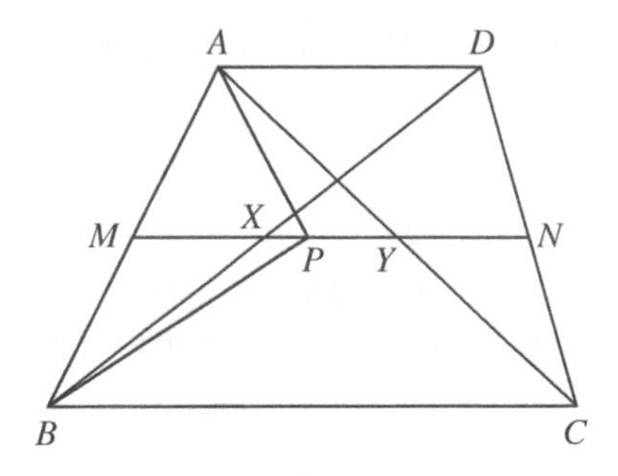

Fig. 7.1

Since

$$\overrightarrow{PA}+\overrightarrow{PC}=2\overrightarrow{PY},\quad \overrightarrow{PB}+\overrightarrow{PD}=2\overrightarrow{PX},$$

combining the given conditions, we know that $\overrightarrow{PY}+2020\overrightarrow{PX}=\overrightarrow{0}$. Hence, point P lies on segment XY, and $PX=\dfrac{1}{2021}$. Let the distance from A to MN be h. By the area formula, we can get

$$\begin{aligned}\frac{t}{s}&=\frac{S_{\triangle PAB}}{S_{ABCD}}=\frac{PM\cdot h}{MN\cdot 2h}=\frac{PM}{2MN}\\&=\frac{1+\dfrac{1}{2021}}{2\times 3}=\frac{337}{2021}.\end{aligned}$$

□

8 Given quintic polynomial $f(x)$ with its leading coefficient being 1, it satisfies $f(n)=8n$, $n=1,2,\ldots,5$. Then the coefficient of the term of degree 1 of $f(x)$ is________.

Solution Let $f(x)=g(x)-8x$, and then $g(x)$ is also a quintic polynomial with its leading coefficient being 1. And there is

$$g(n)=f(n)-8n=0,\quad n=1,2,\ldots,5.$$

Hence, $g(x)$ has 5 real roots, namely, $1,2,\ldots,5$. Therefore,

$$g(x)=(x-1)(x-2)\cdots(x-5),$$

and thus,

$$f(x)=(x-1)(x-2)\cdots(x-5)+8x.$$

Consequently, the coefficient of the term of degree 1 of $f(x)$ is

$$\left(1+\frac{1}{2}+\frac{1}{3}+\frac{1}{4}+\frac{1}{5}\right)\cdot 5!+8=282.$$

Part II Word Problems (16 marks for Question 9, 20 marks for Question 10 and 11, and then 56 marks in total)

9 (16 marks) In ellipse Γ, A is an endpoint of the major axis, B is an endpoint of the minor axis, and F_1, F_2 are the foci. If $\overrightarrow{AF_1}\cdot\overrightarrow{AF_2}+\overrightarrow{BF_1}\cdot\overrightarrow{BF_2}=0$, then find the value of $\tan\angle ABF_1\cdot\tan\angle ABF_2$.

Solution By symmetry, suppose the equation of Γ is $\frac{x^2}{a^2}+\frac{y^2}{b^2}=1(a>b>0)$, and $A(a,0)$, $B(0,b)$, $F_1(-c,0)$, $F_2(c,0)$, where $c=\sqrt{a^2-b^2}$.

By the given conditions, we know that

$$\begin{aligned}\overrightarrow{AF_1}\cdot\overrightarrow{AF_2}+\overrightarrow{BF_1}\cdot\overrightarrow{BF_2}&=(-c-a)(c-a)+(-c^2+b^2)\\&=a^2+b^2-2c^2=0.\end{aligned}$$

Thus, $a^2+b^2-2c^2=-a^2+3b^2=0$, and hence $a=\sqrt{3}b, c=\sqrt{2}b$.

Let O be the origin of the coordinates, and then

$$\tan\angle ABO=\frac{a}{b}=\sqrt{3},$$

$$\tan\angle OBF_1=\tan\angle OBF_2=\frac{c}{b}=\sqrt{2}.$$

Therefore,

$$\begin{aligned}&\tan\angle ABF_1\cdot\tan\angle ABF_2\\&=\tan(\angle ABO+\angle OBF_1)\cdot\tan(\angle ABO-\angle OBF_1)\\&=\frac{\sqrt{3}+\sqrt{2}}{1-\sqrt{3}\cdot\sqrt{2}}\cdot\frac{\sqrt{3}-\sqrt{2}}{1+\sqrt{3}\cdot\sqrt{2}}=-\frac{1}{5}.\end{aligned}$$

□

10 (20 marks) Suppose positive real numbers a, b, and c satisfy $a^2+4b^2+9c^2=4b+12c-2$. Find the minimum of $\frac{1}{a}+\frac{2}{b}+\frac{3}{c}$.

Solution By the given condition, we have

$$a^2+(2b-1)^2+(3c-2)^2=3.$$

By making use of the Cauchy inequality, we get

$$3[a^2+(2b-1)^2+(3c-2)^2]\geq(a+2b-1+3c-2)^2,$$

namely, $(a+2b+3c-3)^2\leq 9$. Therefore,

$$a+2b+3c\leq 6.$$

Again, by the Cauchy inequality we get

$$\left(\frac{1}{a}+\frac{2}{b}+\frac{3}{c}\right)(a+2b+3c)\geq(1+2+3)^2.$$

Thus,

$$\frac{1}{a}+\frac{2}{b}+\frac{3}{c}\geq\frac{36}{a+2b+3c}\geq 6,$$

and the equal sign holds when $a=b=c=1$.

Therefore, the minimum value of $\frac{1}{a}+\frac{2}{b}+\frac{3}{c}$ is 6. □

11 (20 marks) Let the general term of sequence $\{a_n\}$ be

$$a_n=\frac{1}{\sqrt{5}}\left(\left(\frac{1+\sqrt{5}}{2}\right)^n-\left(\frac{1-\sqrt{5}}{2}\right)^n\right),\quad n=1,2,\ldots.$$

Prove that there exist infinite positive integers m such that $a_{m+4}a_m-1$ are perfect squares.

Solution Denote $q_1=\frac{1+\sqrt{5}}{2}$, $q_2=\frac{1-\sqrt{5}}{2}$, and then $q_1+q_2=1$, $q_1q_2=-1$. Thus,

$$a_n=\frac{1}{\sqrt{5}}(q_1^n-q_2^n),\quad n=1,2,\ldots.$$

Hence $a_1=1$, $a_2=1$. Also note that $q_i+1=q_i^2 (i=1,2)$, and there is

$$\begin{aligned}a_{n+1}+a_n&=\frac{1}{\sqrt{5}}(q_1^{n+1}-q_2^{n+1})+\frac{1}{\sqrt{5}}(q_1^n-q_2^n)\\&=\frac{1}{\sqrt{5}}(q_1^n(q_1+1)-q_2^n(q_2+1))\\&=\frac{1}{\sqrt{5}}(q_1^{n+2}-q_2^{n+2}),\end{aligned}$$

namely,

$$a_{n+2}=a_{n+1}+a_n,\quad n=1,2,\ldots.$$

It is easy to know that each term of the sequence $\{a_n\}$ is a positive integer.

It is easy to calculate that $q_1^4+q_2^4=7$, and hence

$$\begin{aligned}a_{2n+3}a_{2n-1}-1&=\frac{1}{\sqrt{5}}(q_1^{2n+3}-q_2^{2n+3})\cdot\frac{1}{\sqrt{5}}(q_1^{2n-1}-q_2^{2n-1})-1\\&=\frac{1}{5}(q_1^{4n+2}+q_2^{4n+2}-(q_1q_2)^{2n-1}q_1^4-(q_1q_2)^{2n-1}q_2^4)-1\end{aligned}$$

$$
\begin{aligned}
&= \frac{1}{5}(q_1^{4n+2} + q_2^{4n+2} + q_1^4 + q_2^4) - 1 \\
&= \frac{1}{5}(q_1^{4n+2} + q_2^{4n+2} + 7) - 1 \\
&= \frac{1}{5}(q_1^{4n+2} + q_2^{4n+2} + 2) \\
&= \left[\frac{1}{\sqrt{5}}(q_1^{2n+1} + q_2^{2n+1})\right]^2 \\
&= a_{2n+1}^2.
\end{aligned}
$$

Therefore, for any positive integer n, $a_{2n+3}a_{2n-1} - 1$ is a perfect square. Hence, $a_{m+4}a_m - 1$ are perfect squares for all positive odd numbers m.

China Mathematical Competition (First Round)

2021

The name of the 2021 China Mathematical Competition was changed to the 2021 China Mathematical Olympiad (First Round) and the 2021 China Mathematical Competition. The competition was held on September 12. Due to the pandemic, Jiangsu, Henan, Fujian, and Inner Mongolia did not participate in the September 12 exam. Paper B was used in Ningxia, Qinghai, Xinjiang, and Tibet, and Paper A was used in the other 23 provinces, municipalities, and autonomous regions.

On October 6, Jiangsu, Henan and Inner Mongolia conducted their examinations, with Jiangsu and Henan using Paper A1 and Inner Mongolia using Paper B1.

On October 23, Fujian had a supplemental competition, and used Paper A2.

While the scope of the test questions in the first round of the 2020 China Mathematical Competition does not exceed the teaching requirements and content specified in the "General High School Mathematics Curriculum Standards (Experiments)" promulgated by the Ministry of Education of China in 2003, the methods of proposing the questions have been improved. The emphasis placed on to test the students' basic knowledge and skills, and their abilities to integrate and use flexibly of them. Each test paper includes eight fill-in-the-blank questions and three answer questions. The answer time is 80 minutes, and the full score is 120 marks.

The scope of the test questions the 2021 China Mathematical Competition (Second Round) is in line with the International Mathematical Olympiad, with some expanded knowledge, plus a few contents of the Mathematical Competition Syllabus. Each test paper consists of four answer questions, including a plane geometry one, and the answering time is 170 minutes. The full score is 180 marks.

Test Paper A
(8:00 – 9:20; September 12, 2021)

Part I Short-Answer Questions (Questions 1–8, eight marks each)

1 Suppose that arithmetic sequence $\{a_n\}$ satisfies $a_{2021} = a_{20}+a_{21} = 1$. Then the value of a_1 is ________.

Solution Let the common difference of $\{a_n\}$ be d. By the given condition, it follows that $\begin{cases} a_1 + 2020d = 1, \\ 2a_1 + 39d = 1. \end{cases}$ The solution is $a_1 = \dfrac{1981}{4001}$. □

2 Given set $A = \{1, 2, m\}$, m is real. Let $B = \{a^2 \mid a \in A\}$, $C = A \cup B$. If the sum of all the elements of C is 6, then the product of all the elements of C is ________.

Solution By the condition, it is known that 1, 2, 4, m, m^2 (allowing for repetition) are all the elements of C.

Note that when m is real, $1+2+4+m+m^2 > 6$, $1+2+4+m^2 > 6$, so it can only be $C = \{1, 2, 4, m\}$, and $1+2+4+m = 6$. Therefore, $m = -1$, and is tested to be consistent with the question. At this point the product of all the elements of C is $1 \times 2 \times 4 \times (-1) = -8$. □

3 Suppose function $f(x)$ satisfies: for any non-zero real number x, there is

$$f(x) = f(1) \cdot x + \frac{f(2)}{x} - 1.$$

Then the minimum of $f(x)$ on $(0, +\infty)$ is ________.

Solution Let $x = 1, 2$, and we can get $f(1) = f(1) + f(2) - 1$ and $f(2) = 2f(1) + \dfrac{f(2)}{2} - 1$, respectively. The solution is $f(2) = 1$, $f(1) = \dfrac{3}{4}$.

Thus, for $x \neq 0$, there is

$$f(x) = \frac{3}{4}x + \frac{1}{x} - 1.$$

When $x \in (0, +\infty)$, $f(x) \geq 2\sqrt{\frac{3}{4}x \cdot \frac{1}{x}} - 1 = \sqrt{3} - 1$. The equal sign holds when $x = 3$.

Therefore, the minimum of $f(x)$ on $(0, +\infty)$ is $\sqrt{3} - 1$. □

4 Suppose $f(x) = \cos x + \log_2 x$ $(x > 0)$. If positive real number a satisfies $f(a) = f(2a)$, then the value of $f(2a) - f(4a)$ is ________.

Solution By the condition, it follows that $\cos a + \log_2 a = \cos 2a + \log_2 2a = 2\cos^2 a - 1 + 1 + \log_2 a$, so $\cos a = 2\cos^2 a$. Thus, we have $\cos a = 0$ or $\cos a = \frac{1}{2}$, and hence correspondingly $\cos 2a = 2\cos^2 a - 1 = -1$ or $\cos 2a = -\frac{1}{2}$. Therefore,

$$\begin{aligned} f(2a) - f(4a) &= \cos 2a + \log_2 2a - \cos 4a - \log_2 4a \\ &= \cos 2a - 2\cos^2 2a \\ &= \begin{cases} -3, & \text{if } \cos 2a = -1, \\ -1, & \text{if } \cos 2a = -\dfrac{1}{2}. \end{cases} \end{aligned}$$

□

5 In $\triangle ABC$, $AB = 1$, $AC = 2$, $B - C = \frac{2\pi}{3}$. Then the area of $\triangle ABC$ is ________.

Solution By the law of sines, it follows that $\frac{\sin B}{\sin C} = \frac{AC}{AB} = 2$. Since $B - C = \frac{2\pi}{3}$, we have

$$2\sin C = \sin B = \sin\left(C + \frac{2}{3}\pi\right) = -\frac{1}{2}\sin C + \frac{\sqrt{3}}{2}\cos C,$$

namely, $\frac{5}{2}\sin C = \frac{\sqrt{3}}{2}\cos C$, and hence $\tan C = \frac{\sqrt{3}}{5}$.

Denote the area of $\triangle ABC$ as S. Notice that $A = \pi - B - C = \frac{\pi}{3} - 2C$, so

$$S = \frac{1}{2}AB \cdot AC \cdot \sin A = \sin A = \frac{\sqrt{3}}{2}\cos 2C - \frac{1}{2}\sin 2C.$$

Since $\tan C = \frac{\sqrt{3}}{5}$, it follows that $\cos 2C = \frac{1-\tan^2 C}{1+\tan^2 C} = \frac{11}{14}$, $\sin 2C = \frac{2\tan C}{1+\tan^2 C} = \frac{5\sqrt{3}}{14}$. Therefore,

$$S = \frac{\sqrt{3}}{2}\cdot\frac{11}{14} - \frac{1}{2}\cdot\frac{5\sqrt{3}}{14} = \frac{3\sqrt{3}}{14}. \qquad \square$$

6 In a plane rectangular coordinate system xOy, the focus of parabola $\Gamma : y^2 = 2px$ $(p > 0)$ is F. Make a tangent line to Γ passing through point P (different from O) on Γ and it intersects the y-axis at point Q. If $|FP| = 2$, $|FQ| = 1$, then the dot product of vectors $\overrightarrow{OP}$ and $\overrightarrow{OQ}$ is ________.

Solution Let $P\left(\frac{t^2}{2p}, t\right)$ $(t \neq 0)$, and then the equation of the tangent line of Γ is $yt = p\left(x + \frac{t^2}{2p}\right)$.

Let $x = 0$, and we get $yt = \frac{t}{2}$. The coordinates of F are $\left(\frac{p}{2}, 0\right)$, and thus

$$|FP| = \sqrt{\left(\frac{p}{2} - \frac{t^2}{2p}\right)^2 + t^2} = \frac{p}{2} + \frac{t^2}{2p},$$

$$|FQ| = \frac{\sqrt{p^2+t^2}}{2}.$$

Combining $|FP| = 2$, $|FQ| = 1$, we can get $p^2 + t^2 = 4p$ and $p^2 + t^2 = 4$, respectively. Hence, $p = 1$, $t^2 = 3$.

Therefore, $\overrightarrow{OP}\cdot\overrightarrow{OQ} = \frac{t^2}{2} = \frac{3}{2}$. $\square$

7 An even cube dice with six faces is marked with six numbers 1, 2, 3, 4, 5 and 6, respectively. The dice are randomly tossed three times, with each toss being independent from each other, and the resulting dots

are a_1, a_2, a_3 in order. Then the probability of event $|a_1 - a_2| + |a_2 - a_3| + |a_3 - a_1| = 6$ is ________.

Solution Note that $|a_1 - a_2|+|a_2 - a_3|+|a_3 - a_1| = 2 \max\limits_{1\le i\le 3} a_i - 2 \min\limits_{1\le i\le 3} a_i$. Therefore, the three a_1, a_2, a_3 tossed satisfy the condition if and only if the difference between the largest number and the smallest number is 3. This is to say that a_1, a_2, a_3 is a permutation of x, $x+d$, $x+3$, where $x \in \{1, 2, 3\}$ and $d \in \{0, 1, 2, 3\}$.

For each $x \in \{1, 2, 3\}$, when $d = 0$ or $d = 3$, there are 3 different permutations of x, $x+d$, $x+3$ for each case; when $d = 1$ or $d = 2$, there are 6 different permutations of x, $x+d$, $x+3$ for each case.

Therefore, there are $3 \times (2 \times 3 + 2 \times 6) = 54$ cases of dots a_1, a_2, a_3 that satisfy the condition.

Consequently, the desired probability is $\frac{54}{6^3} = \frac{1}{4}$. □

8 Given rational number $r = \frac{p}{q} \in (0, 1)$, p, q are coprime positive integers, and pq divides 3600. The number of such rational numbers r is ________.

Solution Suppose set $\Omega = \left\{ r \,\middle|\, r = \frac{p}{q}, p, q \in \mathbb{N}_+, (p, q) = 1, pq \,\middle|\, 3600 \right\}$.

We consider the reduced fractional form $\frac{p}{q}$ of any element r of Ω. Since the standard factorization of 3600 is $2^4 \times 3^2 \times 5^2$, we can set $p = 2^A \times 3^B \times 5^C$, $q = 2^a \times 3^b \times 5^c$, where $\min\{A, a\} = \min\{B, b\} = \min\{C, c\} = 0$ and $A + a \le 4$, $B + b \le 2$, $C + c \le 2$.

Therefore, there are 9 ways to take such number pair (A, a), 5 ways to take number pair (B, b), and 5 ways to take number pair (C, c).

As a result, the number of the elements of Ω is $|\Omega| = 9 \times 5 \times 5 = 225$.

All the rational numbers satisfying the conditions are $\Omega \cap (0, 1)$. Notice that $r \in \Omega$ if and only if $\frac{1}{r} \in \Omega$. In particular, $1 \in \Omega$. Therefore, the elements in $\Omega \backslash \{1\}$ can be matched into

$$\frac{1}{2}(|\Omega| - 1) = 112$$

pairs according to the product of 1, and each pair has exactly one number belonging to $(0, 1)$, that is, there is exactly one number satisfying the conditions. Thus, the number of desired rational numbers r is 112. □

Part II Word Problems (16 marks for Question 9, 20 marks for Question 10 and 11, and then 56 marks in total)

9 (16 marks) Let complex sequence $\{z_n\}$ satisfy

$$z_1 = \frac{\sqrt{3}}{2}, \quad z_{n+1} = \overline{z_n}(1 + z_n\mathrm{i}) \quad (n = 1, 2, \ldots),$$

where i is the imaginary unit. Find the value of z_{2021}.

Solution For $n \in \mathbb{N}_+$, let $z_n = a_n + b_n\mathrm{i}$ $(a_n, b_n \in \mathbb{R})$. Then

$$\begin{aligned} a_{n+1} + b_{n+1}\mathrm{i} = z_{n+1} &= \overline{z_n}(1 + z_n\mathrm{i}) \\ &= \overline{z_n} + |z_n|^2 \cdot \mathrm{i} \\ &= a_n - b_n\mathrm{i} + (a_n^2 + b_n^2)\mathrm{i}, \end{aligned}$$

and hence $a_{n+1} = a_n$, $b_{n+1} = a_n^2 + b_n^2 - b_n$.

And by $z_1 = \frac{\sqrt{3}}{2}$ we know that $a_1 = \frac{\sqrt{3}}{2}$, $b_1 = 0$, so $a_n = \frac{\sqrt{3}}{2}$, and thus

$$b_{n+1} = b_n^2 - b_n + \frac{3}{4},$$

namely, $b_{n+1} - \frac{1}{2} = b_n^2 - b_n + \frac{1}{4} = \left(b_n - \frac{1}{2}\right)^2$.

Hence, when $n \geq 2$,

$$\begin{aligned} b_n &= \frac{1}{2} + \left(b_1 - \frac{1}{2}\right)^{2^{n-1}} \\ &= \frac{1}{2} + \left(-\frac{1}{2}\right)^{2^{n-1}} \\ &= \frac{1}{2} + \frac{1}{2^{2^{n-1}}}. \end{aligned}$$

Consequently,

$$z_{2021} = a_{2021} + b_{2021}\mathrm{i} = \frac{\sqrt{3}}{2} + \left(\frac{1}{2} + \frac{1}{2^{2^{2020}}}\right)\mathrm{i}.$$

□

10 (20 marks) In the plane rectangular coordinate system, the graph of function $y = \frac{x+1}{|x|+1}$ has three different points lying on line l, and the sum of the abscissas of these three points is 0. Find the range of values of the slope of l.

Solution When $x \geq 0$, $y = 1$; when $x < 0$, $y = \dfrac{x+1}{1-x}$ is strictly increasing about x and less than 1.

Suppose line $l : y = kx + b$, and then the known conditions are equivalent to the fact that equation

$$kx + b = \frac{x+1}{|x|+1} \qquad ①$$

has three different real number solutions $x_1, x_2, x_3 (x_1 < x_2 < x_3)$ satisfying $x_1 + x_2 + x_3 = 0$.

Firstly, there is $k \neq 0$, otherwise the function of l can only be $y = 1$, but then the sum of the abscissas of any three common points of the graph of l and function $y = \dfrac{x+1}{|x|+1}$ must be greater than 0. This is not consistent with the question.

When $x < 0$, Equation ① can be arranged as

$$kx^2 - (k - b - 1)x + 1 - b = 0, \qquad ②$$

and it has at most two negative solutions.

When $x \geq 0$, Equation ① follows as

$$kx + b = 1, \qquad ③$$

and it has at most one non-negative solution.

This shows that Equation ② has two different negative solutions x_1, x_2, where

$$x_1 + x_2 = \frac{k - b - 1}{2};$$

Equation ③ has a non-negative solution $x_3 = \dfrac{1-b}{k}$. By $x_1 + x_2 + x_3 = 0$, we know that $k = 2b$.

And then there is $x_3 = \dfrac{1-b}{2b}$. Since $x_3 \geq 0$, we have $0 < b \leq 1$.

Equation ② becomes $2bx^2 + (1 - b)x + 1 - b = 0$. By discriminant

$$(1 - b)^2 - 4 \cdot 2b(1 - b) = (1 - b)(1 - 9b) > 0$$

and combining $0 < b \leq 1$, we have $0 < b < \dfrac{1}{9}$. (After checking, x_1, x_2 are indeed negative at this point, which is consistent with the question.)

In conclusion, the range of the slope $k = 2b$ of l is $0 < b < \dfrac{2}{9}$. □

11 (20 marks) As shown in Fig. 11.1, the edge length of cube $ABCD-EFGH$ is 2. Take any point P_1 on the incircle of square $ABFE$, take any point P_2 on the incircle of square $BCGF$, and take any point P_3 on the incircle of square $EFGH$. Find the minimum and maximum of $|P_1P_2| + |P_2P_3| + |P_3P_1|$.

Solution To establish the Cartesian coordinate system in three-dimensional space, take the centre of the cube as the origin, and the directions of $\overrightarrow{DA}$, $\overrightarrow{DC}$, $\overrightarrow{DH}$ are the positive directions of x-axis, y-axis and z-axis, respectively.

According to the condition, we can set

$$P_1(1, \cos\alpha_1, \sin\alpha_1), \quad P_2(\sin\alpha_2, 1, \cos\alpha_2), \quad P_3(\cos\alpha_3, \sin\alpha_3, 1).$$

We conventionally assume that P4 = P1 and $\alpha 4 = \alpha 1$. Denote

$$d_i = |P_iP_{i+1}| \quad (i = 1, 2, 3).$$

Then

$$d_i^2 = (1 - \sin\alpha_{i+1})^2 + (1 - \cos\alpha_i)^2 + (\sin\alpha_i - \cos\alpha_{i+1})^2.$$

Denote $f = |P_1P_2| + |P_2P_3| + |P_3P_1|$.

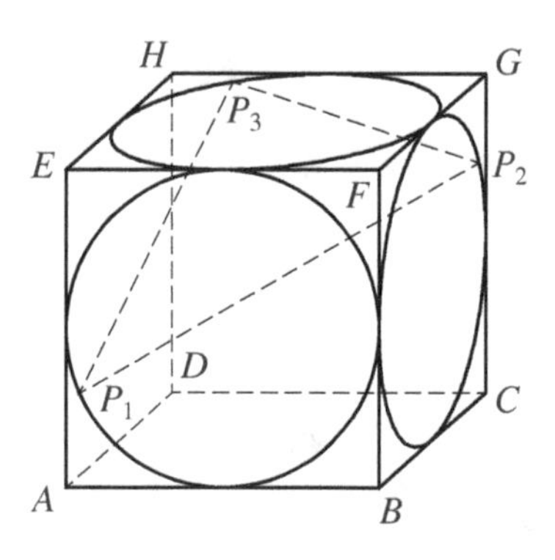

Fig. 11.1

We first find the minimum of f. For $i = 1, 2, 3$, according to ① and the inequality of arithmetic and geometric means we get

$$\begin{aligned} d_i^2 &\geq (1 - \sin\alpha_{i+1})^2 + (1 - \cos\alpha_i)^2 \\ &\geq \frac{1}{2}((1 - \sin\alpha_{i+1}) + (1 - \cos\alpha_i))^2, \end{aligned}$$

and thus $d_i \geq \frac{\sqrt{2}}{2}(2 - \sin\alpha_{i+1} - \cos\alpha_i)$. Therefore,

$$\begin{aligned} f &= d_1 + d_2 + d_3 \\ &\geq \frac{\sqrt{2}}{2}\sum_{i=1}^{3}(2 - \sin\alpha_{i+1} - \cos\alpha_i) \\ &= 3\sqrt{2} - \frac{\sqrt{2}}{2}\sum_{i=1}^{3}(\sin\alpha_i + \cos\alpha_i) \\ &= 3\sqrt{2} - \sum_{i=1}^{3}\sin\left(\alpha_i + \frac{\pi}{4}\right) \\ &\geq 3\sqrt{2} - 3. \end{aligned}$$

When $\alpha_i = \frac{\pi}{4}$ $(i = 1, 2, 3)$, f can take the minimum $3\sqrt{2} - 3$.

Then we will find the maximum of f. By ①, it is clear that

$$d_i^2 = 4 - 2\cos\alpha_i - 2\sin\alpha_{i+1} - 2\sin\alpha_i\cos\alpha_{i+1}.$$

Note that $\sin\alpha_i \geq -1$, $\cos\alpha_i \geq -1$ $(i = 1, 2, 3)$, and then

$$\begin{aligned} \sum_{i=1}^{3} d_i^2 &= 12 - 2\left(\sum_{i=1}^{3}\sin\alpha_{i+1} + \sum_{i=1}^{3}\cos\alpha_i + \sum_{i=1}^{3}\sin\alpha_i\cos\alpha_{i+1}\right) \\ &= 12 - 2\left(\sum_{i=1}^{3}\sin\alpha_i + \sum_{i=1}^{3}\cos\alpha_{i+1} + \sum_{i=1}^{3}\sin\alpha_i\cos\alpha_{i+1}\right) \\ &= 18 - 2\sum_{i=1}^{3}(1 + \sin\alpha_i)(1 + \cos\alpha_{i+1}) \leq 18. \end{aligned}$$

By the Cauchy inequality, we know that $f^2 \leq 3(d_1^2 + d_2^2 + d_3^2) = 54$, so $f \leq 3\sqrt{6}$.

When $\alpha_i = \pi$ $(i = 1, 2, 3)$, f can take the maximum $3\sqrt{6}$.
In conclusion, the minimum of f is $3\sqrt{2} - 3$, and its maximum is $3\sqrt{6}$. □

Test Paper A1
(8:00 – 9:20; October 6, 2021)

Part I Short-Answer Questions (Questions 1–8, eight marks each)

1 Suppose that $A = \{1, 2, 3\}$, $B = \{2x + y \mid x, y \in A, x < y\}$, $B = \{2x + y \mid x, y \in A, x > y\}$. Then the sum of all the elements of $B \cap C$ is ________.

Solution By enumeration, we get $B = \{4, 5, 7\}$, $C = \{5, 7, 8\}$. Thus, $B \cap C = \{5, 7\}$. Therefore, the sum of all the elements of $B \cap C$ is $5 + 7 = 12$. □

2 Given vectors $\vec{a} = (1 + 2^m, 1 - 2^m)$, $\vec{b} = (4^m - 3, 4^m + 5)$, suppose m is real. Then the minimum of the dot product of $\vec{a} \cdot \vec{b}$ is ________.

Solution Let $t = 2m$, and then $\vec{a} = (1+t, 1-t)$, $\vec{b} = (t^2 - 3, t^2 + 5)$. Thus,

$$\begin{aligned}\vec{a} \cdot \vec{b} &= (1 + t)(t^2 - 3) + (1 - t)(t^2 + 5) \\ &= 2(t - 2)^2 - 6 \geq -6.\end{aligned}$$

When $t = 2$, namely, $m = 1$, $\vec{a} \cdot \vec{b}$ takes the minimum -6. □

3 In $\triangle ABC$, $\tan A$ and $\tan B$ are the two roots of equation $x^2 - 10x + 6 = 0$. Then the value of $\cos C$ is ________.

Solution By the condition, we know that $\tan A + \tan B = 10$, $\tan A \tan B = 6$. Thus,

$$\begin{aligned}\tan C &= \tan(\pi - A - B) \\ &= -\tan(A + B) \\ &= -\frac{\tan A + \tan B}{1 - \tan A \tan B} = 2.\end{aligned}$$

Therefore, C is an acute angle, and thus $\cos C = \dfrac{1}{\sqrt{1 + \tan^2 C}} = \dfrac{1}{\sqrt{5}} = \dfrac{\sqrt{5}}{5}$. □

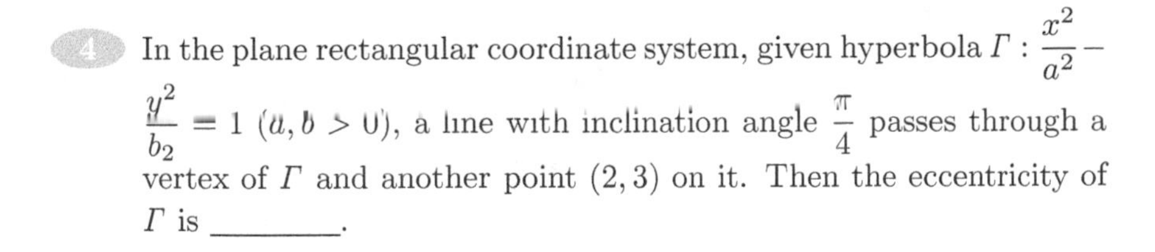

4 In the plane rectangular coordinate system, given hyperbola $\Gamma : \frac{x^2}{a^2} - \frac{y^2}{b_2} = 1$ $(a, b > 0)$, a line with inclination angle $\frac{\pi}{4}$ passes through a vertex of Γ and another point $(2,3)$ on it. Then the eccentricity of Γ is ________.

Solution The slope of the line described in the question is 1 and it passes through point $(2,3)$, so its equation is $y = x + 1$. This line intersects with the x-axis at point $(-1,0)$, and thus $(-1,0)$ is a vertex of Γ. Hence, $a = 1$.

And since point $(2,3)$ is on Γ, we know that $\frac{2^2}{1} - \frac{3^2}{b^2} = 1$, and then $b^2 = 3$.

Denote $c = \sqrt{a^2 + b^2}$. Then the eccentricity of Γ is $\frac{a}{c} = \frac{\sqrt{a^2+b^2}}{a} = 2$. □

5 Let sequence $\{a_n\}$ satisfy $a_1 = 1$, $a_n = \frac{1}{4a_{n-1}} + \frac{1}{n}$ $(n \geq 2)$. Then the value of a_{100} is ________.

Solution By mathematical induction, we can prove that $a_n = \frac{n+1}{2n}$.

When $n = 1$, $a_1 = 1 = \frac{1+1}{2}$, the conclusion is clearly valid.

We assume that the conclusion holds when $n = k$. When $n = k+1$, by the condition and induction hypothesis we have

$$a_{k+1} = \frac{1}{4a_k} + \frac{1}{k+1} = \frac{k}{2(k+1)} + \frac{1}{k+1} = \frac{(k+1)+1}{2(k+1)},$$

that is, the conclusion also holds for $n = k+1$.

Therefore, by the mathematical induction we get $a_n = \frac{n+1}{2n}$. In particular, $a_{100} = \frac{101}{200}$. □

6 Let the side length of the base and height of regular pyramid $P - ABCD$ be equal. Point G is the centroid of face $\triangle PBC$. Then the sine of the angle between line AG and base $ABCD$ is ________.

Solution As shown in Fig. 6.1, take the midpoint M of PM, and then G lies on PM. Take the projections of P, G on base $ABCD$, which are

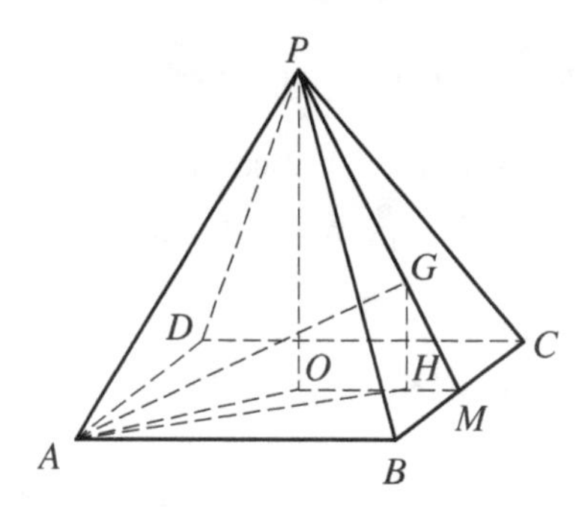

Fig. 6.1

O and H, respectively. Then O is the centre of the base square, H lies on OM, and $\frac{GH}{PO} = \frac{HM}{OM} = \frac{GM}{PM} = \frac{1}{3}$.

For the sake of convenience, let $AB = PO = 6$. Thus, $GH = \frac{PO}{3} = 2$, $OH = \frac{2}{3}OM = 2$. And since $AO = 3\sqrt{2}, \angle AOH = 135°$, we have

$$AH^2 = AO^2 + OH^2 - 2AO \cdot OH \cdot \cos \angle AOH = 34,$$

and thus $AG = \sqrt{AH^2 + GH^2} = \sqrt{38}$.

The angle between line AG and base $ABCD$ is equal to $\angle GAH$. Therefore, the desired sine is

$$\sin \angle GAH = \frac{GH}{AG} = \frac{2}{\sqrt{38}} = \frac{\sqrt{38}}{19}.$$

□

7 Let $a_1, a_2, \ldots, a_{21}$ be a permutation of $1, 2, \ldots, 21$, satisfying

$$|a_{20} - a_{21}| \geq |a_{19} - a_{21}| \geq |a_{18} - a_{21}| \geq \cdots \geq |a_1 - a_{21}|.$$

The number of such permutations is ________.

Solution For a given $k \in \{1, 2, \ldots, 21\}$, consider the number of permutations N_k that satisfy the conditions such that $a_{21} = k$.

When $k \in \{1, 2, \ldots, 11\}$, for $i = 1, 2, \ldots, k-1$, there exist a_{2i-1}, a_{2i} that are permutations of $k - i, k + i$ (if $k = 1$, there exists no such i), and $a_{2j} = j + 1 (2k - 1 \leq j \leq 20)$ (if $k = 11$, there exists no such j), so $N_k = 2^{k-1}$.

Similarly, when $k \in \{12, 13, \ldots, 21\}$, there is $N_k = 2^{21-k}$.
Therefore, the number of such permutations satisfying the condition is

$$\sum_{k=1}^{21} N_k = \sum_{k=1}^{11} 2^{k-1} + \sum_{k=12}^{21} 2^{21-k} = (2^{11} - 1) + (2^{10} - 1) = 3070.$$

□

8 Positive real numbers x, y satisfy the following condition: there exist $a \in [0, x]$, $b \in [0, y]$ such that

$$a^2 + y^2 = 2, \quad b^2 + x^2 = 1, \quad ax + by = 1.$$

Then the maximum of $x + y$ is ________.

Solution In a plane rectangular coordinate system xOy, for positive real number pairs (x, y) that satisfy the condition, take points $L(x, 0), M(x, y), N(0, y)$, and then quadrilateral $OLMN$ is a rectangle. Points P, Q are on sides LM, MN, respectively, as shown in Fig. 8.1.

Since $a^2 + y^2 = 2, b^2 + x^2 = 1, ax + by = 1$, we have

$$|OP| = \sqrt{x^2 + b^2} = 1, \quad |OQ| = \sqrt{a^2 + y^2} = \sqrt{2},$$

$$|PQ| = \sqrt{(a - x)^2 + (b - y)^2}$$

$$= \sqrt{(a^2 + y^2) + (b^2 + x^2) - 2(ax + by)} = 1.$$

Thus, $\triangle OPQ$ is an isosceles right triangle with P as its right-angle vertex.

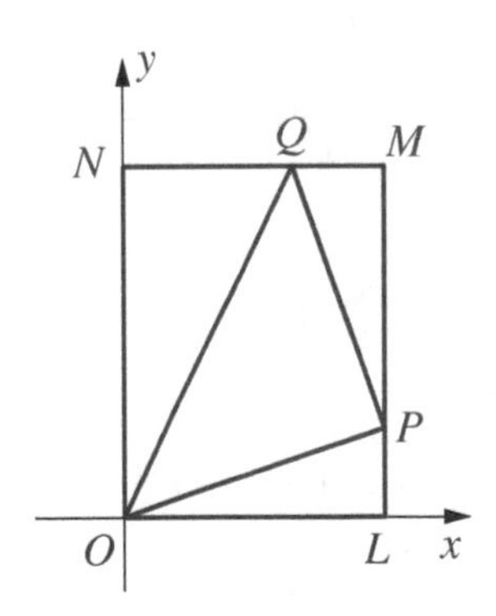

Fig. 8.1

Therefore, we can set $\angle LOP = \theta$, $\angle QON = \frac{\pi}{4} - \theta$, where $0 \leq \theta \leq \frac{\pi}{4}$. Then

$$\begin{aligned} x + y &= |OL| + |ON| \\ &= |OP| \cdot \cos \angle LOP + |OQ| \cdot \cos \angle QON \\ &= \cos\theta + \sqrt{2}\cos\left(\frac{\pi}{4} - \theta\right) = 2\cos\theta + \sin\theta \\ &= \sqrt{5}\sin(\theta + \varphi), \end{aligned}$$

where $\varphi = \arcsin \frac{2\sqrt{5}}{5}$.

When $\theta = \frac{\pi}{2} - \varphi$ $\left(\text{correspondingly, } x = \frac{2\sqrt{5}}{5},\ y = \frac{3\sqrt{5}}{5}\right)$, $x + y$ takes the maximum $\sqrt{5}$. □

Part II Word Problems (16 marks for Question 9, 20 marks for Question 10 and 11, and then 56 marks in total)

9 (16 marks) Given function $f(x) = |2 - \log_3 x|$, positive real numbers a, b, c satisfy $a < b < c$ and $f(a) = 2f(b) = 2f(c)$. Find the value of $\frac{ac}{b}$.

Solution Note that $f(x) = \left|\log_3\left(\frac{x}{9}\right)\right|$ is monotonically decreasing on $(0, 9]$ and monotonically increasing on $[9, +\infty)$.

By the conditions satisfied by a, b, c, we know that $0 < a < b < 9 < c$ and

$$\log_3\left(\frac{9}{a}\right) = 2\log_3\left(\frac{9}{b}\right) = 2\log_3\left(\frac{c}{9}\right).$$

Therefore,

$$\begin{aligned} \log_3\left(\frac{ac}{b}\right) &= \log_3\left(9 \cdot \frac{a}{9} \cdot \frac{9}{b} \cdot \frac{c}{9}\right) \\ &= 2 - \log_3\left(\frac{9}{a}\right) + \log_3\left(\frac{9}{b}\right) + \log_3\left(\frac{c}{9}\right) \\ &= 2, \end{aligned}$$

namely, $\frac{ac}{b} = 3^2 = 9$. □

10 (20 marks) Suppose $a, b \in \mathbb{R}$. If equation

$$(z^2 + az + b)(z^2 + az + 2b) = 0$$

about z has four mutually different complex roots z_1, z_2, z_3, z_4 and their corresponding points in the complex plane are exactly four vertices of a square with side length 1, then find the value of $|z_1| + |z_2| + |z_3| + |z_4|$.

Solution Denote quadratic equations $E_1 : z^2 + az + b = 0$, $E_2 : z^2 + az + 2b = 0$. Let z_1, z_2 be solutions of E_1 and z_3, z_4 be solutions of E_2.

If z_1, z_2, z_3, z_4 are all real numbers, then their corresponding points on the complex plane are all on the real axis, which is not consistent with the question. If z_1, z_2, z_3, z_4 are imaginary numbers, then their corresponding points on the complex plane are all on line $\operatorname{Re} z = -\frac{a}{2}$, which does not fit the question. Therefore, there are two real numbers and two imaginary numbers in z_1, z_2, z_3, z_4.

This shows that discriminant $a^2 - 4b$ of equation E_1 and the discriminant $a^2 - 8b$ of E_2 have different signs.

At this point, there must be $b > 0$ (if $b \leq 0$, then $a^2 - 4b \geq 0$ and $a^2 - 8b \geq 0$, a contradiction), so

$$a^2 - 4b \geq 0 > a^2 - 8b.$$

Hence, $z_{1,2} = \dfrac{-a \pm \sqrt{a^2 - 4b}}{2}$, $z_{3,4} = \dfrac{-a \pm \sqrt{8b - a^2}\mathrm{i}}{2}$.

It is evident that $\dfrac{z_1 + z_2}{2} = \dfrac{z_3 + z_4}{2} = -\dfrac{a}{2}$. Since the side length of the square is 1, there is

$$|z_1 - z_2| = \sqrt{a^2 - 4b} = \sqrt{2},$$
$$|z_3 - z_4| = \sqrt{8b - a^2} = \sqrt{2},$$

namely, $a^2 - 4b = 8b - a^2 = 2$, and the solutions are $a^2 = 6, b = 1$.

Noticing that z_1, z_2 have the same sign and $|z_3| = |z_4|$, we know that

$$\begin{aligned}|z_1| + |z_2| + |z_3| + |z_4| &= |z_1 + z_2| + 2|z_3| \\ &= |-a| + \sqrt{a^2 + (8b - a^2)} \\ &= \sqrt{6} + 2\sqrt{2}.\end{aligned}$$

□

11 (20 marks) As shown in Fig. 11.1, in a plane rectangular coordinate system xOy, the left and right foci of the ellipse $\Gamma : \dfrac{x^2}{2} + y^2 = 1$ are F_1, F_2, respectively. Let P be a point on Γ in the first quadrant, and the extensions of PF_1, PF_2 intersect Γ at points Q_1, Q_2, respectively. Let r_1, r_2 be the radii of the incircles of $\triangle PF_1Q_2$, $\triangle PF_2Q_1$, respectively. Find the maximum of $r_1 - r_2$.

Solution It is easy to find $F_1 = (-1, 0), F_2 = (1, 0)$.

Denote $P(x_0, y_0)$, $Q_1(x_1, y_1)$, $Q_2(x_2, y_2)$. By the given condition, it follows that

$$x_0, y_0 > 0, \quad y_1 < 0, \quad y_2 < 0.$$

By the definition of ellipse we get

$$|PF_1| + |PF_2| = |Q_1F_1| + |Q_1F_2| = |Q_2F_1| + |Q_2F_2| = 2\sqrt{2}.$$

Hence, the perimeters of $\triangle PF_1Q_2$ and $\triangle PF_2Q_1$ are both $l = 4\sqrt{2}$.

And since $|F_1F_2| = 2$,

$$r_1 = \frac{2S_{\triangle PF_1Q_2}}{l} = \frac{(y_0 - y_2) \cdot |F_1F_2|}{l} = \frac{y_0 - y_2}{2\sqrt{2}}.$$

Similarly, we can get $r_2 = \dfrac{y_0 - y_1}{2\sqrt{2}}$, so $r_1 - r_2 = \dfrac{y_1 - y_2}{2\sqrt{2}}$.

In the following, we will first find $y_1 - y_2$.

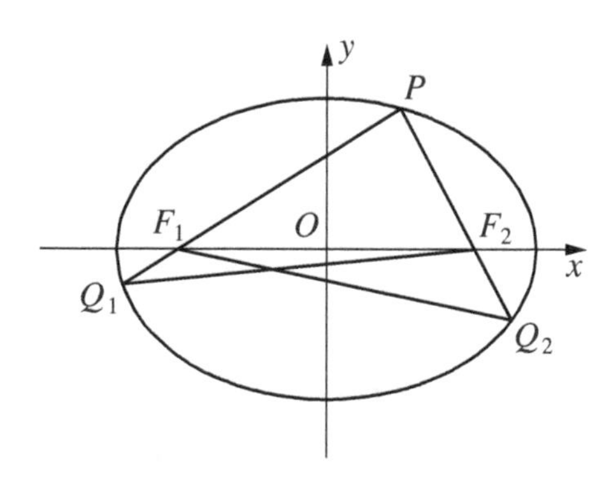

Fig. 11.1

The equation of line PF_1 is $x = \dfrac{(x_0+1)y}{y_0} - 1$. Substituting it into $\dfrac{x^2}{2} + y^2 = 1$ and rearranging it gives

$$\left(\frac{(x_0+1)^2}{2y_0^2}+1\right)y^2 - \frac{x_0+1}{y_0}y - \frac{1}{2} = 0.$$

Multiplying both sides by $2y_0^2$ and noticing that $x_0^2 + 2y_0^2 = 2$, we find that

$$(3+2x_0)y^2 - 2(x_0+1)y_0y - y_0^2 = 0.$$

The two roots of this equation are y_0 and y_1. By Vieta's formulas, we get $y_0y_1 = -\dfrac{y_0^2}{3+2x_0}$. Thus,

$$y_1 = -\frac{y_0}{3+2x_0}.$$

Similarly, we can get $y_2 = -\dfrac{y_0}{3-2x_0}$. Therefore,

$$y_1 - y_2 = \frac{y_0}{3-2x_0} - \frac{y_0}{3+2x_0} = \frac{4x_0y_0}{9-4x_0^2}.$$

Since $9 - 4x_0^2 = \dfrac{1}{2}x_0^2 + 9y_0^2 \geq 2\sqrt{\dfrac{1}{2}x_0^2 \cdot 9y_0^2} = 3\sqrt{2}x_0y_0$, there is

$$r_1 - r_2 = \frac{y_1-y_2}{2\sqrt{2}} = \frac{\sqrt{2}x_0y_0}{9-4x_0^2} \leq \frac{\sqrt{2}x_0y_0}{3\sqrt{2}x_0y_0} = \frac{1}{3},$$

where the equality sign holds when $\dfrac{1}{2}x_0^2 = 9y_0^2$ is required. Accordingly, $x_0 = \dfrac{3\sqrt{5}}{5}$, $y_0 = \dfrac{\sqrt{10}}{10}$.

Therefore, the maximum of $r_1 - r_2$ is $y_0 = \dfrac{1}{3}$. □

Test Paper A2
(8:00 – 9:20; October 23, 2021)

Part I Short-Answer Questions (Questions 1–8, eight marks each)

1. Suppose that geometric sequence $\{a_n\}$ satisfies $a_1 - a_2 = 3$, $a_1 - a_3 = 2$. Then the common ratio of $\{a_n\}$ is ________.

Solution Let the common ratio of $\{a_n\}$ be q. Then

$$a_1(1-q) = a_1 - a_2 = 3,$$
$$a_1(1-q^2) = a_1 - a_3 = 2,$$

and thus $1+q = \dfrac{a_1(1-q^2)}{a_1(1-q)} = \dfrac{2}{3}$. Therefore, $q = -\dfrac{1}{3}$. □

2 The maximum of $f(x) = 2\sin^2 x - \tan^2 x$ is ________.

Solution

$$\begin{aligned} f(x) &= 2(1-\cos^2 x) - \frac{1-\cos^2 x}{\cos^2 x} \\ &= 3 - \left(2\cos^2 x + \frac{1}{\cos^2 x}\right) \\ &\le 3 - 2\sqrt{2\cos^2 x \cdot \frac{1}{\cos^2 x}} = 3 - 2\sqrt{2}, \end{aligned}$$

When $2\cos^2 x = \dfrac{1}{\cos^2 x}$ $\left(\text{e.g., take } x = \arccos \dfrac{1}{\sqrt[4]{2}}\right)$, $f(x)$ takes the maximum $3 - 2\sqrt{2}$. □

3 If the height of a cone is 5 and the lateral surface area is 30π, then the volume of this cone is ________.

Solution Let the base radius of this cone be r and its slant height be l.

By the condition, there is $\pi r l = 30\pi$. Note that $l = \sqrt{r^2+5^2}$, and thus $r\sqrt{r^2+5^2} = rl = 30$. Squaring both sides yields $r^2(r^2+25) = 900$. In other words, we have $(r^2+45)(r^2-20) = 0$. Hence $r^2 = 20$.

Therefore, the volume of this cone is $\dfrac{1}{3}\pi r^2 \cdot 5 = \dfrac{100}{3}\pi$. □

4 Randomly select three vertices from the six vertices of a regular hexagon with side length 1. Then the probability that two of the three vertices are at a distance of $\sqrt{3}$ is ________.

Solution We will show that "two of the three vertices selected are at a distance of $\sqrt{3}$" is a certain event. The regular hexagon with edge length 1 is denoted by $A_1A_2A_3A_4A_5A_6$. If there exist two adjacent vertices are taken out of the three vertices selected, it may be set as A_1, A_2. Note that $A_1A_3 = A_1A_5 = A_2A_4 = A_2A_6 = \sqrt{3}$, so the third vertex must be at a distance of $\sqrt{3}$ from one of the vertices A_1, A_2. If any two of the three vertices selected are not adjacent to each other, it can only be A_1, A_3, A_5 or A_2, A_4, A_6. At this point, there are clearly two vertices with a distance of $\sqrt{3}$.

Therefore, the desired probability is 1. □

5 Complex numbers $z_1, z_2, \ldots, z_{100}$ satisfy $z_1 = 3+2\mathrm{i}$, $z_{n+1} = \overline{z_n} \cdot \mathrm{i}^n$, $(n = 1, 2, \ldots, 99)$, with i as the imaginary unit. Then the value of $z_{99} + z_{100}$ is ________.

Solution By the given conditions, we have

$$z_{n+2} = \overline{z_{n+1}} \cdot \mathrm{i}^{n+1} = \overline{\overline{z_n} \cdot \mathrm{i}^n} \cdot \mathrm{i}^{n+1} = z_n \mathrm{i} \quad (n = 1, 2, \ldots, 98).$$

And since $z_1 = 3 + 2\mathrm{i}$, $z_{99} = z_i \mathrm{i}^{49} = z_1 \mathrm{i} = -2 + 3\mathrm{i}$. Therefore,

$$\begin{aligned} z_{99} + z_{100} &= z_{99} + \overline{z_{99}} \cdot \mathrm{i}^{99} \\ &= (-2+3\mathrm{i}) + (-2-3\mathrm{i})(-\mathrm{i}) \\ &= -5 + 5\mathrm{i}. \end{aligned}$$ □

6 Function $f(x)$ with domain $\mathbb{R}$ satisfies: when $x \in [0,1)$, $f(x) = 2^x - x$, and for any real number x, there is $f(x) + f(x+1) = 1$. Denote $a = \log_2 3$, and then the value of expression $f(a) + f(2a) + f(3a)$ is ________.

Solution By the conditions, we know that $f(x+n) = 1 - f(x)$ when n is odd and $f(x+n) = f(x)$ when n is even.

Note that $a = \log_2 3 \in [1,2)$, $2a = \log_2 9 \in [3,4)$, $3a = \log_2 27 \in [4,5)$. Therefore,

$$\begin{aligned} f(a) + f(2a) + f(3a) &= 1 - f(a-1) + 1 - f(2a-3) + f(3a-4) \\ &= 2 - f\left(\log_2 \frac{3}{2}\right) - f\left(\log_2 \frac{9}{8}\right) + f\left(\log_2 \frac{27}{16}\right) \\ &= 2 - \left(\frac{3}{2} - \log_2 \frac{3}{2}\right) - \left(\frac{9}{8} - \log_2 \frac{9}{8}\right) \\ &\quad + \left(\frac{27}{16} - \log_2 \frac{27}{16}\right) \\ &= \frac{17}{16} + \left(\log_2 \frac{3}{2} + \log_2 \frac{9}{8} - \log_2 \frac{27}{16}\right) = \frac{17}{16}. \end{aligned}$$ □

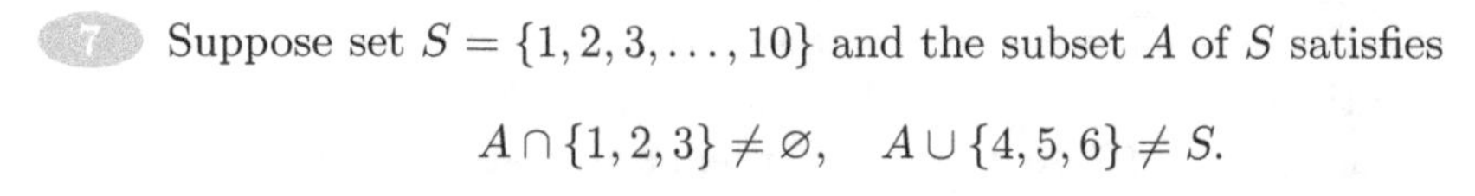

7 Suppose set $S = \{1, 2, 3, \ldots, 10\}$ and the subset A of S satisfies

$$A \cap \{1, 2, 3\} \neq \varnothing, \quad A \cup \{4, 5, 6\} \neq S.$$

The number of such subsets is ________.

Solution First, we will find the number N_1 of subsets A of S such that $A \cap \{1, 2, 3\} \neq \varnothing$ holds.

There are $2^3 - 1 = 7$ ways of selecting at least one element in $1, 2, 3$, while for each number in $4, 5, \ldots, 10$ there are two choices (selected or not selected). Therefore, $N_1 = 7 \times 2^7 = 896$.

Furthermore, we need to deduct the number of selections N_2 which makes $A \cup \{4, 5, 6\} = S$ hold. Of these selections of A, 1, 2, 3, 7, 8, 9, 10 are all selected. And each of 4, 5, 6 has two choices (selected or not selected), so $N_2 = 2^3 = 8$.

Therefore, the number of subsets A satisfying the conditions is $N_1 - N_2 = 896 - 8 = 888$. □

8 In a plane rectangular coordinate system xOy, Γ_1 is a unit circle centred at $(2, 1)$ and Γ_2 is a unit circle centred at $(10, 11)$. Make a line l through the origin O such that l has two intersections with each of Γ_1 and Γ_2, dividing Γ_2 and Γ_2 into four arcs, and two of these four arcs are of equal length. The sum of the slopes of all the lines l satisfying the conditions is ________.

Solution Denote the centres $(2, 1), (10, 11)$ of the two circles Γ_1, Γ_2 as T_1, T_2, respectively.

If l passes through T_1 or T_2, then l bisects the circumference of Γ_1 or that of Γ_2, which yields two equal arcs. The possible slopes of l at this point are $k_1 = k_{OT_1} = \frac{1}{2}$ or $k_2 = k_{OT_2} = \frac{11}{10}$.

If l neither passes through T_1 nor T_2, then Γ_1 and Γ_2 are both divided into two arcs of unequal length by l. And since Γ_1 and Γ_2 are equal circles, the two arcs divided in Γ_1 are equal to the two arcs divided in Γ_2, respectively. This implies that l is parallel to T_1T_2 or passes through its midpoint $M(6, 6)$. The possible slopes of l at this point are $k_3 = k_{T_1T_2} = \frac{5}{4}$ or $k_4 = k_{OM} = 1$.

After checking, line $y = k_1 x$ has no intersection with circle Γ_2, which does not fit the question; when $i = 2, 3, 4$, line $y = k_i x$ has two intersections with each of Γ_1, Γ_2, which is consistent with the question. Therefore, the sum of the slopes of all the lines l satisfying the conditions is

$$k_2 + k_3 + k_4 = \frac{11}{10} + \frac{5}{4} + 1 = \frac{67}{20}.$$ □

Part II Word Problems (16 marks for Question 9, 20 marks for Question 10 and 11, and then 56 marks in total)

9 (16 marks) It is known that $\triangle ABC$ satisfies $AB = 1$, $AC = 2$ and $\cos B + \sin C = 1$. Find the length of side BC.

Solution Denote $a = BC$, $b = AC$, $c = AB$, so $b = 2$, $c = 1$.

By the law of sines, we have $\frac{\sin B}{\sin C} = \frac{b}{c} = 2$, namely, $\sin B = 2\sin C$. And since $\cos B = 1 - \sin C$, there is

$$(2\sin C)^2 + (1 - \sin C)^2 = \sin^2 B + \cos^2 B = 1,$$

and simplifying it gives $5\sin^2 C - 2\sin C = 0$. And because $\sin C \neq 0$, $\sin C = \frac{2}{5}$.

And then we have $\cos B = 1 - \sin C = \frac{3}{5}$.

By the law of cosines, we have $\cos B = \frac{a^2 + c^2 - b^2}{2ac}$, and thus $\frac{3}{5} = \frac{a^2 - 3}{2a}$, i.e.,

$$a^2 - \frac{6}{5}a - 3 = 0.$$

Since $a > 0$, we get the solution $a = \frac{3 + 2\sqrt{21}}{5}$, namely the length of side BC is $\frac{3 + 2\sqrt{21}}{5}$. □

10 (20 marks) In a plane rectangular coordinate system xOy, given ellipse

$$\Gamma : \frac{x^2}{a^2} + \frac{y^2}{b^2} = 1 \quad (a > b > 0),$$

let A be one of the vertices of its major axis, B be one of the vertices of its minor axis, and F be one of its foci. It is known that there exist two points P and Q on Γ that are symmetric about O such that

$$\overrightarrow{FP} \cdot \overrightarrow{FQ} + \overrightarrow{FA} \cdot \overrightarrow{FB} = |AB|^2.$$

(1) Prove that the focal point F lies on the extension of AO;
(2) find the range of the eccentricity of Γ.

Solution By symmetry, it is useful to set $A(a,0)$, $B(0,b)$. Denote $c = |OF| = \sqrt{a^2-b^2}$ and the eccentricity $e = \dfrac{c}{a}$.

Let $|OP| = r$. By the properties of ellipse, we know that the range of r is $[b,a]$.

Note that O is the midpoint of PQ, and thus

$$\overrightarrow{FP}\cdot\overrightarrow{FQ} = (\overrightarrow{FO}+\overrightarrow{OP})\cdot(\overrightarrow{FO}-\overrightarrow{OP}) = \overrightarrow{FO}^2 - \overrightarrow{OP}^2 = c^2 - r^2. \quad ①$$

(1) Suppose the focal point F is $(c,0)$, and then

$$\overrightarrow{FA}\cdot\overrightarrow{FB} = (a-c,0)\cdot(-c,b) = c^2 - ac < 0.$$

Thus

$$\overrightarrow{FP}\cdot\overrightarrow{FQ} + \overrightarrow{FA}\cdot\overrightarrow{FB} < c^2 - r^2 < a^2 + b^2 = |AB|^2,$$

and this is not consistent with the conditions. Hence F is $(-c,0)$, which shows that F lies on the extension of AO.

(2) From $F(-c,0)$, we know that

$$\overrightarrow{FA}\cdot\overrightarrow{FB} = (a+c,0)\cdot(c,b) = c^2 + ac.$$

Combining with ① gives

$$\begin{aligned}\overrightarrow{FP}\cdot\overrightarrow{FQ} + \overrightarrow{FA}\cdot\overrightarrow{FB} &= (c^2-r^2)+(c^2+ac)\\ &= 2c^2+ac-r^2,\end{aligned}$$

so $2c^2+ac-r^2 = a^2+b^2$, i.e.,

$$r^2 = 2c^2+ac-a^2-b^2 = 3c^2+ac-2a^2.$$

Note that $r\in[b,a]$, and thus $3c^2+ac-2a^2 = r^2 \in \left[a^2-c^2, a^2\right]$.

The above equation is equivalent to $\dfrac{a^2-c^2}{a^2} \le \dfrac{3c^2+ac-2a^2}{a^2} \le 1$, i.e.,

$$1-e^2 \le 3e^2+e-2 \le 1.$$

Combining $e\in[0,1]$, we obtain the range of the eccentricity of Γ, namely, $e\in\left[\dfrac{3}{4}, \dfrac{-1+\sqrt{37}}{6}\right]$. □

11 (20 marks) Let a, b be real numbers and function $f(x) = x^3+ax^2+bx$. If there exist three real numbers x_1, x_2, x_3 satisfying $x_1+1\le x_2\le x_3-1$ and $f(x_1)=f(x_2)=f(x_3)$. Find the minimum of $|a|+2|b|$.

Solution For function $f(x)$ and real numbers x_1, x_2, x_3 satisfying the conditions, let $f(x_1) = f(x_2) = f(x_3) = c$, and then x_1, x_2, x_3 are the three real roots of cubic equation $x^3 + ax^2 + bx - c = 0$. By Vieta's formulas, we know that

$$a = -(x_1 + x_2 + x_3), \quad b = x_1x_2 + x_2x_3 + x_3x_1.$$

Further, by the condition we know that $x_2 - x_1 \geq 1$, $x_3 - x_2 \geq 1$, $x_3 - x_1 \geq 2$, and thus

$$\begin{aligned}
a^2 - 3b &= x_1^2 + x_2^2 + x_3^2 - x_1x_2 - x_2x_3 - x_3x_1 \\
&= \frac{1}{2}((x_1 - x_2)^2 + (x_2 - x_3)^2 + (x_3 - x_1)^2) \\
&\geq \frac{1}{2}(1 + 1 + 4) = 3,
\end{aligned}$$

i.e., $b \leq \dfrac{a^2}{3} - 1$.

If $|a| \geq \sqrt{3}$, then $|a| + 2|b| \geq |a| \geq \sqrt{3}$.

If $0 \leq |a| < \sqrt{3}$, then $b \leq \dfrac{a^2}{3} - 1 < 0$. Notice that at this point

$$|a| - \frac{3}{4} < \sqrt{3} - \frac{3}{4},$$

$$\begin{aligned}
|a| + 2|b| &\geq |a| + \frac{2}{3}(3 - a^2) \\
&= -\frac{2}{3}\left(|a| - \frac{3}{4}\right)^2 + \frac{19}{8} \\
&> -\frac{2}{3}\left(\sqrt{3} - \frac{3}{4}\right)^2 + \frac{19}{8} = \sqrt{3}.
\end{aligned}$$

The above shows that $|a| + 2|b| \geq \sqrt{3}$.

On the other hand, when $a = \sqrt{3}, b = 0$, real numbers $x_1 = -1 - \dfrac{\sqrt{3}}{3}$, $x_2 = -\dfrac{\sqrt{3}}{3}$, $x_3 = 1 - \dfrac{\sqrt{3}}{3}$ satisfy the condition because the values of F_1, F_2, F_3 are all $\dfrac{2\sqrt{3}}{9}$. At this point $|a| + 2|b| = \sqrt{3}$.

In conclusion, the minimum of $|a| + 2|b|$ is $\sqrt{3}$. □

Remark Another proof of inequality $b \leq \frac{a^2}{3} - 1$ is as follows.

Since $f(x)$ is a cubic function, the coefficient of its x^3 term is 1, and $f(x_1) = f(x_2) = f(x_3)$. We can let

$$f(x) = (x - x_1)(x - x_2)(x - x_3) + f(x_1).$$

And by $x_1 + 1 < x_2 \leq x_3 - 1$ it follows that $x_2 - 1 \in [x_1, x_2), x_2 + 1 \in (x_2, x_3]$, and thus

$$f(x_2 - 1) \geq f(x_1) \geq f(x_2 + 1).$$

Therefore, $f(x_2 + 1) - f(x_2 - 1) \leq 0$. By simplification, we get $1 + 3x_2^2 + a \cdot 2x_2 + b \leq 0$. Consequently,

$$b \leq -3x_2^2 - a \cdot 2x_2 - 1 = -3\left(x_2 + \frac{a}{3}\right)^2 + \frac{a^2}{3} - 1 \leq \frac{a^2}{3} - 1.$$

Test Paper B
(8:00 – 9:20; September 12, 2021)

Part I Short-Answer Questions (Questions 1–8, eight marks each)

1. Given arithmetic sequence $\{a_n\}$ with common difference $d \neq 0$ and $a_{2021} = a_{20} + a_{21}$, then the value of $\frac{a_1}{d}$ is ________.

Solution By the conditions, we have $a_1 + 2020d = a_1 + 19d + a_1 + 20d$. And since $d \neq 0$, $\frac{a_1}{d} = 1981$. □

2. Suppose m is a real number, and complex numbers $z_1 = 1 + 2\mathrm{i}$, $z_2 = m + 3\mathrm{i}$, where i is the imaginary unit. If $z_1 \cdot \overline{z_2}$ is purely imaginary, then the value of $|z_1 + z_2|$ is ________.

Solution Since $z_1 \cdot \overline{z_2} = (1 + 2\mathrm{i})(m - 3\mathrm{i}) = m + 6 + (2m - 3)\mathrm{i}$ is purely imaginary, we get $m = -6$. Therefore, $|z_1 + z_2| = |-5 + 5\mathrm{i}| = 5\sqrt{2}$. □

3. The range of $y = \sin^2 x + \sqrt{3} \sin x \cos x$ when $\frac{\pi}{4} \leq x \leq \frac{\pi}{2}$ is ________.

Solution

$$\begin{aligned} y &= \sin^2 x + \sqrt{3}\sin x \cos x \\ &= \frac{1-\cos 2x}{2} + \frac{\sqrt{3}}{2}\sin 2x \\ &= \sin\left(2x - \frac{\pi}{6}\right) + \frac{1}{2}. \end{aligned}$$

When $\frac{\pi}{4} \le x \le \frac{\pi}{2}$, there is $\frac{\pi}{3} \le 2x - \frac{\pi}{6} \le \frac{5\pi}{6}$, and thus the range of $\sin\left(2x - \frac{\pi}{6}\right)$ is $\left[\frac{1}{2}, 1\right]$. Therefore, the range of $y = \sin^2 x + \sqrt{3}\sin x \cos x$ is $\left[1, \frac{3}{2}\right]$. □

4 Suppose the domain of function $f(x)$ is $D = (-\infty, 0) \cup (0, +\infty)$ and there is $f(x) = \dfrac{f(1)\cdot x^2 + f(2)\cdot x - 1}{x}$ for any $x \in D$. Then the sum of all the zeros of $f(x)$ is ________.

Solution Let x_1, x_2 and we get

$$\begin{aligned} f(1) &= f(1) + f(2) - 1, \\ f(2) &= 2f(1) + f(2) - \frac{1}{2}, \end{aligned}$$

and the solutions are $f(2) = 1$, $f(1) = \frac{1}{4}$. Therefore,

$$f(x) = \frac{1}{x}\cdot\left(\frac{1}{4}x^2 + x - 1\right) \quad (x \ne 0).$$

Let $f(x) = 0$ and we get $x = -2 \pm 2\sqrt{2}$, so the sum of all the zeros of $f(x) = 0$ is -4. □

5 Suppose $a, b, c > 1$ and $(a^2 b)^{\log_a c} = a \cdot (ac)^{\log_a b}$ is satisfied. Then the value of $\log_c(ab)$ is ________.

Solution Taking the logarithm of the original equation with respect to an arbitrary base a on both sides, we get

$$\log_a c \cdot (2 + \log_a b) = 1 + \log_a b \cdot (1 + \log_a c).$$

Simplifying the above equation gives $2\log_a c = 1 + \log_a b$. Therefore, $c^2 = ab$, and then $\log_c(ab) = \log_c c^2 = 2$. □

6 In $\triangle ABC$, $AB = 1$, $AC = 2$ and $\cos B = 2\sin C$. Then the length of side BC is ________.

Solution By the law of sines, we have $\dfrac{\sin B}{\sin C} = \dfrac{AC}{AB} = 2$. Thus,

$$\sin B = 2\sin C = \cos B,$$

namely, $\tan B = 1$. Therefore, $B = \dfrac{\pi}{4}$.

Let $BC = a > 0$. By the law of cosines, we get $4 = 1 + a^2 - 2a \cdot \dfrac{\sqrt{2}}{2}$. Consequently, $a = \dfrac{\sqrt{2}+\sqrt{14}}{2}$. □

7 In a plane rectangular coordinate system xOy, the graph of parabola $y = ax^2 - 3x + 3$ $(a \neq 0)$ and that of parabola $y^2 = 2px$ $(p > 0)$ are symmetric with respect to line $y = x + m$. Then the product of real numbers a, p, m is ________.

Solution For any point (x_0, y_0) on parabola $y = ax^2 - 3x + 3$ $(a \neq 0)$, there is

$$y_0 = ax_0^2 - 3x_0 + 3. \quad ①$$

Suppose the symmetric point of (x_0, y_0) with respect to line $y = x + m$ is (x_1, y_1).

By $\dfrac{y_1 + y_0}{2} = \dfrac{x_1 + x_0}{2} + m$, $x_1 + y_1 = x_0 + y_0$, it follows that

$$x_1 = y_0 - m, \quad y_1 = x_0 + m.$$

Since (x_1, y_1) lies on parabola $y^2 = 2px$, there is $(x_0+m)^2 = 2p(y_0-m)$. This is equivalent to

$$y_0 = \frac{1}{2p}x_0^2 + \frac{m}{p}x_0 + \frac{m^2}{2p} + m. \quad ②$$

Due to the arbitrary taking of (x_0, y_0), comparing ① and ② gives

$$\frac{1}{2p} = a, \quad \frac{m}{p} = -3, \quad \frac{m^2}{2p} + m = 3.$$

Therefore, we have $3 = \frac{m}{p} \cdot \frac{m}{2} + m = -3 \cdot \frac{m}{2} + m = -\frac{m}{2}$, and its solution is $m = -6$. Hence, in order, we can get $p = 2$, $a = \frac{1}{4}$, satisfying $a \neq 0$ and $p > 0$. Consequently, $apm = 2 \cdot \frac{1}{4} \cdot (-6) = -3$. □

8 Let $a_1, a_2, \ldots, a_{10}$ be a random permutation of $1, 2, \ldots, 10$. Then the probability of both 9 and 12 appearing in the 9 numbers $a_1a_2, a_2a_3, \ldots, a_9a_{10}$ is ________.

Solution A random permutation satisfies the requirement if and only if in this permutation, numbers 1 and 9 are adjacent, and at least one pair of numbers $2, 6$ and $3, 4$ is adjacent. Now we evaluate the number of such permutations N. Suppose the number of permutations $2, 6$ being adjacent is N_1, the number of permutations $3, 4$ being adjacent is N_2, and the number of permutations both $2, 6$ and $3, 4$ being adjacent is N_3.

To determine N_1, we first "bind" the numbers $1, 9$, and $2, 6$. Then arrange them with the remaining 6 numbers, and there are 8! ways. Considering also the order of 1, 9 and that of 2, 6, we get

$$N_1 = 2^2 \times 8!.$$

Similarly, we can find $N_2 = 2^2 \times 8!$, $N_3 = 2^3 \times 7!$.

By the inclusion-exclusion principle, $N = N_1 + N_2 - N_3 = 8 \times (8! - 7!) = 7 \times 8!$.

Therefore, the desired probability is $\frac{N}{10!} = \frac{7}{90}$. □

Part II Word Problems (16 marks for Question 9, 20 marks for Question 10 and 11, and then 56 marks in total)

9 (16 marks) Sequence $\{a_n\}$ satisfies $a_1 = a_2 = a_3$. Let

$$b_n = a_n + a_{n+1} + a_{n+2} (n \in \mathbb{N}_+).$$

If $\{b_n\}$ is a geometric sequence with common ratio 3, find the value of a_{100}.

Solution By the condition, we know that $b_n = b_1 \cdot 3^{n-1} = 3^n (n \in \mathbb{N}_+)$. Thus,

$$a_{n+3} - a_n = b_{n+1} - b_n = 3^{n+1} - 3^n = 2 \cdot 3^n \quad (n \in \mathbb{N}_+).$$

Therefore,

$$\begin{aligned} a_{100} &= a_1 + \sum_{k=1}^{33}(a_{3k+1} - a_{3k-2}) \\ &= 1 + \sum_{k=1}^{33} 2 \cdot 3^{3k-2} \\ &= 1 + 6 \cdot \frac{27^{33} - 1}{27 - 1} \\ &= 1 + \frac{3}{13}(3^{99} - 1) = \frac{3^{100} + 10}{13}. \end{aligned}$$

□

10 (20 marks) In a plane rectangular coordinate system xOy, the graph of function $y = \frac{1}{|x|}$ is Γ. Let points P, Q on Γ satisfy: P is in the first quadrant, Q is in the second quadrant, and line PQ is tangent to the part of Γ in the second quadrant at point Q. Find the minimum of $|PQ|$.

Solution When $x > 0$, $y = \frac{1}{x}$. When $x < 0$, $y = -\frac{1}{x}$, and its corresponding derivative is $y' = -\frac{1}{x^2}$.

Suppose $Q\left(-a, \frac{1}{a}\right)$, where $a > 0$. By the condition, the slope of PQ is $y'|_{x=-a} = \frac{1}{a^2}$.

The equation of line PQ is $y = \frac{1}{a^2}(x + a) + \frac{1}{a} = \frac{x + 2a}{a^2}$.

Combining the above equation with $y = \frac{1}{x}(x > 0)$ yields $x^2 + 2ax - a^2 = 0$, and thus we know the abscissa $x_P = (\sqrt{2} - 1)a$ of point P, with the negative root discarded. Therefore,

$$\begin{aligned} |PQ| &= \sqrt{1 + \left(\frac{1}{a^2}\right)^2} \cdot |x_P - x_Q| \\ &= \sqrt{1 + \frac{1}{a^4}} \cdot \sqrt{2}a \\ &\geq \sqrt{2\sqrt{1 \cdot \frac{1}{a^4}}} \cdot \sqrt{2}a = 2. \end{aligned}$$

When $a = 1$, namely, $Q(-1, 1)$, the minimum of $|PQ|$ is 2. □

11 (20 marks) In regular pyramid $P - A_1A_2\cdots A_n$ $(n \geq 3)$, O is the centre of the regular n-sided polygon $A_1A_2\cdots A_n$ base and B is the midpoint of edge A_1A_n.

(1) Prove that $PO^2 \sin\dfrac{\pi}{n} + PA_1^2 \cos^2\dfrac{\pi}{n} = PB^2$;

(2) For regular pyramid $P - A_1A_2\cdots A_n$, let the angle formed by the lateral edge and the base be α and the angle formed by the lateral face and the base be β. Try to determine the magnitude relation between $\dfrac{1}{n}\sum\limits_{i=1}^{n}\cos\angle A_iPB$ and $\sin\alpha\sin\beta$, and then prove it.

Solution (1) Since PO is perpendicular to base $A_1A_2\cdots A_n$, it follows that $\angle POA_1 = \angle POB = 90°$.

Let $OA_1 = r$, and then $OB = OA_1 \cdot \cos\angle A_1OB = r\cos\dfrac{\pi}{n}$. Hence,

$$r^2 + PO^2 = PA_1^2, \quad r^2\cos^2\frac{\pi}{n} + PO^2 = PB^2.$$

Eliminating r^2 gives $PO^2\left(1 - \cos^2\dfrac{\pi}{n}\right) = PB^2 - PA_1^2\cos^2\dfrac{\pi}{n}$, namely,

$$PO^2\sin\frac{\pi}{n} + PA_1^2\cos^2\frac{\pi}{n} = PB^2.$$

(2) By the given condition, it follows that $\overrightarrow{PO}\cdot\overrightarrow{OA_i} = 0$ $(i = 1, 2, \ldots, n)$, $\overrightarrow{PO}\cdot\overrightarrow{OB} = 0$.

Suppose the length of the lateral edge of the regular pyramid is l. Then

$$\begin{aligned}
l\cdot|PB|\cdot\sum_{i=1}^{n}\cos\angle A_iPB &= \sum_{i=1}^{n}|PA_i|\cdot|PB|\cdot\cos\angle A_iPB \\
&= \sum_{i=1}^{n}\overrightarrow{PA_i}\cdot\overrightarrow{PB} \\
&= \sum_{i=1}^{n}(\overrightarrow{PO} + \overrightarrow{OA_i})\cdot(\overrightarrow{PO} + \overrightarrow{OB}) \\
&= \sum_{i=1}^{n}(\overrightarrow{PO}^2 + \overrightarrow{OA_i}\cdot\overrightarrow{OB}) \\
&= n\overrightarrow{PO}^2 + \overrightarrow{OB}\cdot\sum_{i=1}^{n}\overrightarrow{OA_i} \\
&= n\cdot|PO|^2.
\end{aligned}$$

The last step in the above equation makes use of $\vec{s}=\sum\limits_{i=1}^{n}\overrightarrow{OA_i}=\vec{0}$. (Since O is the centre of the regular n-sided polygon $A_1A_2\cdots A_n$ and each $\overrightarrow{OA_i}$ $(i=1,2,\ldots,n)$ is still a permutation of these vectors after $\frac{2\pi}{n}$ anticlockwise rotation, their sum vector $\vec{s}$ is invariant after the rotation, and thus $\vec{s}$ can only be the zero vector.) Hence,

$$\begin{aligned}\frac{1}{n}\sum_{i=1}^{n}\cos\angle A_iPB &= \frac{|PO|^2}{l\cdot|PB|}=\frac{PO}{l}\cdot\frac{PO}{PB}\\ &= \sin\angle PA_1O\cdot\sin\angle PBO. \qquad ①\end{aligned}$$

It is clear that $\angle PA_1O$ is the angle formed by the lateral edge and the base. And by $OB\perp A_1A_n$, $PB\perp A_1A_n$, we know that $\angle PBO$ is the angle formed by the lateral face and the base. Thus, $\angle PA_1O=\alpha$, $\angle PBO=\beta$.

Therefore, by ①, we obtain that $\frac{1}{n}\sum\limits_{i=1}^{n}\cos\angle A_iPB$ and $\sin\alpha\sin\beta$ are equal. □

Test Paper B1
(8:00 – 9:20; October 6, 2021)

Part I Short-Answer Questions (Questions 1–8, eight marks each)

1 Suppose $f(x)$ is an odd function with domain $\mathbb{R}$. If $f(1)=2$, $f(2)=3$, then the value of $f(f(-1))$ is ________.

Solution By the given conditions we have $f(-1)=-f(1)=-2$. Therefore,

$$f(f(-1))=f(-2)=-f(2)=-3. \qquad □$$

2 Let m be a real number. If the real and imaginary parts of complex $z=1+\mathrm{i}+\frac{m}{1+\mathrm{i}}$, with i being the imaginary unit, are greater than zero, then the range of m is ________.

Solution After calculation, we get $z=1+\mathrm{i}+\frac{(1-\mathrm{i})m}{2}=\frac{2+m}{2}+\frac{2-m}{2}\mathrm{i}$.

By the condition, it follows that $\frac{2+m}{2}>0$, $\frac{2-m}{2}>0$, and then we find the solution is $-2<m<2$. □

3 Suppose $A = \{1, 2, 3\}$, $B = \{4x - y \mid x, y \in A\}$, $C = \{4x + y \mid x, y \in A\}$. Then the sum of all the elements of $B \cap C$ is ________.

Solution When x and y take all the elements of A, respectively, $4x - y$ gets exactly the values 1, 2, 3, 5, 6, 7, 9, 10, 11, and $4x + y$ gets exactly the values 5, 6, 7, 9, 10, 11, 13, 14, 15.

Hence, $B \cap C = \{5, 6, 7, 9, 10, 11\}$. Then the sum of all the elements of $B \cap C$ is 18. □

4 In a plane rectangular coordinate system xOy, given parabola $\Gamma : y^2 = 2px$ $(p > 0)$, a line with inclination angle $\frac{\pi}{4}$ intersects Γ at point $P(3, 2)$ and another point Q. Then the area of $\triangle OPQ$ is ________.

Solution Since point $P(3, 2)$ is on Γ, we have $2p = \frac{4}{3}$.

The slope of the mentioned line is 1 and it passes through point $P(3, 2)$. Therefore, its equation is $y = x - 1$, and it passes through point $A(1, 0)$ on the x-axis. Substitute $x = y + 1$ into $y^2 = \frac{4}{3}x$, eliminating x and arranging it gives

$$3y^2 - 4y - 4 = 0.$$

The solution is $y_1 = 2, y_2 = -\frac{2}{3}$. Further, we can get

$$S_{\triangle OPQ} = S_{\triangle OAP} + S_{\triangle OAQ} = \frac{1}{2}|OA|\,(|y_1| + |y_2|) = \frac{4}{3}. \qquad \square$$

5 Sequence $\{a_n\}$ satisfies $a_1 = 2$ and $a_{n+1} = (n+1)a_n - n, n = 1, 2, \ldots$. Then the general term formula of $\{a_n\}$ is ________.

Solution By the condition, we have $a_{n+1} - 1 = (n+1)(a_n - 1)$. Therefore,

$$\begin{aligned} a_n - 1 &= n(a_{n-1} - 1) = n(n-1)(a_{n-2} - 1) = \cdots \\ &= n(n-1)\cdots 2(a_1 - 1) = n!, \end{aligned}$$

namely, $a_n = n! + 1$. □

6 In right square pyramid $P{-}ABCD$, point G is the centroid of the lateral face $\triangle PBC$. Let V_1, V_2 be the volumes of tetrahedrons $PABG$, $PADG$, respectively. Then the value of $\frac{V_1}{V_2}$ is ________.

Solution As shown in Fig. 6.1, take the midpoint M of BC, and then G lies on the median PM of $\triangle PBC$.

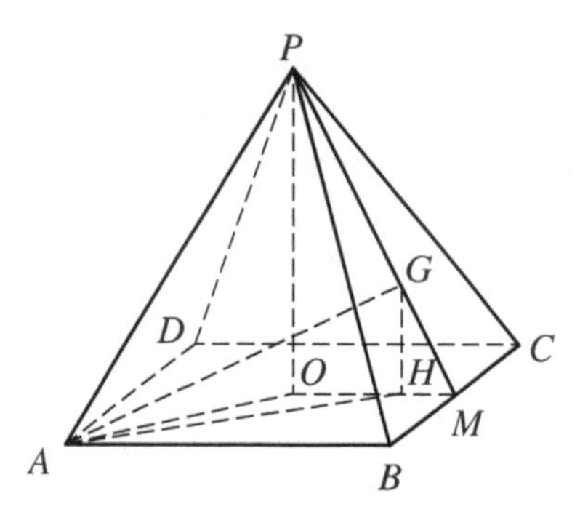

Fig. 6.1

Note that

$$V_1 = V_{G-PAB} = \frac{PG}{PM} \cdot V_{M-PAB},$$

$$V_2 = V_{G-PAD} = \frac{PG}{PM} \cdot V_{M-PAD}.$$

Therefore,

$$\frac{V_1}{V_2} = \frac{V_{M-PAB}}{V_{M-PAD}} = \frac{V_{P-MAB}}{V_{P-MAD}} = \frac{S_{\triangle MAB}}{S_{\triangle MAD}} = \frac{1}{2}. \qquad \square$$

7 Suppose $\alpha, \beta \geq 0$, $\alpha + \beta \leq 2\pi$. Then the minimum of $\sin\alpha + 2\cos\beta$ is ________.

Solution When $0 \leq \alpha \leq \pi$, $\sin\alpha + 2\cos\beta \geq 0 + 2\cdot(-1) = -2$.

When $\pi < \alpha \leq 2\pi$, there is $0 \leq \beta \leq 2\pi - \alpha < \pi$. At this point, as β gets bigger, $\cos\beta$ gets smaller. Therefore,

$$\begin{aligned}\sin\alpha + 2\cos\beta &\geq \sin\alpha + 2\cos(2\pi - \alpha)\\ &= \sin\alpha + 2\cos\alpha\\ &= \sqrt{5}\sin(\alpha + \varphi),\end{aligned}$$

where $\varphi = \arcsin\dfrac{2\sqrt{5}}{5}$.

When $\alpha = \dfrac{3\pi}{2} - \varphi$, $\beta = 2\pi - \alpha = \dfrac{\pi}{2} + \varphi$, $\sin\alpha + 2\cos\beta$ gets the minimum $\sqrt{5}$. $\square$

8 Let $a_1, a_2, \ldots, a_{10}$ be a permutation of $1, 2, \ldots, 21$, satisfying

$$|a_{20} - a_{21}| \geq |a_{19} - a_{21}| \geq |a_{18} - a_{21}| \geq \cdots \geq |a_1 - a_{21}|$$

The number of such permutations is ________.

Solution For a given $k \in \{1, 2, \ldots, 21\}$, we consider the number of permutations N_k satisfying the condition such that $a_{21} = k$.

When $k \in \{1, 2, \ldots, 11\}$, for $i = 1, 2, \ldots, k-1$, a_{2i-1}, a_{2i} are permutations of $k-i, k+i$ (if $k = 1$, there is no such i), and $a_j = j+1$ $(2k-1 \leq j \leq 20)$ (if $k = 11$, there is no such j). Therefore, $N_k = 2^{k-1}$.

Similarly, when $k \in \{12, 13, \ldots, 21\}$, there is $N_k = 2^{21-k}$.

Therefore, the number of permutations satisfying the condition is

$$\begin{aligned}\sum_{k=1}^{21} N_k &= \sum_{k=1}^{11} 2^{k-1} + \sum_{k=12}^{21} 2^{21-k} \\ &= (2^{11} - 1) + (2^{10} - 1). \\ &= 3070.\end{aligned}$$

□

Part II Word Problems (16 marks for Question 9, 20 marks for Question 10 and 11, and then 56 marks in total)

9 (16 marks) Suppose the included angle between non-zero vectors $\vec{a}$ and $\vec{b}$ in the plane is $\frac{\pi}{3}$. If $|\vec{a}|$, $|\vec{b}|$, $|\vec{a}+\vec{b}|$ form arithmetic sequence in order, find the value of $|\vec{a}| : |\vec{b}| : |\vec{a}+\vec{b}|$.

Solution Denote $s = |\vec{a}|$, $t = |\vec{b}|$, and then $s, t > 0$. Note that the included angle between $\vec{a}$ and $\vec{b}$ is $\frac{\pi}{3}$, and we have

$$\begin{aligned}|\vec{a}+\vec{b}|^2 &= \vec{a}^2 + (\vec{b})^2 + 2\vec{a}\cdot\vec{b} \\ &= s^2 + t^2 + 2st\cos\frac{\pi}{3} \\ &= s^2 + t^2 + st.\end{aligned}$$

By the condition, we know that $s, t, \sqrt{s^2+t^2+st}$ form arithmetic sequence in order. Then there is $\sqrt{s^2+t^2+st} = 2t - s$. Squaring and arranging the above equation gives $5st - 3t^2 = 0$. And since $t \neq 0$, it follows that $5s = 3t$, i.e., $s : t = 3 : 5$.

Therefore, $|\vec{a}| : |\vec{b}| : |\vec{a}+\vec{b}| = 3 : 5 : 7$. □

10 (20 marks) Given function $f(x) = |2 - \log_3 x|$, positive real numbers a, b, c satisfy $a < b < c$ and $f(a) = 2f(b) = 2f(c)$. Find the minimum of $\frac{ac}{b}$.

Solution Notice that $f(x) = \left|\log_3\left(\frac{x}{9}\right)\right|$ is monotonically decreasing on $(0, 9]$ and monotonically increasing on $[9, +\infty)$.

By the conditions satisfied by a, b, c, we have $0 < a < b < 9 < c$ and

$$\log_3\left(\frac{9}{a}\right) = 2\log_3\left(\frac{9}{b}\right) = 2\log_3\left(\frac{c}{9}\right).$$

Therefore,

$$\log_3\left(\frac{ac}{b}\right) = \log_3\left(9 \cdot \frac{a}{9} \cdot \frac{9}{b} \cdot \frac{c}{9}\right) = 2 - \log_3\left(\frac{9}{a}\right) + \log_3\left(\frac{9}{b}\right) + \log_3\left(\frac{c}{9}\right) = 2,$$

namely, $\frac{ac}{b} = 3^2 = 9$. □

11 (20 marks) As shown in Fig. 11.1, in a plane rectangular coordinate system xOy, the left and right foci of ellipse $\Gamma : \frac{x^2}{2} + y^2 = 1$ are F_1, F_2, respectively. Let P be a point on Γ in the first quadrant and the extensions of PF_1, PF_2 intersect Γ at points $Q_1(x_1, y_1)$, $Q_2(x_2, y_2)$, respectively.

Find the maximum of $y_1 - y_2$.

Solution As shown in Fig. 11.1, we find $F_1(-1, 0)$, $Q_2(1, 0)$.

Denote $P(x_0, y_0)$. By the condition, it follows that $x_0, y_0 > 0$, $y_1 < 0$, $y_2 < 0$.

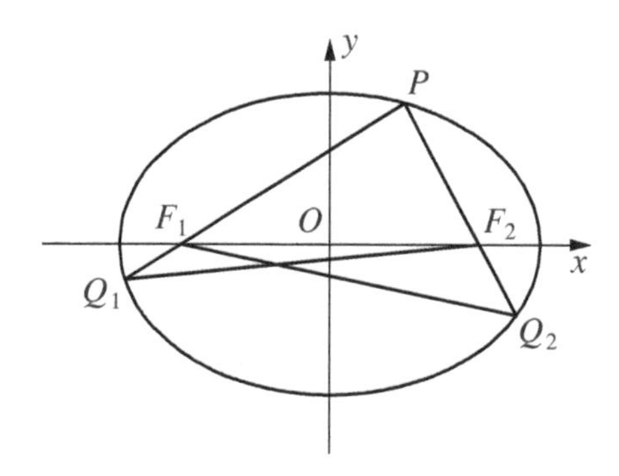

Fig. 11.1

The equation of line PF_1 is $x = \dfrac{(x_0+1)\,y}{y_0} - 1$. Substituting it into $\dfrac{x^2}{2} + y^2 = 1$ and organizing it yields

$$\left(\frac{(x_0+1)^2}{2y_0^2}+1\right)y^2 - \frac{x_0+1}{y_0}y - \frac{1}{2} = 0.$$

Multiplying both sides by $2y_0^2$ and noting that $x_0^2 + 2y_0^2 = 2$, we get

$$(3+2x_0)y^2 - 2(x_0+1)y_0 y - y_0^2 = 0.$$

The two roots of this equation are y_0, y_1. By Vieta's formulas, we get $y_0 y_1 = -\dfrac{y_0^2}{3+2x_0}$. Thus,

$$y_1 = -\frac{y_0}{3+2x_0}.$$

Similarly, we can get $y_2 = -\dfrac{y_0}{3-2x_0}$. Therefore,

$$y_1 - y_2 = \frac{y_0}{3-2x_0} - \frac{y_0}{3+2x_0} = \frac{4x_0 y_0}{9-4x_0^2}.$$

Since $9 - 4x_0^2 = \dfrac{1}{2}x_0^2 + 9y_0^2 \geq 2\sqrt{\dfrac{1}{2}x_0^2 \cdot 9y_0^2} = 3\sqrt{2}x_0 y_0$, it follows that

$$y_1 - y_2 \leq \frac{4x_0 y_0}{3\sqrt{2}x_0 y_0} = \frac{2\sqrt{2}}{3},$$

where the equal sign holds when $\dfrac{1}{2}x_0^2 = 9y_0^2$ is required, and accordingly $x_0 = \dfrac{3\sqrt{5}}{5}$, $y_0 = \dfrac{\sqrt{10}}{10}$.

Therefore, the maximum of $y_1 - y_2$ is $\dfrac{2\sqrt{2}}{3}$. □

China Mathematical Competition (Second Round)

2020

Test Paper A
(9:40 – 12:30; September 13, 2020)

1 (40 marks) As shown in Fig. 1.1, in isosceles $\triangle ABC$, $AB = BC$, and I is its incentre. M is the midpoint of BI. P lies on side AC, satisfying $AP = 3PC$. Point H on the extension of PI satisfy $MH \perp PH$. Q is the midpoint of the minor arc of the circumcircle of $\triangle ABC$. Prove that $BH \perp QH$.

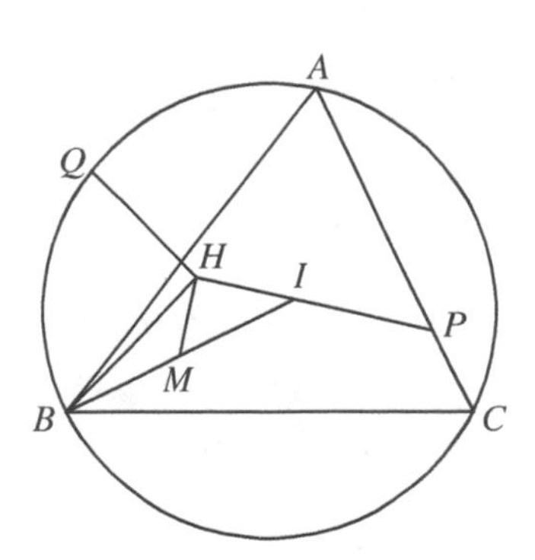

Fig. 1.1

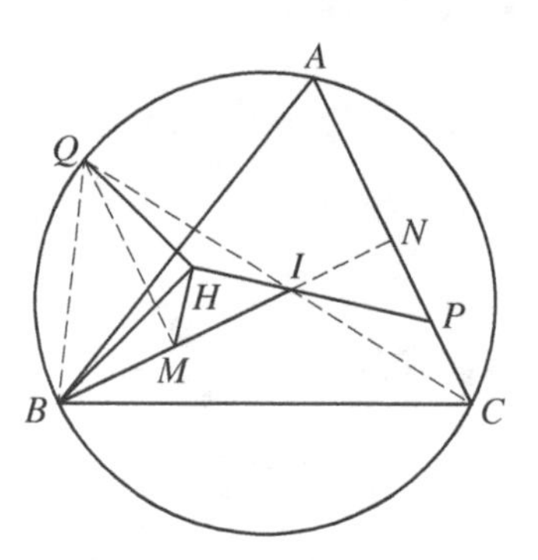

Fig. 1.2

Solution As shown in Fig. 1.2, take the midpoint N of AC. By $AP = 3PC$, we know that P is the midpoint of NC. It is easy to find that points B, I, N are collinear and $\angle INC = 90°$.

Since I is the incentre $\triangle ABC$, we see that CI passes through point Q and

$$\begin{aligned}\angle QIB &= \angle IBC + \angle ICB = \angle ABI + \angle ACQ \\ &= \angle ABI + \angle ABQ = \angle QBI.\end{aligned}$$

And since M is the midpoint of BI, we have $QM \perp BI$. And then $QM // CN$.

We shall consider $\triangle HMQ$ and $\triangle HIB$. Since $MH \perp PH$, there is $\angle HMQ = 90° - \angle HMI = \angle HIB$.

And since $\angle IHM = \angle INP = 90°$, $\dfrac{HM}{HI} = \dfrac{NP}{NI}$. Therefore,

$$\frac{HM}{HI} = \frac{NP}{NI} = \frac{1}{2} \cdot \frac{NC}{NI} = \frac{1}{2} \cdot \frac{MQ}{MI} = \frac{MQ}{IB}.$$

Hence, $\triangle HMQ \backsim \triangle HIB$, and then $\angle HQM = \angle HBI$.

And thus points H, M, B, Q are concyclic. Therefore, there is $\angle BHQ = \angle BMQ = 90°$, namely, $BH \perp QH$. □

2 (40 marks) Given positive integer $n \geq 3$, let $a_1, a_2, \ldots, a_{2n}$, $b_1, b_2, \ldots, b_{2n}$ be $4n$ non-negative real numbers, satisfying

$$a_1 + a_2 + \cdots + a_{2n} = b_1 + b_2 + \cdots + b_{2n} > 0.$$

And for any $i = 1, 2, \ldots, 2n$, there is $a_i a_{i+2} \geq b_i + b_{i+1}$, where $a_{2n+1} = a_1$, $a_{2n+2} = a_2$, $b_{2n+1} = b_1$. Find the minimum of $a_1 + a_2 + \cdots + a_{2n}$.

Solution Denote $S = a_1 + a_2 + \cdots + a_{2n} = b_1 + b_2 + \cdots + b_{2n}$.

Without loss of generality, suppose $T = a_1 + a_3 + \cdots + a_{2n-1} \leq \frac{S}{2}$.

For $n = 3$, since

$$T^2 - 3 \cdot \sum_{k=1}^{3} a_{2k-1}a_{2k+1} = \frac{1}{2}\left((a_1 - a_3)^2 + (a_3 - a_5)^2 + (a_5 - a_1)^2\right) \geq 0,$$

by combining the given conditions we can find

$$\frac{S^2}{4} \geq T^2 \geq 3 \cdot \sum_{k=1}^{3} a_{2k-1}a_{2k+1} \geq 3 \cdot \sum_{k=1}^{3} (b_{2k-1} + b_{2k}) = 3S.$$

And since $S > 0$, there is $S \geq 12$.

S takes the minimum when $a_i = b_i = 2$ $(1 \leq i \leq 6)$.

When $n \geq 4$, on one hand we have

$$\sum_{k=1}^{n} a_{2k-1}a_{2k+1} \geq \sum_{k=1}^{n} (b_{2k-1} + b_{2k}) = S.$$

On the other hand, if n is even, then

$$\sum_{k=1}^{n} a_{2k-1}a_{2k+1} \leq (a_1 + a_5 + \cdots + a_{2n-3})(a_3 + a_7 + \cdots + a_{2n-1}) \leq \frac{T^2}{4}.$$

The first inequality is due to the fact that each term of $(a_1 + a_5 + \cdots + a_{2n-3})(a_3 + a_7 + \cdots + a_{2n-1})$ is nonnegative after expansion and contains terms $a_{2k-1}a_{2k+1}$ $(1 \leq k \leq n)$. The second inequality makes use of the inequality of arithmetic and geometric means.

If n is odd, we suppose $a_1 \leq a_3$, and then

$$\begin{aligned}\sum_{k=1}^{n} a_{2k-1}a_{2k+1} &\leq \left(\sum_{k=1}^{n-1} a_{2k-1}a_{2k+1}\right) + a_{2n-1}a_3 \\ &\leq (a_1 + a_5 + \cdots + a_{2n-1})(a_3 + a_7 + \cdots + a_{2n-3}) \\ &\leq \frac{T^2}{4}.\end{aligned}$$

Thus, there is always $S \leq \sum_{k=1}^{n} a_{2k-1}a_{2k+1} \leq \frac{T^2}{4} \leq \frac{S^2}{16}$. And since $S > 0$, $S \geq 16$.

When $a_1 = a_2 = a_3 = a_4 = 4$, $a_i = 0$ $(5 \leq i \leq 2n)$, $b_1 = 0$, $b_2 = 16$, $b_i = 0$ $(3 \leq i \leq 2n)$, S takes the minimum of 16.

To sum up, when $n = 3$, the minimum of S is 12; when $n \geq 4$, the minimum of S is 16. □

3 (50 marks) Let $a_1 = 1$, $a_2 = 2$, $a_n = 2a_{n-1} + a_{n-2}$, $n = 3, 4, \ldots$. Prove that for integer $n \geq 5$, a_n must have a prime factor that is congruent to 1 modulo 4.

Solution Denote $\alpha = 1 + \sqrt{2}$, $\beta = 1 - \sqrt{2}$. Then it is easy to find $a_n = \frac{\alpha^n - \beta^n}{\alpha - \beta}$.

Let $b_n = \frac{\alpha^n + \beta^n}{2}$. Then sequence $\{b_n\}$ satisfies

$$b_n = 2b_{n-1} + b_{n-2} \quad (n \geq 3). \tag{1}$$

Since $b_1 = 1$ and $b_2 = 3$ are both integers, by ① and the mathematical induction, we know that each term of $\{b_n\}$ is an integer.

From $\left(\frac{\alpha^n + \beta^n}{2}\right)^2 - \left(\frac{\alpha - \beta}{2}\right)^2 \left(\frac{\alpha^n - \beta^n}{\alpha - \beta}\right)^2 = (\alpha\beta)^n$, it is obvious that

$$b_n^2 - 2a_n^2 = (-1)^n \quad (n \geq 1). \tag{2}$$

When $n > 1$ is odd, since a_1 is odd, by the recurrence relation of $\{a_n\}$ and the mathematical induction, we know that a_n is odd greater than 1, so a_n has an odd prime factor p. From ②, we get $b_n^2 \equiv -1 \pmod{p}$, so

$$b_n^{p-1} \equiv (-1)^{\frac{p-1}{2}} \pmod{p}.$$

The above equation shows that $(p, b_n) = 1$. Hence, by Fermat's Little Theorem, we get $b_n^{p-1} \equiv 1 \pmod{p}$, and thus

$$(-1)^{\frac{p-1}{2}} \equiv 1 \pmod{p}.$$

Since $p > 2$, there must be $(-1)^{\frac{p-1}{2}} = 1$, and thus $p \equiv 1 \pmod{4}$.

On the other hand, for positive integers m, n, if $m \mid n$, let $n = km$, and then

$$a_n = \frac{\alpha^n - \beta^n}{\alpha - \beta} = \frac{\alpha^m - \beta^m}{\alpha - \beta} \cdot \left(\alpha^{(k-1)m} + \alpha^{(k-2)m}\beta^m + \cdots + \alpha^m\beta^{(k-2)m} + \beta^{(k-1)m}\right)$$

$$= \begin{cases} a_m \cdot \sum_{i=0}^{l-1} (\alpha\beta)^{im}\left(\alpha^{(2l-1-2i)m} + \beta^{(2l-1-2i)m}\right), & k = 2l, \\ a_m \cdot \left(\sum_{i=0}^{l-1} (\alpha\beta)^{im}\left(\alpha^{(2l-2i)m} + \beta^{(2l-2i)m}\right) + (\alpha\beta)^{lm}\right), & k = 2l+1. \end{cases}$$

Since $\alpha^s + \beta^s = 2b_s$ is integer (for positive integers s) and $\alpha\beta = -1$ is integer, the above equation indicates that a_n is equal to the product of a_m and an integer, and hence $a_m \mid a_n$.

Accordingly, if n has odd factor m greater than 1, then by the conclusion proved above a_m has prime factor $p \equiv 1 \pmod 4$. And since $a_m \mid a_n$, $p \mid a_n$, namely, a_n also has a prime factor that is congruent to 1 modulo 4.

At last, if n has no odd factor greater than 1, then n is an integer power of 2.

Let $n = 2^l$ ($l \geq 3$). Since $a_8 = 408 = 24 \times 17$ has prime factor 17 that is congruent to 1 modulo 4, by $8 \mid 2^l$ we have $a_8 \mid a_{2^l}$ for $l \geq 4$, and thus a_{2^l} also has prime factor 17. The proof is completed. □

4 (50 marks) Given a convex 20-sided polygon P. Dividing P into 18 triangles by its 17 non-intersecting diagonals in the interior, the resulting graph is called a triangulation graph of P. For any triangulation graph T of P, the 20 edges of P and the added 17 diagonals are both called the edges of T. The set of any 10 edges of T without common endpoints between any two edges is called a perfect matching of T. When T takes all the triangulation graphs of P, find the maximum number of perfect matchings of T.

Solution We consider the general problem by replacing 20-sided polygon with $2n$-sided polygon.

For a diagonal of convex $2n$-sided polygon P, if there are an odd number of vertices of P on its both sides, it is called odd chord, otherwise it is called even chord. First, we should notice the following basic fact:

For any triangulation graph T of P, the perfect matching of T does not have odd chords. $(*)$

Suppose there was an odd chord e_1 in a perfect matching. Since a perfect matching of T gives a pairwise division of the set of vertices of P, and there are an odd number of vertices on each side of e_1, in this perfect matching there must be another edge e_2 of T, with endpoints on each side of e_1, respectively. And since P is a convex polygon, e_1 and e_2 intersect in the interior of P. This contradicts the fact that T is a triangulation graph.

Denote $f(T)$ as the number of perfect matchings of T. Let $F_1 = 1$, $F_2 = 2$, and $F_{k+2} = F_{k+1} + F_k$ for $k \geq 2$, which is the Fibonacci sequence.

In the following we will prove by induction on n. If T is any triangulation graph of a convex $2n$-sided polygon, then $f(T) \leq F_n$.

Let $P = A_1A_2 \cdots A_{2n}$ be a convex $2n$-sided polygon. There are exactly two ways to choose n edges from the $2n$ edges of P to construct a perfect matching, namely, $A_1A_2, A_3A_4, \ldots, A_{2n-1}A_{2n}$ or $A_2A_3, A_4A_5, \ldots, A_{2n-2}A_{2n-1}, A_{2n}A_1$.

When $n = 2$, the triangulation graph T of convex quadrilateral P has no even chords, so the perfect matching of T can only be done by using the edges of P. Therefore, $f(T) = 2 = F_2$.

When $n = 3$, the triangulation graph T of convex hexagon P has at most one even chord. If T has no even chord, as above it is known that $f(T) = 2$. If T contains one even chord, we may let it be A_1A_4, and the perfect matching of A_1A_4 is unique. The other two edges can only be A_2A_3, A_5A_6, and then $f(T) = 3$. In a word, $f(T) \leq 3 = F_3$.

The conclusion holds for $n = 2, 3$. Suppose $n \geq 4$ and the conclusion holds for all cases less than n. Consider a triangulation graph T of convex $2n$-sided polygon $P = A_1A_2 \cdots A_{2n}$. If T has no even chords, then as above it follows that $f(T) = 2$.

For even chord e, denote the smaller value of the number of vertices of P in both sides of e as $w(e)$. If T contains even chords, take one of the even chords e so that $w(e)$ reaches the minimum. Let $w(e) = 2k$, and we may set e to be $A_{2n}A_{2k+1}$, so that each $A_i (i = 1, 2, \ldots, 2k)$ cannot give rise to an even chord.

In fact, suppose A_iA_j is an even chord, and if $j \in \{2k+2, 2k+3, \ldots, 2n-1\}$, then A_iA_j intersects e in the interior of P, a contradiction. If $j \in \{1, 2, \ldots, 2k+1, 2n\}$, then $w(A_iA_j) < 2k$, which contradicts the minimality of $w(e)$.

Furthermore, by $(*)$, we know that there are no odd chords in perfect matching, so A_iA_j can only be paired with its neighboring vertices.

In particular, A_1 can only be paired with A_2 or A_{2n}. The following is divided into two cases.

Case 1: Choose edge A_1A_2, then edges $A_3A_4, \ldots, A_{2k-1}A_{2k}$ must be selected. Note that there are $2k, 2n-2k-2$ vertices on each side of $A_{2n}A_{2k+1}$, respectively. And there is $2n-2k-2 \geq w(A_{2n}A_{2k+1}) = 2k$. Since $n \geq 4$, we have $2n-2k \geq 6$. In convex $2n-2k$-sided polygon P_1, the edges of T give the triangulation graph T_1 of P_1. Then pick $n-k$ edges $e_1, e_2, \ldots, e_{n-k}$ in T, which together with $A_1A_2, A_3A_4, \ldots, A_{2k-1}A_{2k}$ forms a perfect matching of T if and only if e1 is a perfect matching of T.

Case 2: Choose edge A_1A_{2n}, then edges $A_2A_3, \ldots, A_{2k}A_{2k+1}$ must be selected. In convex $2n-2k-2$-sided polygon $P_2 = A_{2k+2}A_{2k+3}\cdots A_{2n-1}$, construct the following triangulation graph T_2. For $2k+2 \leq i < j \leq 2n-1$, if segment A_iA_j is an edge of T, it will also be an edge of T_2. Since these edges do not intersect each other in the interior, some more diagonals of P_2 can be added appropriately to obtain a triangulation graph T_2 which contains all the edges of T between vertices $A_{2k+2}, A_{2k+3}, \ldots, A_{2n-1}$. Hence for every perfect matching of T containing edges $A_{2n}A_1, A_2A_3, \ldots, A_{2k}A_{2k+1}$, the remaining edges must be a perfect matching of T_2. Therefore, the number of perfect matchings of T in *Case 2* does not exceed $f(T_2)$.

By the induction hypothesis, we obtain $f(T_1) \leq F_{n-k}, f(T_2) \leq F_{n-k-1}$. Combining the above two cases and $k \geq 1$, there is

$$f(T) \leq f(T_1) + f(T_2) \leq F_{n-k} + F_{n-k-1} = F_{n-k+1} \leq F_n.$$

The following shows that the equal sign is valid. Consider the triangulation graph $\triangle_n$ of convex $2n$-sided polygon $A_1A_2\cdots A_{2n}$: add diagonals $A_2A_{2n}, A_{2n}A_3, A_3A_{2n-1}, A_{2n-1}A_4, A_4A_{2n-2}, \ldots, A_{n+3}A_n, A_nA_{n+2}$. Repeating the previous process of demonstration yields $f(\triangle_2) = 2$, $f(\triangle_3) = 3$. For $\triangle_n$, $n \geq 4$, consider the even chord A_nA_3. In case 1, using A_1A_2, since the triangulation graph of convex $2n-2$-sided polygon $A_3A_4\cdots A_{2n}$ is exactly $\triangle_{n-1}$, there are $f(\triangle_{n-1})$ perfect matchings of T at this point. In case 2, using A_1A_{2n}, since the edges of T in convex $2n-4$-sided polygon $A_4A_5\cdots A_{2n-1}$ form exactly the triangulation graph $\triangle_{n-1}$, there is no need to add any diagonals, and the number of perfect matchings of T in this case is exactly $f(\triangle_{n-2})$. Hence, for $n \geq 4$, there is

$$f(\triangle_n) = f(\triangle_{n-1}) + f(\triangle_{n-2}).$$

By mathematical induction, we have $f(\triangle_n) = F_n$. The conclusion is proved.

Therefore, for convex 20-sided polygon P, the maximum value of $f(T)$ is equal to $F_{10} = 89$. □

Test Paper B
(9:40 – 12:30; September 13, 2020)

1 (40 marks) As shown in Fig. 1.1, A, B, C, D, E are five points in order on circle Ω, satisfying $\overparen{ABC} = \overparen{BCD} = \overparen{CDE}$. Points P, Q are on segments AD, BE, respectively, and P lies on segment CQ. Prove that $\angle PAQ = \angle PEQ$.

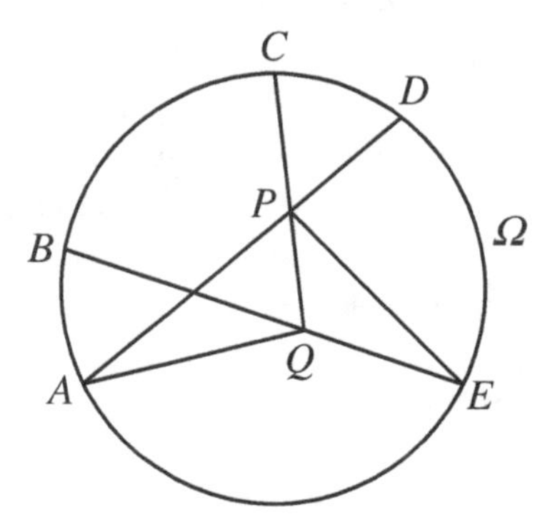

Fig. 1.1

Solution As shown in Fig. 1.2, let S be the intersection of AD and BE, and the extension of CQ intersects circle Ω at point T.

Noticing that $\overparen{ABC} = \overparen{BCD} = \overparen{CDE}$, we can let the central angles of $\overparen{AB}$, $\overparen{CD}$, be both α, and the central angles of $\overparen{BC}$, $\overparen{DE}$, be both β.

Therefore,

$$\angle ATQ = \angle ATC = \alpha + \beta,$$

$$\angle PTE = \angle CTE = \alpha + \beta,$$

$$\angle PSQ = \angle BDA + \angle DBE = \alpha + \beta.$$

By $\angle ATQ = \angle PSQ$ we see points S, A, T, Q are concyclic. And from $\angle PTE = \angle PSQ$ we know points P, S, T, E are concyclic.

Consequently, $\angle PAQ = \angle PEQ$. □

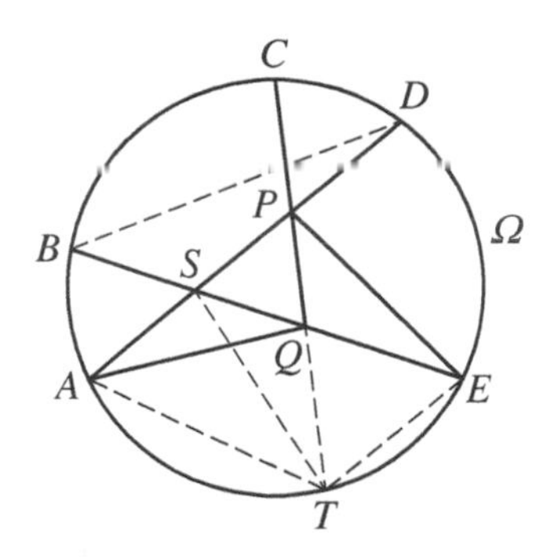

Fig. 1.2

2 (40 marks) Given set $A = \{1, 2, \ldots, 19\}$, does there exist non-empty subsets S_1, S_2 of set A satisfying the following conditions?

(1) $S_1 \cap S_2 = \varnothing$, $S_1 \cup S_2 = A$;
(2) S_1 and S_2 both have at least four elements;
(3) the sum of all elements of S_1 is equal to the product of all elements of S_2.

Prove your conclusion.

Solution The answer is definitely yes.

Let $S_2 = \{1, 2, x, y\}, 2 < x < y \leq 19$. Then

$$1 + 2 + \cdots + 19 - 1 - 2 - x - y = 2xy,$$

and thus $2xy + x + y = 187$.

Therefore,

$$(2x + 1)(2y + 1) = 375 = 15 \times 25,$$

so $x = 7, y = 12$ is also a solution set.

Hence, by taking $S_1 = \{3, 4, 5, 6, 7, 8, 10, 11, 13, 14, 15, 16, 17, 18, 19\}$ and $S_2 = \{1, 2, 7, 12\}$, then S_1 and S_2 satisfy the conditions. □

Remark Give a direct example and verify it can get 40 marks.

3 (50 marks) Given integer $n \geq 2$, let $a_1, a_2, \ldots, a_n, b_1, b_2, \ldots, b_n > 0$, satisfying

$$a_1 + a_2 + \cdots + a_n = b_1 + b_2 + \cdots + b_n$$

and there is always $a_i a_j \geq b_i + b_j$ for any i, j $(1 \leq i < j \leq n)$. Find the minimum value of $a_1 + a_2 + \cdots + a_n$.

Solution Denote $S = a_1 + a_2 + \cdots + a_n = b_1 + b_2 + \cdots + b_n$.

By the given conditions, we can find

$$\sum_{1\le i<j\le n} a_i a_j \ge \sum_{1\le i<j\le n} (b_i + b_j) = (n-1)S.$$

Furthermore, since $\sum\limits_{1\le i<j\le n} a_i a_j \ge \sum\limits_{1\le i<j\le n} (b_i + b_j) = \frac{n-1}{2} \cdot \sum\limits_{i=1}^{n} a_i^2$,

$$S^2 = \left(\sum_{i=1}^{n} a_i\right)^2 = \sum_{i=1}^{n} a_i^2 + 2 \sum_{1\le i<j\le n} a_i a_j$$
$$\ge \left(\frac{2}{n-1} + 2\right) \sum_{1\le i<j\le n} a_i a_j \ge 2nS.$$

Note that $S > 0$, so $S \ge 2n$.

On the other hand, when $a_i = b_i = 2 (i = 1, 2, \ldots, n)$, the condition is satisfied and $S = 2n$.

To sum up, the minimum value of $S = a_1 + a_2 + \cdots + a_n$ is $2n$. □

4 (50 marks) Let a, b be positive integers not greater than 12 satisfying that there exists constant C such that $a^n + b^{n+9} \equiv C \pmod{13}$ holds for any positive integer n. Find all ordered pairs (a, b) that satisfy the conditions.

Solution 1 By the given conditions, we know that for any positive integer n, there is

$$a^n + b^{n+9} \equiv a^{n+3} + b^{n+12} \pmod{13}. \quad ①$$

Note that 13 is a prime number and a, b are coprime with 13. By Fermat's Little Theorem it is known that

$$a^{12} \equiv b^{12} \equiv 1 \pmod{13}.$$

Therefore, take $n = 12$ in ① and the simplification gives $1 + b^9 \equiv a^3 + 1 \pmod{13}$, and thus

$$b^9 \equiv a^3 \pmod{13}.$$

Substituting it into ①, we get $a^n + a^3 b^n \equiv a^{n+3} + b^{n+12} \equiv a^{n+3} + b^n \pmod{13}$, namely,

$$(a^n - b^n)(1 - a^3) \equiv 0 \pmod{13}. \quad ②$$

□

In the following, we discuss it in two cases.

Case 1: If $a^3 \equiv 1 \pmod{13}$, then $b^3 \equiv a^3b^3 \equiv b^{12} \equiv 1 \pmod{13}$.

Also, since $a, b \in \{1, 2, \ldots, 12\}$, it is clear by checking that $a, b \in \{1, 3, 9\}$.

At this point $a^n + b^{n+9} \equiv a^n + b^n \pmod{13}$. By the condition it follows that

$$a + b \equiv a^3 + b^3 \equiv 2 \pmod{13},$$

and hence it can only be $a = b = 1$.

After checking, when $(a, b) = (1, 1)$, for any positive integer n, $a^n + b^{n+9}$ is congruent to 2 modulo 13. Since it is a constant, the condition is satisfied.

Case 2: If $a^3 \not\equiv 1 \pmod{13}$, then by ② we know that for any positive integer n, there is

$$a^n \equiv b^n \pmod{13}.$$

In particular, $a \equiv b \pmod{13}$, so $a = b$. Therefore, $a^3 \equiv b^9 = a^9 \pmod{13}$, i.e.,

$$a^3(a^3 - 1)(a^3 + 1) \equiv 0 \pmod{13},$$

and thus $a^3 \equiv -1 \pmod{13}$. By checking $a \equiv \pm1, \pm2, \ldots, \pm6 \pmod{13}$, we know that $a = 4, 10, 12$.

It is checked that when $(a, b) = (4, 4), (10, 10), (12, 12)$, there is

$$a \equiv \pm1, \pm2, \ldots, \pm6 \pmod{13}$$

for any positive integer n, satisfying the condition.

Combining *Case 1* and *Case 2*, the desired ordered pairs (a, b) are $(1, 1), (4, 4), (10, 10), (12, 12)$.

Solution 2 By the given conditions, we know that for any positive integer n, there is

$$(a^n + b^{n+9})(a^{n+2} + b^{n+11}) \equiv (a^{n+1} + b^{n+10})^2 \pmod{13}.$$

Simplification gives $a^nb^{n+11} + a^{n+2}b^{n+9} \equiv 2a^{n+1}b^{n+10} \pmod{13}$, namely,

$$a^nb^{n+9}(a - b)^2 \equiv 0 \pmod{13}.$$

Since 13 is a prime number, $a, b \in \{1, 2, \ldots, 12\}$. Hence $13 \mid (a - b)^2$, and thus $a = b$.

Therefore, when n varies, the remainder of $a^n + b^{n+9} = a^n(1 + a^9)$ modulo 13 is a constant.

When $1 + a^9 \not\equiv 0 \pmod{13}$, by the above equation we know that the remainder of a^n modulo 13 is a constant. In particular, there is $a^2 \equiv a \pmod{13}$, so $a = 1$.

When $1 + a^9 \equiv 0 \pmod{13}$, by Fermat's Little Theorem it is known that $a^{12} \equiv 1 \pmod{13}$. Thus,

$$a^3 \equiv a^3 \cdot (-a^9) \equiv -a^{12} \equiv -1 \pmod{13}.$$

By checking $a \equiv \pm 1, \pm 2, \ldots, \pm 6 \pmod{13}$, we know that $a = 4, 10, 12$.

To sum up, the desired ordered pairs (a, b) are $(1, 1), (4, 4), (10, 10), (12, 12)$. □

China Mathematical Competition (Second Round)

2021

Test Paper A

(9:40 – 12:30; September 12, 2021)

1 (40 marks) Given positive integer k ($k \geq 2$) and k non-zero real numbers $a_1, a_2, \ldots, a_k$, prove that there are at most finite k-element integer arrays $(n_1, n_2, \ldots, n_k)$ satisfying that $n_1, n_2, \ldots, n_k$ are pairwise distinct and

$$a_1 \cdot n_1! + a_2 \cdot n_2! + \cdots + a_k \cdot n_k! = 0.$$

Solution Take a certain positive integer $N \geq \dfrac{|a_1| + |a_2| + \cdots + |a_k|}{\min\limits_{1 \leq i \leq k} |a_i|}$ (note that $a_1, a_2, \ldots, a_k \neq 0$).

We will show that when positive integers $n_1, n_2, \ldots, n_k$ satisfy the condition, there must be $\max\limits_{1 \leq i \leq k} n_i \leq N$.

We assume that this does not hold and set $\max\limits_{1 \leq i \leq k} n_i = n_1 > N$. For $i = 2, \ldots, k$, since positive integer $n_i < n_1$, it follows that

$$n_i! \leq (n_1 - 1)! = \frac{n_1!}{n_1} < \frac{n_1!}{N}.$$

Hence,

$$|a_2 \cdot n_2! + \cdots + a_k \cdot n_k!| \leq \sum_{i=2}^{k} |a_i| \cdot n_i! < \frac{n_1!}{N} \sum_{i=2}^{k} |a_i| < \frac{n_1!}{N} \sum_{i=1}^{k} |a_i|$$

$$\leq \min_{1\leq i\leq k} |a_i| \cdot n_1! \leqslant |a_1| \cdot n_1!.$$

However, $|a_2 \cdot n_2! + \cdots + a_k \cdot n_k!| = |-a_1 \cdot n_1!| = |a_1| \cdot n_1!$, a contradiction.

Therefore, there are at most N^k sets of positive integers $(n_1, n_2, \ldots, n_k)$ that satisfy the condition. The proof is completed. □

2 (40 marks) As shown in Fig. 2.1, in $\triangle ABC$, M is the midpoint of side AC. D, E are two points on the tangent of the circumcircle of $\triangle ABC$ at point A, satisfying $MD//AB$ and A is the midpoint of segment DE. The circle passing through points A, B, E intersects side AC at point P. The circle passing through points A, D, P intersects the extension of DM at point Q.

Prove that $\angle BCQ = \angle BAC$.

Solution As shown in Fig. 2.2, take the midpoint N of side BC. Then D, M, Q, N are collinear and $MN//AB$.

By the alternate segment theorem, there is $\angle DAM = \angle CBA = \angle CNM$. And since $\angle AMD = \angle NMC$, it follows that $\triangle AMD \backsim \triangle NMC$. Therefore,

$$\frac{NM}{NC} = \frac{AM}{AD}. \qquad ①$$

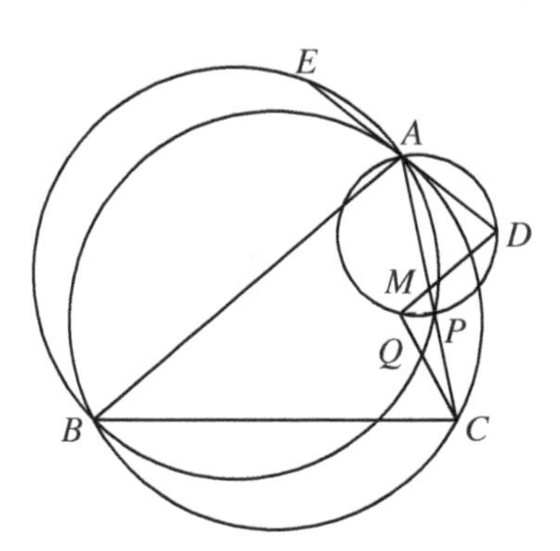

Fig. 2.1

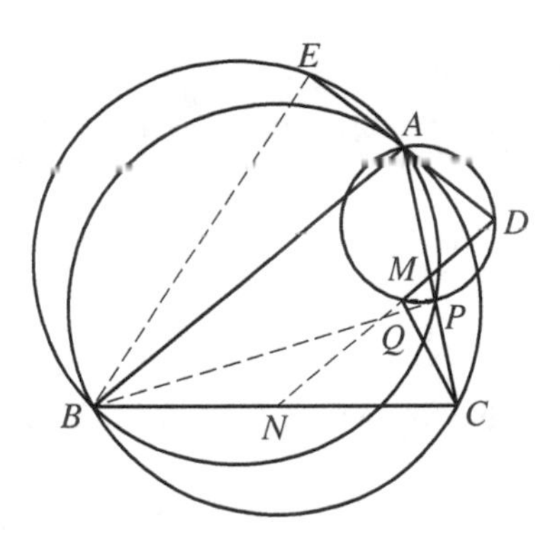

Fig. 2.2

By the given condition, we know that points A, D, P, Q are concyclic. Hence $\angle APQ = \angle ADQ = \angle ADM = \angle ACB$, and thus $PQ // BC$. Therefore,

$$\frac{NQ}{NM} = \frac{CP}{CM}. \qquad ②$$

Combining ①, ② and $AD = AE$ yields

$$\frac{NQ}{NC} = \frac{NQ}{NM} \cdot \frac{NM}{NC} = \frac{CP}{CM} \cdot \frac{AM}{AD} = \frac{CP}{AD} = \frac{CP}{AE},$$

namely,

$$\frac{NQ}{NC} = \frac{CP}{AE}. \qquad ③$$

By the alternate segment theorem, we have $\angle BAE = \angle BCA = \angle BCP$. And since A, P, B, E are concyclic, it follows that $\angle BEA = \angle BPC$. Therefore, $\triangle BAE \backsim \triangle BCP$, and thus $\frac{CP}{AE} = \frac{BC}{BA}$. Combining ③ gives

$$\frac{NQ}{NC} = \frac{BC}{BA}.$$

Since $MN // AB$, $\angle CNQ = \angle ABC$. Hence, $\triangle CNQ \backsim \triangle ABC$. Therefore, $\angle NCQ = \angle BAC$, namely, $\angle BCQ = \angle BAC$. □

3 (50 marks) Suppose integer $n \geq 4$. Prove that if n divides $2^n - 2$, then $\frac{2^n - 2}{n}$ is composite.

Solution Denote the integer $\dfrac{2^n-2}{n}$ as y.

If n is odd, then since 2^n-2 is even, we know that y is even. Also as $n \geq 4$, it follows that $\dfrac{2^n-2}{n} > 2$ and thus y is composite.

In the following, we consider the case where n is even, and let $n = 2m(m > 1)$.

Since $y = \dfrac{2^{2m}-2}{2m} = \dfrac{2^{2m-1}-1}{m}$ is an integer, m is odd.

Let δ be the multiplicate order of 2 modulo m. Then $\delta < m$, and $\delta \mid 2m-1$ since $m \mid 2^{2^{m-1}} - 1$.

Let $2m-1 = \delta r$. From $\delta < m < 2m-1$ we know that $r > 1$.

(1) If $m \neq 2^\delta - 1$, $m < 2^\delta - 1$ due to $m \mid 2^\delta - 1$. At this point, there is

$$y = \frac{2^{2m-1}-1}{m} = \frac{2^{\delta r}-1}{m} = \frac{2^{\delta r}-1}{2^\delta - 1} \cdot \frac{2^\delta - 1}{m}.$$

And since $r > 1$, it is the product of two integers greater than 1, and thus it is composite.

(2) If $m = 2^\delta - 1$, then $2(2^\delta - 1) - 1 = 2m - 1 = \delta r$. By $m > 1$ we know that $\delta > 1$. Hence,

$$r = \frac{2^{\delta+1}-3}{\delta} > \delta. \qquad ①$$

Since $2^\delta - 1 \,|\, 2^{\delta r} - 1, 2^r - 1 \,|\, 2^{\delta r} - 1$, $2^\delta - 1$ is the multiple of $[2^\delta - 1, 2^r - 1]$, namely,

$$\frac{(2^\delta - 1)(2^r - 1)}{(2^\delta - 1, 2^r - 1)} \,\Big|\, 2^{\delta r} - 1.$$

Note that $(2^\delta - 1, 2^r - 1) = 2^{(\delta, r)} - 1$, and thus $(2^\delta - 1)(2^r - 1)(2^{\delta r} - 1)$ $(2^{(\delta, r)} - 1)$. Therefore,

$$y = \frac{2^{2m-1}-1}{m} = \frac{2^{\delta r}-1}{2^\delta - 1} = \frac{(2^{\delta r}-1)(2^{(\delta,r)}-1)}{(2^\delta - 1)(2^r - 1)} \cdot \frac{2^r - 1}{2^{(\delta,r)}-1} \qquad ②$$

is the product of two integers.

Because $r > \delta$ (see ①), $\dfrac{2^r - 1}{2^{(\delta,r)} - 1} \geq \dfrac{2^r - 1}{2^\delta - 1} > 1$. And since $\delta \geq 2$, it follows that

$$\begin{aligned}\frac{(2^{\delta r} - 1)(2^{(\delta,r)} - 1)}{(2^\delta - 1)(2^r - 1)} &\geq \frac{(2^{2r} - 1) \cdot 1}{(2^\delta - 1)(2^r - 1)} \\ &> \frac{2^{2r} - 1}{(2^r - 1)(2^r - 1)} \\ &= \frac{2^r + 1}{2^r - 1} > 1.\end{aligned}$$

Therefore, Equation ② shows that y is the product of two integers greater than 1, and thus it is composite.

In summary, the conclusion is confirmed. □

4 (50 marks) Find the smallest positive number c with the following property: for any integer $n \geq 4$ and set $A \subseteq \{1, 2, \ldots, n\}$, if $|A| > cn$, then there exists function $f : A \to \{1, -1\}$ that satisfies

$$\left| \sum_{a \in A} f(a) \cdot a \right| \leq 1.$$

Solution The desired smallest possible number is $c = \dfrac{2}{3}$.

First, when $n = 6$ and $A = \{1, 4, 5, 6\}$, there exists no f that satisfies the requirement, because the sum of the elements of A is 16 and A cannot be divided into a union of two subsets with sum of elements being both 8. At this point $|A| = \dfrac{2}{3}n$, so $c < \dfrac{2}{3}$ does not have the properties described in the question.

In the following, we will prove that $c = \dfrac{2}{3}$ satisfies the requirement. That is, when $|A| > \dfrac{2}{3}n$, there exists f that satisfies the condition.

Lemma *Let $x_1, x_2, \ldots, x_m$ be positive integers, whose sum is s and $s < 2m$. Then for any integer $x \in [0, s]$, there exists index set $I \subseteq \{1, 2, \ldots, m\}$ that satisfies $\sum_{i \in I} x_i = x$.* (Summation over the empty index set is considered to be zero).

Proof of lemma We prove by induction on m. When $m = 1$, it can only be $x_1 = s = 1$, and the conclusion is clearly valid.

Suppose $m > 1$ and the conclusion holds for $m - 1$. We may set $m = 1$, and then

$$x_1 + x_2 + \cdots + x_{m-1} \le \frac{m-1}{m} \cdot (x_1 + x_2 + \cdots + x_m)$$
$$< \frac{m-1}{m} \cdot 2m = 2(m-1). \quad ①$$

And since $x_1 + x_2 + \cdots + x_{m-1} \ge m - 1$, it follows that

$$x_m \le m \le 1 + x_1 + x_2 + \cdots + x_{m-1}. \quad ②$$

For any integer $x \in [0, s]$, if $x \le x_1 + x_2 + \cdots + x_{m-1}$, by ① and the induction hypothesis there exists index set $I \subseteq \{1, \ldots, m-1\}$ such that $\sum\limits_{i \in I} x_i = x$. If

$$x \ge 1 + x_1 + x_2 + \cdots + x_{m-1},$$

then using the induction hypothesis on $x - x_m$ (by ② $x - x_m \ge 0$), there exists index set

$$I \subseteq \{1, \ldots, m-1\}$$

such that $\sum\limits_{i \in I} x_i = x - x_m$. At this point, index set

$$I' = I \cup \{m\} \subseteq \{1, 2, \ldots, m\}$$

satisfies $\sum\limits_{i \in I'} x_i = x$. The lemma is proven.

Let us return to the original problem. Noting that $n \ge 4$, we discuss it in the following two cases.

(1) $|A|$ is even and let $|A| = 2m$. The elements of A from smallest to largest are denoted as

$$a_1 < b_1 < a_2 < b_2 < \cdots < a_m < b_m.$$

Let $x_i = b_i - a_i > 0$, $1 \le i \le m$, so that

$$s = \sum_{i=1}^{m} x_i = (b_m - a_1) - \sum_{i=1}^{m-1} (a_{i+1} - b_i) \le n - 1 - (m-1) = n - m < 2m.$$

The above equation makes use of $2m = |A| > \frac{2}{3}n$. Hence, $x_1, x_2, \ldots, x_m$ satisfy the conditions of the lemma.

By taking $x = \left[\frac{s}{2}\right] \in [0, s]$ and making use of the lemma, we know that there exists $I \subseteq \{1, 2, \ldots, m\}$ such that $\sum_{i \in I} x_i = \left[\frac{s}{2}\right]$. Let

$$\varepsilon_i = \begin{cases} 1, & i \in I, \\ -1, & i \in \{1, 2, \ldots, n\} \backslash I, \end{cases}$$

and then

$$\sum_{i=1}^{m} \varepsilon_i (b_i - a_i) = \sum_{i=1}^{m} \varepsilon_i x_i = \left[\frac{s}{2}\right] - \left(s - \left[\frac{s}{2}\right]\right) = 2\left[\frac{s}{2}\right] - s \in \{0, -1\}.$$

Therefore, the conclusion is valid. (We only need to let $f(a_i) = -\varepsilon_i$, $f(b_i) = \varepsilon_i$.)

(2) $|A|$ is odd. Let $|A| = 2m + 1$, and then $m \geq 1$. The elements of A from smallest to largest are denoted as

$$a < a_1 < b_1 < \cdots < a_m < b_m.$$

Let $x_i = b_i - a_i > 0$, $1 \leq i \leq m$. As in case (1), we have $s = x_1 + x_2 + \cdots + x_m < 2m$, and obviously there is $s \geq m$. Since $2m + 1 = |A| > \frac{2}{3}n$, $n \leq 3m + 1$. And thus

$$a \leq n - 2m \leq m + 1 \leq s + 1.$$

Since $x_1, x_2, \ldots, x_m$ satisfies the conditions of the lemma, by using the lemma for $x_1, x_2, \ldots, x_m$ and

$$x = \left[\frac{a + s}{2}\right] \in [0, s],$$

there exists $I \subseteq \{1, 2, \ldots, m\}$ such that $\sum_{i \in I} x_i = x = \left[\frac{a+s}{2}\right]$. Let

$$\varepsilon_i = \begin{cases} 1, & i \in I, \\ -1, & i \in \{1, 2, \ldots, n\} \backslash I, \end{cases}$$

and then

$$\begin{aligned} \left| -a + \sum_{i=1}^{m} \varepsilon_i x_i \right| &= \left| -a + \sum_{i \in I} x_i - \sum_{i \notin I} x_i \right| = |-a + x - (s - x)| \\ &= \left| 2\left[\frac{a+s}{2}\right] - (a + s) \right| \leq 1. \end{aligned}$$

Therefore, the conclusion is valid. (For $i \in I$, we only need to let $f(a_i) = 1$, $f(b_i) = -1$; for $i \in \{1, 2, \ldots, n\} \backslash I$, we let $f(a_i) = 1$, $f(b_i) = -1$ and $f(a) = -1$.) $\square$

Test Paper A1
(9:40 – 12:30; October 6, 2021)

1 (40 marks) As shown in Fig. 1.1, in $\triangle ABC$, $AB > AC$. Two points X, Y in $\triangle ABC$ are on the bisector of $\angle BAC$ and satisfy $\angle ABX = \angle ACY$. Let the extension of BX and segment CY intersect at point P. The circumcircle ω_1 of $\triangle BPY$ and the circumcircle ω_2 of $\triangle CPX$ intersect at P and another point Q. Prove that points A, P, Q are collinear.

Solution By $\angle BAX = \angle CAY$, $\angle ABX = \angle ACY$, we know that $\triangle ABX \backsim \triangle ACY$. Therefore,

$$\frac{AB}{AC} = \frac{AX}{AY}. \tag{1}$$

As shown in Fig. 1.2, extend AX and it intersects ω_1, ω_2 at U, V, respectively. Then

$$\angle AUB = \angle YUB = \angle YPB = \angle YPX = \angle XVC = \angle AVC,$$

and thus $\triangle ABU \backsim \triangle ACV$. Therefore,

$$\frac{AB}{AC} = \frac{AU}{AV}. \tag{2}$$

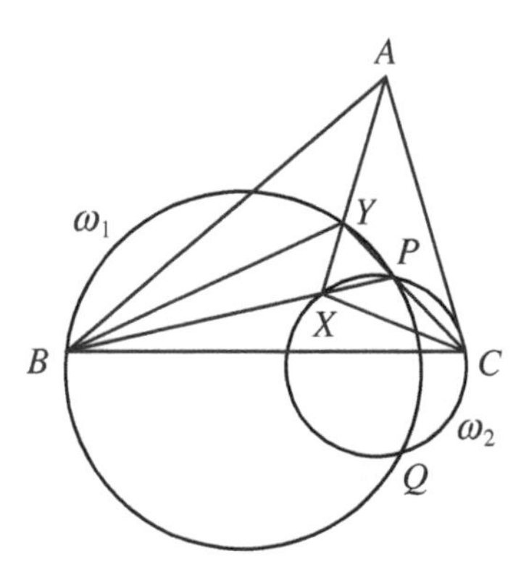

Fig. 1.1

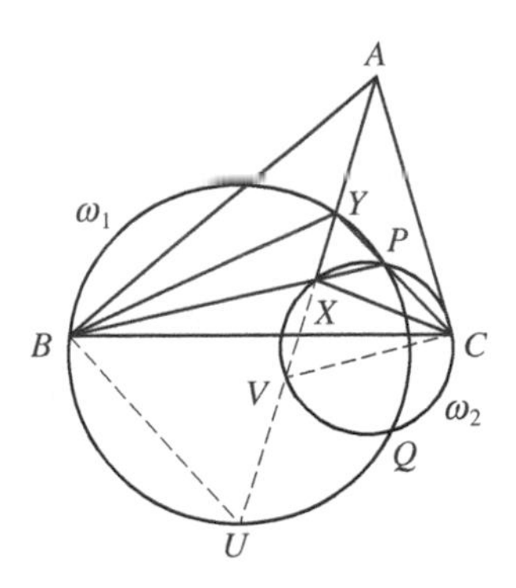

Fig. 1.2

From ① and ②, we can obtain $\dfrac{AX}{AY} = \dfrac{AU}{AV}$, i.e., $AU \cdot AY = AV \cdot AX$.

The two sides of the above equation are the circle powers of point A to circles ω_1, ω_2, respectively. This implies that A is on the radical axis (i.e., line PQ) of circles ω_1, ω_2. In other words, points A, P, Q are collinear. □

2 (40 marks) Find the necessary and sufficient conditions satisfied by positive integers a, b, n ($n \geq 2$) such that there exists one-to-one map $f : S \to S$ from set S to itself that satisfies: for any $x \in S$, x and $f(x)$ are always coprime.

Solution Let a, b, n be consistent with the question. If $(a, b) = d > 1$, then d divides all the elements of S. Hence, for any $f : S \to S$, taking an $x \in S$, we have $(x, f(x)) \geq d > 1$, a contradiction.

Therefore, there must be $(a, b) = 1$.

If n is odd and a is even, then ab is odd since $(a, b) = 1$. Then S contains $\dfrac{n+1}{2}$ even numbers and $\dfrac{n-1}{2}$ odd numbers. Assume that there exists $(a, b) = 1$ that meets the requirement, and then for each even number $x \in S$, $f(x)$ must be odd since $(x, f(x)) = 1$. Hence, S contains at least $\dfrac{n+1}{2}$ odd numbers, which contradicts the previous result.

Thus we obtain the necessary condition that a, b, n need to satisfy: $(a, b) = 1$ and a is odd when n is odd.

In the following we will show that this condition is also sufficient.

We assume that this condition is valid. First note that for any $t = 0, 1, \ldots, n-2$, there is

$$(a + bt, a + b(t+1)) = (a + bt, b) = (a, b) = 1.$$

When n is even, let

$$f(a+bt)=a+b(t+1),$$
$$f(a+b(t+1))=a+bt(t=0,2,\ldots,n-2),$$

and f defined in this way meets the requirements.

When n is odd, let $f(a)=a+b$, $f(a+b)=a+2b$, $f(a+2b)=a$ and set

$$f(a+bt)=a+b(t+1),$$
$$f(a+b(t+1))=a+bt(t=3,5,\ldots,n-1).$$

Since a is odd at this point, $(a+2b,a)=(a,2b)=(a,b)=1$, so f defined in this way meets the requirement.

To sum up, the desired necessary and sufficient conditions are: $(a,b)=1$ and a is odd when n is odd. □

3 (50 marks) Let positive real number sequences $\{a_n\}$, $\{b_n\}$ satisfy: for any integer $n \geq 101$, there is

$$a_n=\sqrt{\frac{1}{100}\sum_{j=1}^{100} b_{n-j}^2}, \quad b_n=\sqrt{\frac{1}{100}\sum_{j=1}^{100} a_{n-j}^2}.$$

Prove that there exists positive integer m such that $|a_m-b_m|<0.001$.

Solution 1 Let $c_n=a_n^2-b_n^2$, $n=1,2,\ldots$. Then for $n \geq 101$, there is

$$c_n=a_n^2-b_n^2=\frac{1}{100}\left(\sum_{i=1}^{100} b_{n-j}^2-\sum_{i=1}^{100} a_{n-j}^2\right)=-\frac{1}{100}\sum_{i=1}^{100} c_{n-j}. \quad ①$$

If there exists a $c_n=0$, then the conclusion obviously holds. In the following, we assume that all $c_n \neq 0$.

We first prove a lemma.

Lemma *There exists constant $\lambda \in (0,1)$ with the following property: if positive integer n and positive real number M satisfy $|c_n|, |c_{n+1}|, \ldots, |c_{n+99}| \leq M$, then there exists integer $k > n+99$ such that $|c_k|, |c_{k+1}|, \ldots, |c_{k+99}| \leq \lambda M$.*

Proof of the lemma (1) If not all of $c_n, c_{n+1}, \dots, c_{n+99}$ are of the same sign, then by ① we know that

$$|c_{n+100}| \leq \frac{99}{100} M = \lambda_1 M,$$

where $\lambda_1 = \dfrac{99}{100} \in (0,1)$. Repeatedly using ① and the triangle inequality, we get

$$|c_{n+101}| \leq \frac{1}{100}(99 + \lambda_1)M = \lambda_2 M,$$

$$|c_{n+102}| \leq \frac{1}{100}(99 + \lambda_1 + \lambda_2)M = \lambda_3 M,$$

$$\vdots$$

$$|c_{n+199}| \leq \frac{1}{100}(1 + \lambda_1 + \lambda_2 + \cdots + \lambda_{99})M = \lambda_{100} M,$$

where

$$\lambda_2 = \frac{1}{100}(99 + \lambda_1) \in (0,1),$$

$$\lambda_3 = \frac{1}{100}(98 + \lambda_1 + \lambda_2) \in (0,1),$$

$$\vdots$$

$$\lambda_{100} = \frac{1}{100}(1 + \lambda_1 + \lambda_2 + \cdots + \lambda_{99}) \in (0,1).$$

It is sufficient to take $k = n + 100$, $\lambda = \max\{\lambda_1, \lambda_2, \dots, \lambda_{100}\} \in (0,1)$. (Note that λ here is a constant independent of n, M.)

(2) If $c_n, c_{n+1}, \dots, c_{n+99}$ are of the same sign, then by ① we know that c_{n+100} and $c_n, c_{n+1}, \dots, c_{n+99}$ are not of the same sign. And there is $|c_{n+100}| \leq M$. Therefore, by the conclusion of case (1), there is

$$|c_{n+101}|, |c_{n+102}|, \dots, |c_{n+200}| \leq \lambda M$$

for the λ determined in (1).

It is sufficient to take $k = n + 101$. The proof of the lemma is complete.

Let us return to the original question. Let $M = \max\{|c_1|, |c_2|, \dots, |c_{100}|\}$, and then repeatedly using the above lemma, we know that for any positive integer t, there exists positive integer m such that $|c_m| \leq \lambda^t M$.

By $0 < \lambda < 1$, we can find $\lim_{t\to\infty} \lambda^t M = 0$. Therefore, there exists a positive integer m such that $|c_m| < (0.001)^2$. And since

$$|c_m| = |a_m^2 - b_m^2| = |(a_m - b_m)(a_m + b_m)| > |a_m - b_m|^2,$$

it follows that $|a_m - b_m| < 0.001$. □

Solution 2 Let $c_n = a_n^2 - b_n^2$, $n = 1, 2, \ldots$. Then for $n \geq 101$, there is

$$c_n = a_n^2 - b_n^2 = \frac{1}{100}\left(\sum_{j=1}^{100} b_{n-j}^2 - \sum_{j=1}^{100} a_{n-j}^2\right) = -\frac{1}{100}\sum_{j=1}^{100} c_{n-j}. \qquad ①$$

Sequence $\{c_n\}$ is a 100th order linear homogeneous recursive sequence with constant coefficients and its characteristic equation is

$$100x^{100} + x^{99} + x^{98} + \cdots + x + 1 = 0. \qquad ②$$

We will prove that the roots of the above characteristic equations are all complex numbers with modulus less than 1. Let z be a root, and if $|z| \geq 1$, ② can be rewritten as

$$100 = -\frac{1}{z} - \frac{1}{z^2} - \cdots - \frac{1}{z^{100}}.$$

Taking the modulus on both sides and using the triangle inequality, we get

$$100 = \left|\frac{1}{z} + \frac{1}{z^2} + \cdots + \frac{1}{z^{100}}\right| \leq \frac{1}{|z|} + \frac{1}{|z|^2} + \cdots + \frac{1}{|z|^{100}} \leq 100,$$

where the equal sign holds if and only if $|z| = 1$ and the arguments of $z, z^2, \ldots, z^{100}$ are the same, i.e., $z = 1$. But $z = 1$ does not satisfy equation ②, so the roots of equation ② are all complex numbers with modulus less than 1.

Suppose $t_1, t_2, \ldots, t_k$ are all the different complex roots of equation ② with multiplicity $m_1, m_2, \ldots, m_k$, respectively. Then the general term formula of sequence $\{c_n\}$ is

$$c_n = P_1(n)t_1^n + P_2(n)t_2^n + \cdots + P_k(n)t_k^n,$$

where P_i is the complex coefficient polynomial with degree less than m_i.

Since $|t_i| < 1, i = 1, 2, \ldots, k$, there is $\lim_{n \to \infty} P_i(n) t_i^n = 0$. Thus as $n \to \infty$, we have $|c_n| \to 0$. Therefore, there exists positive integer m such that $|c_m| < (0.001)^2$. In addition, since

$$|c_m| = |c_m^2 - b_m^2| = |(a_m - b_m)(a_m + b_m)| \geq |a_m - b_m|^2,$$

it follows that $|a_m - b_m| < 0.001$. □

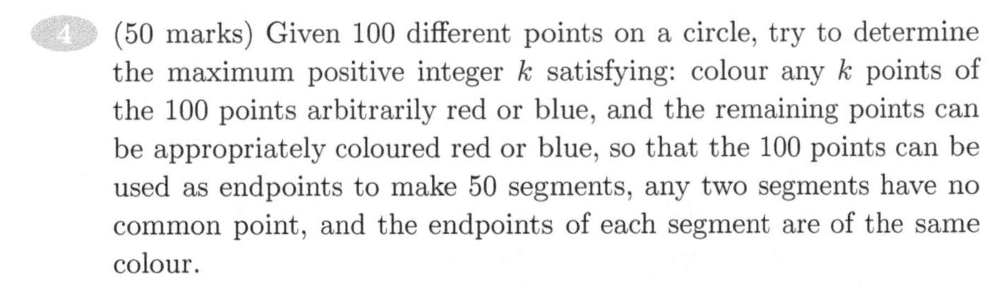

4 (50 marks) Given 100 different points on a circle, try to determine the maximum positive integer k satisfying: colour any k points of the 100 points arbitrarily red or blue, and the remaining points can be appropriately coloured red or blue, so that the 100 points can be used as endpoints to make 50 segments, any two segments have no common point, and the endpoints of each segment are of the same colour.

Solution The answer is 50.

Replace the 100 points with $2m$ points, denoted as $A_1, A_2, \ldots, A_{2m}$ in clockwise order, and colour each point red or blue. If it is possible to pair these $2m$ points with m segments, satisfying the conditions stated in the question, then we say the colouring method is wonderful.

Denote $S = \{A_1, A_3, \ldots, A_{2m-1}\}$, $T = \{A_2, A_4, \ldots, A_{2m}\}$.

Lemma *Colour the $2m$ points red or blue. The colouring method is wonderful if and only if the number of the red points in S is equal to the red points in T.*

Proof of the lemma We first prove the necessity. Assuming that there exist pairwise connections that satisfy the requirement, then the red points in S must be paired with the red points in T (Because if both endpoints of a segment l are the points in S, then there must be an odd number of points on each arc on both sides of l, and there must be a segment that intersects segment l). Similarly, the red points in T must be paired with the red points in S. Therefore, the required pairwise connections give a one-to-one correspondence between the red points in S and the red points in T.

Then we prove the sufficiency. We prove by induction on m. When the number of the red points in S is equal to the number of the red points in T, there exists a pairwise connection that satisfies the requirement. When $m = 1$, it is known by the assumption that the two points are of the same colour and a segment can be connected as required.

Assume that $m \geq 2$ and that the conclusion holds for $m-1$. Consider the case of m. At this time, there exist two adjacent points that are of the same colour, otherwise red and blue are separated. Thus, all points in S are of one colour and all points in T are of another colour, which is not consistent with the assumption. We may suppose A_{2m-1}, A_{2m} are of the same colour. (Otherwise, the subscripts can be re-labeled, at which point S, T remain unchanged or are exchanged, but this does not affect the conditions). Connect segment $A_{2m-1}A_{2m}$. The remaining $2m-2$ points satisfy the conditions in the induction hypothesis, and any two of them will not intersect $A_{2m-1}A_{2m}$. By the induction hypothesis, the $m-1$ segments can be paired to meet the requirements, and then adding the $A_{2m-1}A_{2m}$ segment gives a paired connection of $2m$ points to meet the requirements.

We return to the original question. If $k \geq 51$, all the 50 points in S can be coloured red, and the $k-50$ points in T can be coloured blue. No matter how the remaining points are coloured, the necessary sufficient condition in the lemma cannot be obtained. Therefore, $k \geq 51$ does not satisfy the requirement.

Suppose $k = 50$. Assume that 50 points are arbitrarily coloured, and there are a, b points coloured red in S, T, respectively. Let $a \geq b$, and then there are at least $a-b$ points in T uncoloured. (Otherwise, if the number of uncoloured points in T is less than $a-b$, then the number of coloured points is not less than $a+(50-(a-b)+1) = 51+b > 50$, a contradiction.)

To sum up, the desired maximum of k is equal to 50. □

Test Paper A2
(9:40 – 12:30; October 23, 2021)

1. (40 marks) As shown in Fig. 1.1, in acute $\triangle ABC$, $AB > AC$. M is the midpoint of minor arc $\overset{\frown}{BC}$ of the circumcircle Ω of $\triangle ABC$, and K is the intersection of the exterior angle bisector of $\angle BAC$ and the extension of BC. Take a point D (different from A) on the line passing through point A and perpendicular to BC such that $DM = AM$. Let the circumcircle of $\triangle ADK$ intersect circle Ω at point A and another point T.

   Prove that AT bisects segment BC.

Solution 1 As shown in Fig. 1.2, extend DM and it intersects BC at point X. Let A be the midpoint of minor arc $\overset{\frown}{BC}$ of circle Ω, and then points N, A, K are collinear.

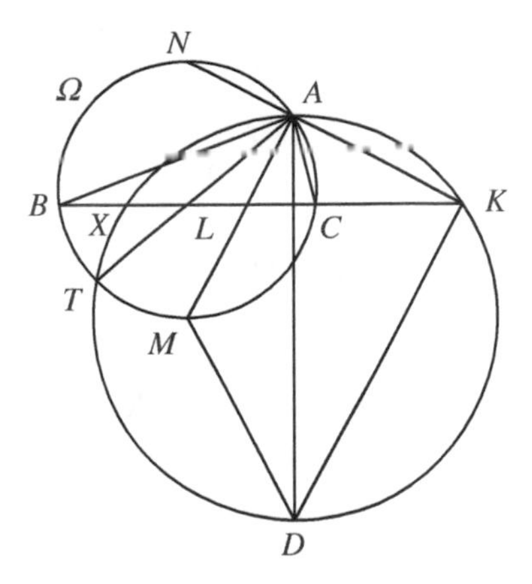

Fig. 1.1

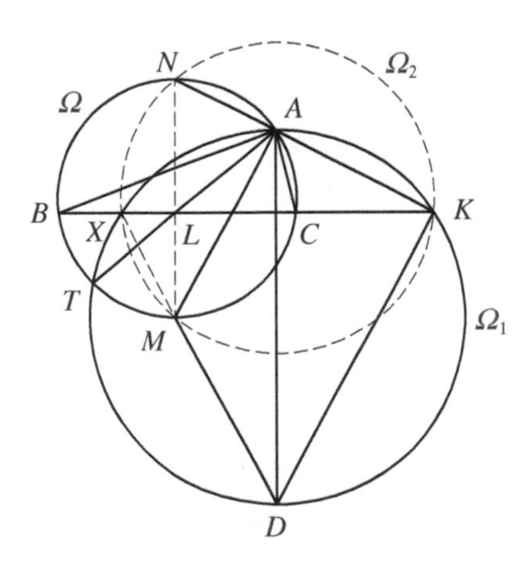

Fig. 1.2

Suppose L is the midpoint of BC. It is easy to find that AM is the angle bisector of $\angle BAC$, and thus $AM \perp AK$. And since $AD \perp BC, AM = DM$, we have

$$\angle AKX = 90^\circ - \angle DAK = \angle MAD = \angle ADM = \angle ADX.$$

Therefore, points A, K, D, X are concyclic, and hence points A, K, D, T, X are concyclic. We denote the circle as Ω_1.

Note that $MN//AD$ since both MN and AD are perpendicular to BC. Therefore, we have

$$\angle NMX = \angle ADX = \angle AKX = \angle NKX.$$

Hence, points N, K, M, X are concyclic. Denote the circle as Ω_2.

Applying Monge's theorem to circles Ω, Ω_1, Ω_2, we know that AT (radical axis of Ω, Ω_1), MN (radical axis of Ω, Ω_2) and XK (radical axis of Ω_1, Ω_2) are concurrent. Since MN and XK intersect at the point L, it follows that AT passes through point L, namely, AT bisects segment BC. □

Solution 2 As shown in Fig. 1.3, extend DM and it intersects BC at point X.

It is easy to find that AM is the angle bisector of $\angle BAC$, and thus $AM \perp AK$. And since $AD \perp BC, AM = DM$, we get

$$\angle AKX = 90° - \angle DAK = \angle MAD = \angle ADM = \angle ADX.$$

Therefore, points A, K, D, X are concyclic.

Take the midpoint L of BC. Extend AL and it intersects circle Ω at point T'.

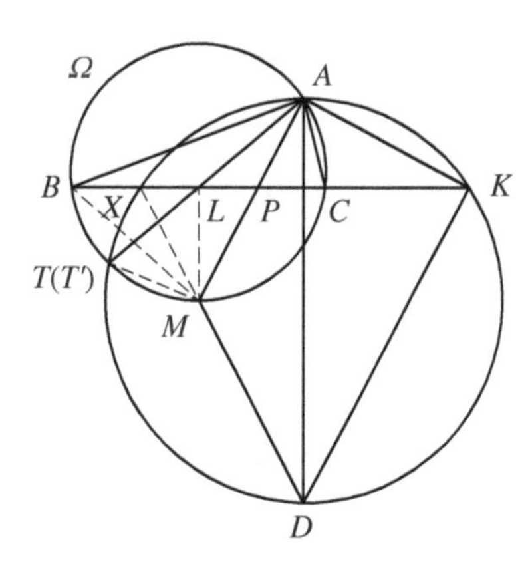

Fig. 1.3

Suppose AM and BC intersect at point P, and then

$$\begin{aligned}\angle LXM &= 90^\circ - \angle MDA = 90^\circ - \angle MAD \\ &= \angle APC = \angle ABC + \angle BAM \\ &= \angle ABC + \angle CAM = \angle AT'C + \angle CT'M \\ &= \angle AT'M + \angle LT'M.\end{aligned}$$

Therefore, points M, L, X, T' are concyclic.

In addition, there is $ML//AD$ since both ML and AD are perpendicular to BC. Therefore, we have

$$\angle AT'X = \angle LT'X = \angle LMX = \angle ADX.$$

and hence points A, D, T', X are concyclic.

Consequently, points A, K, D, T', X are concyclic. Hence, T' lies on the circumcircle of $\triangle AKD$, and thus T and T' coincide.

Therefore, AT passes through the midpoint L of BC, which means that AT bisects segment BC. □

2 (40 marks) Try to determine the largest positive number $M > 1$ with the following property: for any 10 different real numbers chosen from the interval $[1, M]$, three of them can be selected, denoted from smallest to largest by $a < b < c$, such that the quadratic equation $ax^2 + bx + c = 0$ has no real roots.

Solution $M_{\max} = 4^{255}$.

We first prove $M = 4^{255}$ satisfies the property stated in the question. Let $a_1 < a_2 < \cdots < a_{10}$ be any 10 different real numbers chosen from $[1, M]$. We use proof by contradiction. Suppose that for any $1 \le i < j < k \le 10$, the equation $a_i x^2 + a_j x + a_k = 0$ has real roots, so $a_j^2 - 4a_i a_k \ge 0$, i.e.,

$$\frac{a_j}{a_i} \ge 4 \cdot \frac{a_k}{a_j}, \quad 1 \le i < j < k \le 10. \qquad ①$$

We prove that for $1 \le k \le 9$, there is

$$\frac{a_{10}}{a_k} > 4^{29-k} - k. \qquad ②$$

For $k = 9, 8, 7, \ldots, 1$, we prove ② in turn. When $k = 9$, since $a_{10} > a_9$, we have $\frac{a_{10}}{a_9} > 1$ and the conclusion holds. Suppose that for some $2 \leq k \leq 9$, $\frac{a_{10}}{a_k} > 4^{2^{9-k}-1}$ holds. Then from ① we have

$$\frac{a_k}{a_{k-1}} \geq 4\frac{a_{10}}{a_k} > 4 \times 4^{2^{9-k}-1} = 4^{2^{9-k}},$$

and thus

$$\frac{a_{10}}{a_{k-1}} = \frac{a_{10}}{a_k} \cdot \frac{a_k}{a_{k-1}} > 4^{2^{9-k}-1} \times 4^{2^{9-k}} = 4^{2^{9-(k-1)}-1}.$$

Therefore, ② holds for all $1 \leq k \leq 9$. In particular, when $k = 1$, there is $\frac{a_{10}}{a_1} > 4^{2^8-1} = 4^{255}$.

However, since $a_1, a_{10} \in [1, M]$, it follows that $\frac{a_{10}}{a_1} \leq M = 4^{255}$, a contradiction! The proof by contradiction shows that the assumption is not true.

In addition, if $M > 4^{255}$, we can set $M = 4^{255}\lambda^{256}$, $\lambda > 1$. We take $a_1, a_2, \ldots, a_{10} \in [1, M]$ as follows,

$$a_{10} = M,$$
$$a_k = \frac{a_{10}}{4^{2^{2^{9-k}-1}}\lambda^{2^{9-k}}}, \quad 1 \leq k \leq 9.$$

so $1 = a_1 < a_2 < \cdots < a_9 < a_{10} = M$. For any $1 \leq i < j < k \leq 10$, there is

$$a_j^2 - 4a_ia_k \geq a_j^2 - 4a_{j-1}a_{10} = \left(\frac{a_{10}}{4^{2^{9-j}-1}\lambda^{2^{9-j}}}\right)^2 - 4 \cdot \frac{a_{10}}{4^{2^{10-j}-1}\lambda^{2^{10-j}}} \cdot a_{10} = 0.$$

Therefore, the equation $a_ix^2 + a_jx + a_k = 0$ has real roots. Therefore, $M > 4^{255}$ does not have the property stated in the question.

In summary, we get the desired largest positive number $M_{\max} = 4^{255}$.

□

3 (50 marks) Given integer $n \geq 2$, let non-negative real numbers $a_1, a_2, \ldots, a_n$ satisfy

$$a_1 \geq a_2 \geq \cdots \geq a_n, a_1 + a_2 + \cdots + a_n = n.$$

Find the minimum of $a_1 + a_1a_2 + a_1a_2a_3 + \cdots + a_1a_2 \cdots a_n$.

Solution For integer $m \geq 1$, we denote

$$f_m(x_1, x_2, \ldots, x_m) = x_1 + x_1x_2 + \cdots + x_1x_2\cdots x_m.$$

We prove the following lemma by induction on m.

Lemma *If the average of non-negative real numbers $x_1 \geq x_2 \geq \cdots \geq x_m$ is A and $A \leq 1$, then there is*

$$f_m(x_1, x_2, \ldots, x_m) \geq f_m(A, A, \ldots, A).$$

Proof of the lemma The conclusion clearly holds for $m = 1$. Assuming that the conclusion holds for m, we consider the case of $m+1$.

Let the average of non-negative real numbers $x_1 \geq x_2 \geq \cdots \geq x_{m+1}$ be $A \leq 1$.

Since $x_1 \geq A$, the average of $x_2, x_3, \ldots, x_{m+1}$, denoted by B, is not greater than A, and thus $B \leq A \leq 1$. By the induction hypothesis, we have

$$\begin{aligned} f_{m+1}(x_1, x_2, \ldots, x_{m+1}) &= x_1(1 + f_m(x_2, x_3, \ldots, x_{m+1})) \\ &\geq x_1(1 + f_m(B, B, \ldots, B)) \\ &= ((m+1)A - mB)(1 + B + B^2 + \cdots + B^m). \end{aligned}$$

In the following, we prove that

$$((m+1)A - mB)(1 + B + B^2 + \cdots + B^m) \geq f_{m+1}(A, A, \ldots, A).$$

In fact,

$$\begin{aligned} &((m+1)A - mB)(1 + B + B^2 + \cdots + B^m) - f_{m+1}(A, A, \ldots, A) \\ &\quad = ((m+1)A - mB)(1 + B + B^2 + \cdots + B^m) \\ &\quad\quad - A(1 + A + A^2 + \cdots + A^m) \\ &\quad = m(A - B)(1 + B + B^2 + \cdots + B^m) \\ &\quad\quad + A(B + B^2 + \cdots + B^m - A - A^2 - \cdots - A^m) \end{aligned}$$

$$\begin{aligned}
&= (A-B)(m(1+B+B^2+\cdots+B^m) \\
&\quad - A(1+(A+B)+(A^2+AB+B^2) \\
&\quad + \cdots + (A^{m-1}+A^{m-2}B+\cdots+B^{m-1})) \\
&\geq (A-B)(m(1+B+B^2+\cdots+B^m)-1-(1+B)-(1+B+B^2) \\
&\quad - \cdots - (1+B+\cdots+B^{m-1}) \text{ (Here } A \leq 1 \text{ is used)} \\
&= (A-B)(B+2B^2+3B^3+\cdots+mB^m) \\
&\geq 0.
\end{aligned}$$

The lemma is proven.

Since the average of non-negative real numbers $a_1 \geq a_2 \geq \cdots \geq a_n$ in the original question is 1, by the lemma we have

$$f_n(a_1, a_2, \ldots, a_n) \geq f_n(1, 1, \ldots, 1) = n.$$

Therefore, the desired minimum is equal to n. □

4 (50 marks) k consecutive positive integers are written down in ascending order to form a positive integer, which is called a k-consecutive number. For example, writing down 99, 100, 101 in order gives 99100101, which is a 3-consecutive number.

Prove that for any positive integers N, k, there exists a k-consecutive number that is divisible by N.

Solution N can be uniquely written as $N = N_1N_2$, where N_1 is coprime with 10 and N_2 has no prime factors other than 2 and 5. Since 10 and N_1 are coprime, by Euler's theorem, it follows that $10^{\varphi(N_1)} \equiv 1 \pmod{N_1}$.

Take a sufficiently large positive integer t and denote $m = t\varphi(N_1)$ such that

$$10^m - 10^{m-1} = 9 \cdot 10^{m-1} > N + k.$$

Assume that $x, x+1, x+2, \ldots, x+(k-1)$ are k consecutive m-digit numbers, i.e.,

$$10^{m-1} \leq x \leq x+(k-1) < 10^m,$$

and then the k-consecutive number obtained after writing down $x, x+1$, $x+2, \ldots, x+(k-1)$ in order is

$$\begin{aligned}M &= x \cdot 10^{(k-1)m} + (x+1)10^{(k-2)m} + \cdots + (x+k-2)10^m + (x+k-1) \\ &= x\sum_{j=1}^{k} 10^{(k-j)m} + \sum_{j=1}^{k-1} j \cdot 10^{(k-1-j)m} \\ &= Ax + B,\end{aligned}$$

where $A = \sum\limits_{j=1}^{k} 10^{(k-j)m}$, $B = \sum\limits_{j=1}^{k-1} j \cdot 10^{(k-1-j)m}$.

Note that $10^m \equiv 1 \pmod{N_1}$, and thus

$$A \equiv k \pmod{N_1},$$
$$B \equiv \frac{1}{2}k(k-1) \pmod{N_1}.$$

Since N_1 does not contain the prime factor 2, we have

$$(A, N_1) = (k, N_1) \mid (k(k-1), N_1) = \left(\frac{1}{2}k(k-1), N_1\right) = (B, N_1).$$

In addition, since $A \equiv 1 \pmod{10}$, it follows that A and N_2 are coprime, and thus $(A, N) = (A, N_1)$. Hence, (A, N) divides (B, N_1) and, of course, divides (B, N). Therefore, the congruence equation $Ax + B \equiv 0 \pmod{N}$ has a solution.

Consequently, there exists $10^{m-1} \le x < 10^{m-1} + N$, satisfying $N \mid Ax + B$. Also, since

$$10^m - 10^{m-1} > N + k,$$

there is

$$x + (k-1) < 10^{m-1} + N + (k-1) < 10^{m-1} + N + k < 10^m.$$

Thus, $x, x+1, x+2, \ldots, x+(k-1)$ are all m-digit numbers. Therefore, the k-consecutive number obtained by writing down these k numbers in ascending order is indeed M and is divisible by N. The conclusion is proved. $\square$

Test Paper B
(9:40 – 12:30; September 12, 2021)

1. (40 marks) As shown in Fig. 1.1, I is the incentre of $\triangle ABC$, $AB > AC$. Points P, Q are the projections of I on sides AB, AC, respectively. Line PQ intersects the circumcircle of $\triangle ABC$ at points X, Y (P is between X and Y). Given that points B, I, P, X are concyclic, prove that points C, I, Q, Y are concyclic.

Solution As shown in Fig. 1.2, denote the circumcircle of $\triangle ABC$ as ω.

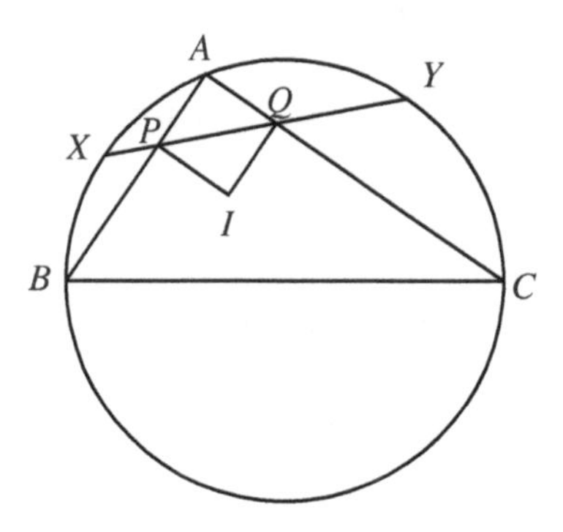

Fig. 1.1

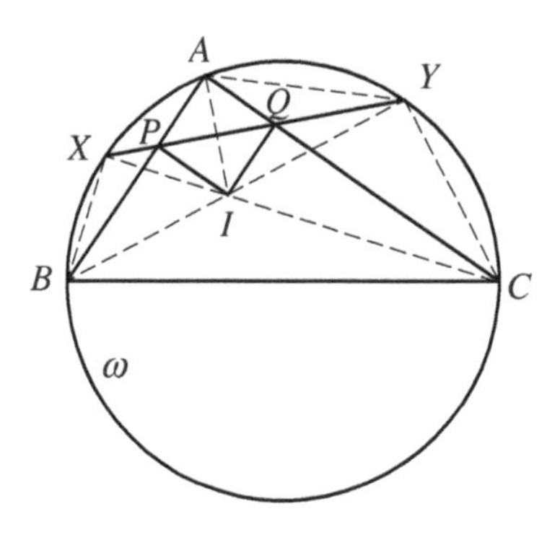

Fig. 1.2

Since $\angle API = \angle AQI = 90°$, points A, P, I, Q are concyclic. Therefore,

$$\begin{aligned}\angle BPQ &= \angle BPI + \angle IPQ \\ &= 90° + \angle IAQ \\ &= 90° + \frac{1}{2}\angle BAC \\ &= \angle BIC.\end{aligned}$$

And since points B, I, P, X are concyclic, we have $\angle BPX = \angle BIX$, and thus

$$\angle BIX + \angle BIC = \angle BPX + \angle BPQ = 180°.$$

Hence, points C, I, X are collinear.

Since B, I, P, X are concyclic, there is $\angle BXI = \angle BPI = 90°$, and thus $BX \perp IX$, i.e., $BX \perp CX$. Hence, $\angle BAC = \angle BXC = 90°$.

Therefore, quadrilateral $APIQ$ is a square, and PQ is the perpendicular bisector of segment AI.

Let Y' be the midpoint of $\overset{\frown}{AC}$. Then by the well-known theorem of the incentre, we have $Y'A = Y'I$. Thus, Y' is the intersection of the perpendicular bisector of AI and circle ω. In addition, since line PQ intersects circle ω at two points X, Y and Y' is obviously different from X, so Y' and Y coincide. Therefore, Y is the midpoint of $\overset{\frown}{AC}$.

Consequently, points B, I, Y are collinear. Therefore, $\angle IYC = \angle BYC = \angle BAC = 90° = \angle IQC$, and thus points C, I, Q, Y are concyclic. □

2 (40 marks) Find the maximum positive integer n such that there exist 8 integers x_1, x_2, x_3, x_4 and y_1, y_2, y_3, y_4 satisfying

$$\{0, 1, \ldots, n\} \subseteq \{|x_i - x_j| \mid 1 \leq i < j \leq 4\} \cup \{|y_i - y_j| \mid 1 \leq i < j \leq 4\}.$$

Solution Let n meet the requirement in the question. Then integers $x_1, x_2, x_3, x_4, y_1, y_2, y_3, y_4$ satisfy that $0, 1, \ldots, n$ all belong to set $X \cup Y$, where $X = \{|x_i - x_j| \mid 1 \leq i < j \leq 4\}$, $Y = \{|y_i - y_j| \mid 1 \leq i < j \leq 4\}$.

Note that $0 \in X \cup Y$. We may set $0 \in X$. Then there must be two numbers equal in x_1, x_2, x_3, x_4, and we may set $x_1 = x_2$. Then $X = \{0\} \cup \{|x_i - x_j| \mid 2 \le i < j \le 4\}$, so

$$|X| \le 1 + 3 = 4.$$

And since $|Y| \le \mathrm{C}_4^2 = 6$, it follows that $n+1 \le |X \cup Y| \le |X| + |Y| \le 10$, yielding $n \le 9$.

On the other hand, let $(x_1, x_2, x_3, x_4) = (0, 0, 7, 8)$, $(y_1, y_2, y_3, y_4) = (0, 4, 6, 9)$, and then

$$X = (0, 1, 7, 8), \quad Y = (2, 3, 4, 5, 6, 9),$$

which means that 0, 1, 9 belong to set $X \cup Y$.

In summary, the maximum positive integer n is 9. □

3 (50 marks) Given that $a, b, c, d \in [0, \sqrt[4]{2})$ satisfying $a^3 + b^3 + c^3 + d^3 = 2$, find the minimum value of $\dfrac{a}{\sqrt{2-a^4}} + \dfrac{b}{\sqrt{2-b^4}} + \dfrac{c}{\sqrt{2-c^4}} + \dfrac{d}{\sqrt{2-d^4}}$.

Solution When $a > 0$, we have $\dfrac{a}{\sqrt{2-a^4}} = \dfrac{a^3}{\sqrt{a^4(2-a^4)}} \ge a^3$.

When $a = 0$, $\dfrac{a}{\sqrt{2-a^4}} \ge a^3$ also holds.

Therefore,

$$\frac{a}{\sqrt{2-a^4}} + \frac{b}{\sqrt{2-b^4}} + \frac{c}{\sqrt{2-c^4}} + \frac{d}{\sqrt{2-d^4}} \ge a^3 + b^3 + c^3 + d^3 = 2.$$

The equal sign of the above inequality holds when $a = b = 1$ and $c = d = 0$.

Therefore, the minimum value of $\dfrac{a}{\sqrt{2-a^4}} + \dfrac{b}{\sqrt{2-b^4}} + \dfrac{c}{\sqrt{2-c^4}} + \dfrac{d}{\sqrt{2-d^4}}$ is 2. □

4 (50 marks) Let a be a positive integer. Sequence $\{a_n\}$ satisfies:

$$a_1 = a, \quad a_{n+1} = a_n^2 + 20, \quad n = 1, 2, \ldots.$$

(1) Prove that there exists a positive integer a that is not a cube such that one of the terms in sequence $\{a_n\}$ is a cube;
(2) prove that at most one of the terms in sequence $\{a_n\}$ is a cube.

Solution (1) Note that $14^2 + 20 = 216 = 6^3$, so take $a = 14$, and then $a_2 = a + 20 = 6^3$ is a cube. Therefore, $a = 14$ satisfies the condition.

(2) Suppose that there exists a cube number in $\{a_n\}$ and a_k is the first cube that appears.

Since for any integer m, we have $m \equiv 0, \pm1, \pm2, \pm3, \pm4 \pmod 9$. Thus, $m^3 \equiv 0, \pm1 \pmod 9$, so $a_k \equiv 0, \pm1 \pmod 9$.

Hence,

$$a_{k+1} = a_k^2 + 20 \equiv 2,\ 3 \pmod 9,$$

$$a_{k+2} = a_{k+1}^2 + 20 \equiv 6,\ 2 \pmod 9,$$

$$a_{k+3} = a_{k+2}^2 + 20 \equiv 2,\ 6 \pmod 9,$$

By induction, we know that $a_{k+i} \equiv 2, 6 \pmod 9$, $i = 4, 5, \ldots$. Thus, the terms after a_k are not $0, \pm1 \pmod 9$, so they are not cubes.

Therefore, at most one of the terms in sequence $\{a_n\}$ is a cube. □

Test Paper B1
(9:40 – 12:30; October 6, 2021)

1 (40 marks) Let a, b, c be non-negative real numbers. Denote

$$S = a + 2b + 3c, \quad T = a + b^2 + c^3.$$

(1) Find the minimum of $T - S$.
(2) if $S = 4$, find the maximum of T.

Solution (1) Note that $b, c \geq 0$. By the inequality of arithmetic and geometric means, we get

$$b^2 + 1 \geq 2b, \quad c^3 + 1 + 1 \geq 3\sqrt[3]{c^3 \cdot 1 \cdot 1} = 3c.$$

Thus

$$T - S = (b^2 - 2b) + (c^3 - 3c) \geq -1 - 2 = -3.$$

When a is any non-negative real number and $b = c = 1$, $T - S$ takes the minimum -3.

(2) Since $a, b, c \geq 0$ and $S = a + 2b + 3c = 4$, we know that $0 \leq b \leq 2$, $0 \leq c \leq \frac{3}{4}$. Therefore,

$$2b - b^2 = b(2 - b) \geq 0, \quad 3c - c^3 = c(3 - c^2 \geq 0.$$

Thus,

$$T = a + b^2 + c^3 \leq a + 2b + 3c = S = 4.$$

When $a = 4, b = c = 0$ (or $a = 0, b = 2, c = 0$), T takes the maximum 4. □

2 (40 marks) As shown in Fig. 2.1, in $\triangle ABC$, $AB > AC$. Points X, Y in $\triangle ABC$ are on the bisector of $\angle BAC$, satisfying $\angle ABX = \angle ACY$. Let the extension of BX and segment CY intersect at point P. The circumcircle ω_1 of $\triangle BPY$ and the circumcircle ω_2 of $\triangle CPX$ intersect at P and another point Q.

Prove that points A, P, Q are collinear.

Solution Since $\angle BAX = \angle CAY$, $\angle ABX = \angle ACY$, it follows that $\triangle ABX \backsim \triangle ACY$. Hence,

$$\frac{AB}{AC} = \frac{AX}{AY}. \quad ①$$

As shown in Fig. 2.2, extend AX and it intersects circles ω_1, ω_2 at points U, V, respectively. Then

$$\angle AUB = \angle YUB = \angle YPB = \angle YPX,$$
$$= \angle XVC = \angle AVC,$$

and thus $\triangle ABU \backsim \triangle ACV$. Therefore,

$$\frac{AB}{AC} = \frac{AU}{AV}. \quad ②$$

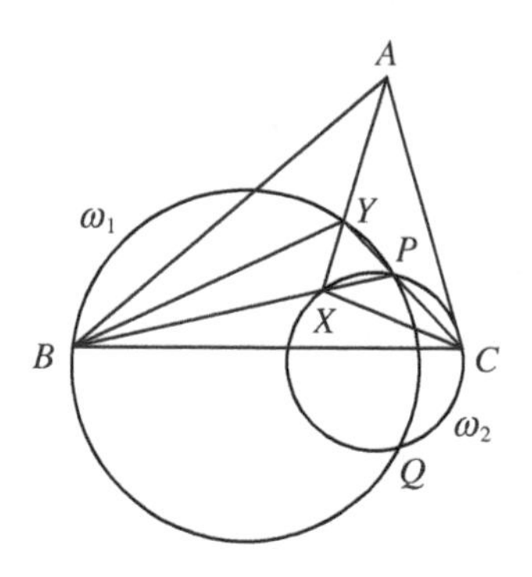

Fig. 2.1

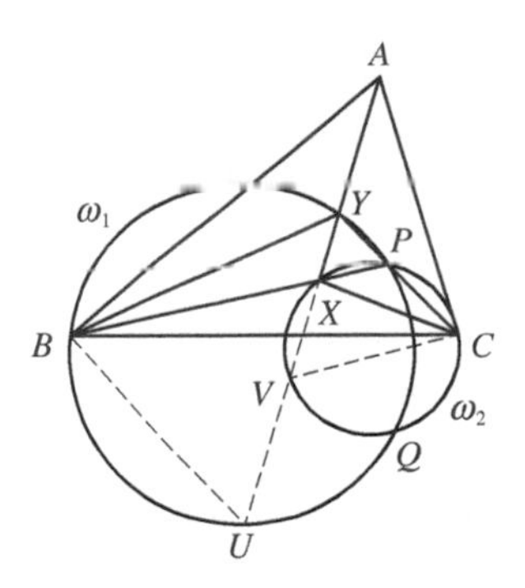

Fig. 2.2

From ① and ②, we get $\dfrac{AX}{AY} = \dfrac{AU}{AV}$, i.e.,

$$AU \cdot AY = AV \cdot AX.$$

The two sides of the above equation are the circle powers of point A to circles ω_1, ω_2, respectively. This implies that A is on the radical axis (i.e., line PQ) of circles ω_1, ω_2. In other words, points A, P, Q are collinear. □

3 (50 marks) Let m, n be integers greater than 1, and n is not a perfect square. If n^2+n+1 is divisible by m, prove that

$$|m-n| > \sqrt{3n} - 2.$$

Solution Let $n = m + k$, and then

$$k^2+k+1 = n^2+n+1+m(m-2n-1) \equiv n^2+n+1 \equiv 0 \pmod{m}.$$

Therefore, k^2+k+1 is divisible by m. Note that k^2+k+1 is a positive integer, so we can set

$$k^2+k+1 = mt, \qquad ①$$

where t is a positive integer.

If $t = 1$, we get $m = k^2+k+1$, so $n = m+k$ is a perfect square, which contradicts the condition. Hence, $t > 1$.

Note that $k^2+k = k(k+1)$ is the product of two adjacent integers, which is even, so k^2+k+1 is odd. From ①, we know that t is odd, so $t \geq 3$.

Therefore, by using ①, we get

$$k^2 + k + 1 \geq 3m = 3(n - k),$$

and thus $3n \leq k^2 + 4k + 1 < (k+2)^2$.

Consequently,

$$|k + 2| > \sqrt{3n},$$

and hence $|k| > \sqrt{3n} - 2$, i.e., $|m - n| > \sqrt{3n} - 2$. □

4 (50 marks) 9 football teams play in a single round-robin tournament (each two teams play once). In each match, the winner gets 3 points, the loser gets 0 point, and both teams get 1 point for a draw. The score of a team is the sum of the points it gets in each match. After the tournament, it is found that the 9 teams have different points from each other. The 9 teams are recorded as $T_1, T_2, \ldots, T_9$ in descending order of points. It is known that team T_1 has a record of 3 wins, 4 draws and 1 loss, and team T_9 has a record of 0 win, 5 draws and 3 losses. Questions:

(1) Is it possible for T_3 to win against T_4?
(2) Is it possible for T_4 to win against T_3?

Solution According to the scoring rules, we know that the scores of T_1, T_9 are $13, 5$, respectively. From the fact that the scores of the 9 teams are different from each other, we know that the scores of $T_1, T_2, \ldots, T_9$ can only be $13, 12, \ldots, 5$, respectively.

Let there be x matches in the tournament that result in a win or loss, and the sum of the score of each team is

$$13 + 12 + \cdots + 6 + 5 = 81.$$

Then $3x + 2(9 \times 8 \div 2 - x) = 81$, so $x = 9$.

Since the score of T_4 is 10 and the score of T_5 is 9, and a team without a win can score at most 8 points, we know that T_4 and T_5 both have wins; since the scores of T_2 and T_3 are more than 10, and a team with no more than one win can score at most 10 points, T_2 and T_3 each have at least two wins. And since T_1 has 3 wins, it follows that $T_1, T_2, \ldots, T_5$ has at least 9 wins. Combining $x = 9$ we know that the number of wins for T_2, T_3, T_4, T_5 is exactly 2, 2, 1, 1, respectively. In particular, the number of wins for T_4 is 1. In addition, since the score of T_4 is 10, the record of T_4 is 1 win, 7 draws and 0 loss. Because T_4 has no losses, it is impossible for T_3 to win against T_4.

On the other hand, it is possible for T_4 to win against T_3. In fact, an example of the full match result for the tournament can be constructed as follows (such examples are not unique).

	T_1	T_2	T_3	T_4	T_5	T_6	T_7	T_8	T_9	win	draw	loss	score
T_1		draw	draw	draw	loss	draw	win	win	win	3	4	1	13
T_2	draw		draw	draw	win	draw	draw	draw	win	2	6	0	12
T_3	draw	draw		loss	draw	draw	draw	win	win	2	5	1	11
T_4	draw	draw	win		draw	draw	draw	draw	draw	1	7	0	10
T_5	win	loss	draw	draw		draw	draw	draw	draw	1	6	1	9
T_6	draw	draw	draw	draw	draw		draw	draw	draw	0	8	0	8
T_7	loss	draw	draw	draw	draw	draw		draw	draw	0	7	1	7
T_8	loss	draw	loss	draw	draw	draw	draw		draw	0	6	2	6
T_9	loss	loss	loss	draw	draw	draw	draw	draw		0	5	3	5

In summary, it is impossible for T_3 to win against T_4, but it is possible for T_4 to win against T_3.

China Mathematical Olympiad

2020 (Changsha, Hunan)

2020 China Mathematical Olympiad (also named the 36th National Mathematics Winter Camp for Middle School Students), was sponsored by the China Mathematical Society and hosted by the Hunan Mathematical Society and the Changsha Changjun Middle School. It was held at the Changsha Changjun Middle School in Changsha from November 22 to 28, 2020, with 32 teams from 31 provinces, municipalities and autonomous regions in China's mainland, Hong Kong Special Administrative Region of China (attending online), including 455 participants.

Two tests were held during the competition, and according to their test scores, 60 students were selected into the national training team. There are 449 senior students from China's Mainland participating the competition, among which 144 ones won the gold medals, 195 ones won the silver medals, and 110 ones won the bronze medals.

The members of the main examination committee of this winter camp are:

Xiong Bin (East China Normal University);
Yu Hongbing (Suzhou University);
Yao Yijun (Fudan University);
Qu Zhenhua (East China Normal University);
Xiao Liang (Peking University);
Yang Shiwu (Peking University);
Ai Yinghua (Tsinghua University);
He Yijie (East China Normal University);

Wang Bin (Chinese Academy of Sciences Academy of Mathematics and Systems Science);
Wang Xinmao (University of Science and Technology of China);
Luo Ye (Hong Kong University);
Li Weigu (Peking University);
Li Ting (Sichuan University);
Fu Yunhao (Southern University of Science and Technology);
Ji Chungang (Nanjing Normal University);
Li Ming (Nankai University);
Zhang Sihui (University of Shanghai for Science and Technology);
Lai Li (Tsinghua University);
Yang Biaogui (Fujian Normal University);
Wu Yuchi (East China Normal University);
Lin Tiangqi (East China Normal University);
Luo Zhenhua (East China Normal University);
Liao Yuxuan (Peking University);
Bai Yaoyuan (Peking University);
Zhang Chengchun (Fudan University).

First Day
(8:00 – 12:30; November 24, 2020)

1 Suppose all the odd terms of sequence $\{z_n\}_{n\geq 1}$ are real and all the even terms are pure imaginary, and for any positive integer k, there is $|z_k z_{k+1}| = 2^k$. For positive integer n, denote $f_n = |z_1 + z_2 + \cdots + z_n|$.

(1) Find the minimum possible value of f_{2020};
(2) find the minimum possible value of $f_{2020} \cdot f_{2021}$.

(Contributed by He Yijie)

Solution (1) Let $z_{2m-1} = a_m$, $z_{2m} = b_m\mathrm{i}$, where $a_m, b_m \in \mathbb{R}$ $(m = 1, 2, \ldots)$. For any positive integer k, there is

$$\frac{|z_{k+2}|}{|z_k|} = \frac{|z_{k+1}z_{k+2}|}{|z_k z_{k+1}|} = \frac{2^{k+1}}{2^k} = 2.$$

Thus, $|a_{m+1}| = 2|a_m|$, $|b_{m+1}| = 2|b_m|$ $(m = 1, 2, \ldots)$.

Denote $a = a_1$, $b = b_1$, so $|a| \cdot |b| = 2$ and

$$|a_m| = 2^{m-1}|a|, \quad |b_m| = 2^{m-1}|b| \quad (m = 1, 2, \ldots).$$

Denote $S = a_1 + a_2 + \cdots + a_{1010}$, $T = b_1 + b_2 + \cdots + b_{1010}$. It is easy to find that

$$\begin{aligned}|S| &\geq |a_{1010}| - |a_{1009}| - |a_{1008}| - \cdots - |a_1| \\ &= (2^{1009} - 2^{1008} - 2^{1007} - \cdots - 1) \cdot |a| = |a|.\end{aligned}$$

Similarly, we can obtain $|T| \geq |b|$. Therefore,

$$f_{2020} = |S + T\mathrm{i}| = \sqrt{S^2 + T^2} \geq \sqrt{a^2 + b^2} \geq \sqrt{2|ab|} = 2.$$

It is easy to verify that the following sequence $\{z_n\}$ satisfies the condition and such that $f_{2020} = 2$, i.e.,

$$z_{2m-1} = \begin{cases} -2^{m-1}\sqrt{2}, & m \leq 1009, \\ 2^{m-1}\sqrt{2}, & m \geq 1010, \end{cases}$$

$$z_{2m} = \begin{cases} -2^{m-1}\sqrt{2}\mathrm{i}, & m \leq 1009, \\ 2^{m-1}\sqrt{2}\mathrm{i}, & m \geq 1010. \end{cases}$$

To sum up, the minimum value of f_{2020} is 2.

(2) By following the notations and conclusions in the solution to question (1), we have

$$\begin{aligned}f_{2020} \cdot f_{2021} &= |S + T\mathrm{i}| \cdot |S + a_{1011} + T\mathrm{i}| \\ &= |(S + T\mathrm{i})(S + a_{1011} - T\mathrm{i})| \\ &\geq |\mathrm{Im}((S + T\mathrm{i})(S + a_{1011} - T\mathrm{i}))| \\ &= |a_{1011}T| = 2^{1010}|a| \cdot |T| \\ &\geq 2^{1010}|a| \cdot |b| = 2^{1011}. \qquad ①\end{aligned}$$

In the following, we construct a sequence such that Formula ① takes the equal sign. For undetermined positive numbers a and $b = \frac{2}{a}$, let

$$z_{2m-1} = \begin{cases} -2^{m-1}a, & m \leq 1010, \\ 2^{m-1}a, & m \geq 1011, \end{cases}$$

$$z_{2m} = \begin{cases} -2^{m-1}b\mathrm{i}, & m \leq 1009, \\ 2^{m-1}b\mathrm{i}, & m \geq 1010. \end{cases}$$

At this point $S = -(2^{1010}-1)a$, $S + a_{1011} = a$, $T = b = \frac{2}{a}$. Since there is already $|T| = |b|$, in order for Formula ① to take the equal sign, it is only necessary that

$$\mathrm{Re}((S+T\mathrm{i})(S+a_{1011}-T\mathrm{i})) = S(S+a_{1011}) + T^2 = 0,$$

namely,

$$-(2^{1010}-1)a^2 + \frac{4}{a^2} = 0.$$

The solution is $a = \sqrt{\frac{2}{\sqrt{2^{1010}-1}}}$.

In summary, the minimum value of $f_{2020} \cdot f_{2021}$ is 2^{1011}. □

2 Given integer $m > 1$, find the smallest positive integer n such that for any integers $a_1, \ldots, a_n$ and $b_1, \ldots, b_n$, there exist integers $x_1, \ldots, x_n$ that satisfy the following two conditions:

(1) at least one of $x_1, \ldots, x_n$ is coprime with m;

(2) $\sum\limits_{i=1}^{n} a_i x_i \equiv \sum\limits_{i=1}^{n} b_i x_i \equiv 0 \pmod{m}$.

(Contributed by Ai Yinghua)

Solution Suppose there are k different prime factors of m. The prime factorization of m is $m = \prod\limits_{i=1}^{k} p_i^{\alpha_i}$, where k is a positive integer, $p_1 < \cdots < p_k$ are prime numbers, and $\alpha_1, \ldots, \alpha_k$ are positive integers. Then the minimum value of the desired n is $2k+1$.

On one hand, consider the following construction. For each $1 \le i \le k$, denote $M_i = \frac{m}{p_i^{\alpha_i}}$. Define ordered set of integers

$$(\bar{a}_1, \ldots, \bar{a}_{2k}) = (M_1, \ldots, M_k,\ 0, \ldots, 0),$$

$$(\bar{b}_1, \ldots, \bar{b}_{2k}) = (0, \ldots, 0,\ M_1, \ldots, M_k).$$

For each positive integer $n \le 2k$, consider $\bar{a}_1, \ldots, \bar{a}_n$ and $\bar{b}_1, \ldots, \bar{b}_n$, and for integers $x_1, \ldots, x_n$ satisfying the congruence

$$\sum_{i=1}^{n} \bar{a}_i x_i = \sum_{i=1}^{n} \bar{b}_i x_i \equiv 0 \pmod{m},$$

we will prove that each x_i is not coprime with m.

If $n < 2k$, then add $x_{n+1} = \cdots = x_{2k} = 0$, which gives

$$\sum_{i=1}^{2k} \bar{a}_i x_i - \sum_{i=1}^{2k} \bar{b}_i x_i \equiv 0 \pmod{m}.$$

Hence, we have

$$\sum_{i=1}^{k} M_i \cdot x_i \equiv 0 \pmod{m},$$

$$\sum_{i=1}^{k} M_i \cdot x_{k+i} \equiv 0 \pmod{m}.$$

Note that for each $j \leq k$, only M_j in $M_1, \ldots, M_k$ is not a multiple of p_j. Combining the above equation it is clear that x_j and x_{k+j} are both multiples of p_j, and thus they are not coprime with m. This proves the above assertion and thus proves that the conditions stated in the question are not satisfied for any positive integer n not exceeding $2k$.

On the other hand, let's prove that $n = 2k + 1$ satisfies the conditions in the question. We first prove the following lemma.

Lemma *Suppose integer $n > 2$, p is a prime number and e is a positive integer. Then for any integers $a_1, \ldots, a_n$ and $b_1, \ldots, b_n$, there exist integers $x_1, \ldots, x_n$ such that*

$$\sum_{j=1}^{n} a_j x_j \equiv \sum_{j=1}^{n} b_j x_j \equiv 0 \pmod{p^e},$$

and except for at most two indices j, all x_j are congruent to 1 modulo p^e.

Proof of lemma We may assume that $a_1, \ldots, a_n, b_1, \ldots, b_n$ are not all 0. It may further be useful to suppose

$$v_p(a_1) = \min\{v_p(a_j) \mid 1 \leq j \leq n\} = d,$$

where $v_p(a)$ denotes the number of factors p in a. We consider a stronger system of congruences

$$\begin{cases} \displaystyle\sum_{j=1}^{n} \frac{a_j}{p^d} \cdot x_j \equiv 0 \pmod{p^e}, \\ \displaystyle\sum_{j=1}^{n} b_j x_j \equiv 0 \pmod{p^e}. \end{cases}$$

Since $\frac{a_1}{p^d}$ and p^e are coprime, x_1 can be expressed in terms of $x_2, \ldots, x_n$ from the previous first equation as

$$x_1 \equiv -\left(\frac{a_1}{p^d}\right)^{-1}\left(\sum_{j=2}^{n}\frac{a_j}{p^d}\cdot x_j\right) \pmod{p^e}, \qquad ①$$

where $\frac{a_1}{p^d}$ denotes the modular inverse of $\frac{a_1}{p^d}$ modulo p^e. Putting this solution into the second equation gives

$$\sum_{j=1}^{n} b_j x_j \equiv 0 \pmod{p^e}.$$

After combining like terms, it becomes

$$\sum_{j=1}^{n} b_j x_j \equiv 0 \pmod{p^e}. \qquad ②$$

If $c_2, \ldots, c_n$ are all zero, then Equation ② is always established. At this point, taking

$$x_2 \equiv \cdots \equiv x_n \equiv 1 \pmod{p^e}$$

and substituting it into Equation ① can find x_1. If $c_2, \ldots, c_n$ are not all zero, similar to the treatment of the first equation, we might set

$$v_p(c_2) = \min\{v_p(c_i) \mid 2 \le i \le n\} = d'.$$

Then equation

$$\sum_{j=2}^{n}\frac{c_j}{p^{d'}}\cdot x_j \equiv 0 \pmod{p^e}$$

is stronger than Equation ② and has solution

$$x_2 \equiv -\left(\frac{c_2}{p^{d'}}\right)^{-1}\left(\sum_{j=3}^{n}\frac{c_j}{p^{d'}}\cdot x_j\right) \pmod{p^e}. \qquad ③$$

Take $x_3 \equiv \cdots \equiv x_n \equiv 1 \pmod{p^e}$, we can solve x_2 from Equation ③, and then find the solution o x_1 by substituting it into Equation ①. This completes the proof of the Lemma.

To return to the original question, for each $i \leq k$, it is known from the Lemma that there exists integers $x_{i,1}, \ldots, x_{i,2k+1}$ such that

$$\sum_{j=1}^{2k+1} a_i x_{i,j} \equiv \sum_{j=0}^{2k+1} b_i x_{i,j} \equiv 0 \pmod{p_i^{a_i}}$$

and there is $x_{i,j} \equiv 1 \pmod{p_i^{\alpha_i}}$ except for at most two indices j. We call such exceptional indices j "bad", and each $1 \leq i \leq k$ produces at most two "bad" indices, so that there are at most $2k$ "bad" indices. Since the total number of indices j is $2k+1$, there is one index t that is not "bad".

For each $j \leq 2k+1$, define x_j as the solutions to the following system of congruences

$$x_j \equiv x_{1,j} \pmod{p_1^{\alpha_1}}, \ldots, x_j \equiv x_{k,j} \pmod{p_k^{\alpha_k}}.$$

By the Chinese Remainder Theorem, all of the above systems of congruences have solutions. The $x_1, \ldots, x_{2k+1}$ thus obtained clearly satisfies

$$\sum_{i=1}^{2k+1} a_i x_i = \sum_{i=1}^{2k+1} b_i x_i \equiv 0 \pmod{m}.$$

For the former not "bad" index, there is

$$x_t \equiv x_{1,t} \equiv 1 \pmod{p_1^{\alpha_1}}, \ldots, x_t \equiv x_{k,t} \equiv 1 \pmod{p_k^{\alpha_k}}.$$

In particular, x_t and m are coprime. This proves that $n = 2k+1$ satisfies the conditions of the question. □

3. Suppose that positive integer n can be divisible by exactly 36 different prime numbers. For $k = 1, 2, \ldots, 5$, let c_k be the number of integers on interval $\left[\frac{(k-1)n}{5}, -\frac{kn}{5}\right]$ that are coprime with n. It is known that $c_1, c_2, \ldots, c_5$ are not all equal. Prove that

$$\sum_{1 \leq i < j \leq 5} (c_i - c_j)^2 \geq 2^{36}.$$

(Contributed by Fu Yunhao)

Solution Suppose that $n = 2$, where $p_1, \ldots, p_{36}$ are different prime numbers and $\alpha_1, \ldots, \alpha_{36}$ are positive integers. Obviously if $\frac{kn}{5}$ $(k = 0, 1, 2, 3, 4, 5)$ are integers, then they are not coprime with n. Since either integers x and $n-x$ are both coprime with n or neither of them are coprime with n, it follows that $c_1 = c_5$, $c_2 = c_4$.

For positive integer x, define function $\mu(x)$ as follows: if x is divisible by the square of some prime number, then $\mu(x) = 0$; if x is the product of t different prime numbers (t can be 0, and at this point $x = 1$), then

$$\mu(x) = (-1)^t.$$

For a factor m of n and positive integer y, the number of multiples of m in $[1, y]$ is exactly $\left[\frac{y}{m}\right]$. So by the inclusion-exclusion principle, the number of integers in $[1, y]$ that are coprime with n is

$$\sum_{m \mid n} \mu(m) \left(\left[\frac{y}{m}\right]\right).$$

Substituting into the problem yields for $k = 0, \ldots, 5$, we have

$$\begin{aligned} c_k &= \sum_{m \mid n} \mu(m) \left(\left[\frac{kn}{5m}\right] - \left[\frac{(k-1)n}{5m}\right]\right) \\ &= \sum_{d \mid n} \mu\left(\frac{n}{d}\right) \left(\left[\frac{kd}{5}\right] - \left[\frac{(k-1)d}{5}\right]\right). \end{aligned}$$

Let $p_1, \ldots, p_{36}$ be all the prime factors of n. Note that the above equation only needs to be summed over d satisfying $\frac{n}{d} \mid p_1 p_2 \cdots p_{36}$. For these d, denote $r_k(d) := \left[\frac{kd}{5}\right] - \left[\frac{(k-1)d}{5}\right]$ and consider the cases that d modulo 5 with different remainders.

$d = 5m$	All $r_k(d)$ are m				
$d = 5m+1$	$r_1(d) = m$	$r_2(d) = m$	$r_3(d) = m$	$r_4(d) = m$	$r_5(d) = m+1$
$d = 5m+2$	$r_1(d) = m$	$r_2(d) = m$	$r_3(d) = m+1$	$r_4(d) = m$	$r_5(d) = m+1$
$d = 5m+3$	$r_1(d) = m$	$r_2(d) = m+1$	$r_3(d) = m$	$r_4(d) = m+1$	$r_5(d) = m+1$
$d = 5m+4$	$r_1(d) = m$	$r_2(d) = m+1$	$r_3(d) = m+1$	$r_4(d) = m+1$	$r_5(d) = m+1$

- If $25 \mid n$, then d that satisfies $\frac{n}{d}$ divides $p_1 p_2 \cdots p_{36}$ are all multiples of 5, so all c_k are the same, a contradiction!
- If there exists some prime factor, say p_1, that is congruent to 1 modulo 5, then d satisfying $p_1 \mid \frac{n}{d}$ and d satisfying $p_1 \nmid \frac{n}{d}$ can be paired by quotient p_1. The remainder of d modulo 5 of each pair is the same, and

$\mu\left(\frac{n}{d}\right)$ is the opposite number of each other, so that all c_k obtained are also exactly the same, a contradiction!

In the following, we consider the case that n does not have a prime factor that is congruent to 1 modulo 5 and $25 \nmid n$. Note that

$$c_2 - c_1 = \sum_{d \mid n,\, d \equiv 3,4 \pmod 5} \mu\left(\frac{n}{d}\right),$$

$$c_3 - c_1 = \sum_{d \mid n,\, d \equiv 2,4 \pmod 5} \mu\left(\frac{n}{d}\right).$$

If $5 \parallel n$, the difference of c_1, c_2, c_3, c_4, c_5 becomes the opposite of the original after replacing n with $\frac{n}{5}$, so $\sum_{1 \le i < j \le 5} (c_i - c_j)^2$ remains invariant. In the following, only the cases of $n = p_1^{\alpha_1} p_2^{\alpha_2} \cdots p_{36}^{\alpha_{36}}$ or $n = p_1^{\alpha_1} p_2^{\alpha_2} \cdots p_{35}^{\alpha_{35}}$ and $p_i \equiv 2, 3, 4 \pmod 5$ are considered. Let $S_i = \sum_{d \mid n,\, d \equiv i \pmod 5} \mu\left(\frac{n}{d}\right)$, then it is easy to know that

$$S_1 + S_2 + S_3 + S_4 = 0$$

and

$$\begin{aligned}
\sum_{1 \le i < j \le 5} (c_i - c_j)^2 &= 4(S_3 + S_4)^2 + 2(S_2 + S_4)^2 + 2(S_2 - S_3)^2 \\
&= (S_3 + S_4 - S_1 - S_2)^2 + \frac{1}{2}(S_2 + S_4 - S_1 - S_3)^2 \\
&\quad + 2(S_2 - S_3)^2 \\
&= \frac{3}{2}(S_1 - S_4)^2 + \frac{7}{2}(S_2 - S_3)^2 + (S_1 - S_4)(S_2 - S_3).
\end{aligned}$$

If we let $Z = S_1 - S_4 + (S_2 - S_3)\mathrm{i} = a + b\mathrm{i}$, then

$$\sum_{1 \le i < j \le 5} (c_i - c_j)^2 = \frac{3}{2}a^2 + \frac{7}{2}b^2 + ab.$$

Let $n \equiv 2^t \pmod 5$, for different indices $j_1, j_2, \ldots, j_s$, if primes $p_{j_1}, p_{j_2}, \ldots, p_{j_s}$ have a_2 numbers congruent to 2 modulo 5, a_3 numbers congruent to 3 modulo 5 and a_4 numbers congruent to 4 modulo 5, then for integer

$$d = \frac{n}{p_{j_1} p_{j_2} \cdots p_{j_s}} \equiv 2^{t-a_2+a_3-2a_4} \pmod 5,$$

we have

$$\mu\left(\frac{n}{d}\right) = (-1)^s = (-1)^{a_2+a_3+a_4}.$$

Since for any integer ℓ there is

$$2^{4\ell+1} \equiv 2 \pmod 5, \quad 2^{4\ell+2} \equiv 4 \pmod 5,$$
$$2^{4\ell+3} \equiv 3 \pmod 5, \quad 2^{4\ell} \equiv 1 \pmod 5,$$
$$\mathrm{i}^{4\ell+1} = \mathrm{i}, \quad \mathrm{i}^{4\ell+2} = -1, \quad \mathrm{i}^{4\ell+3} = -\mathrm{i}, \mathrm{i}^{4\ell} = 1.$$

the contribution of $\mu\left(\frac{n}{d}\right)$ for Z is

$$\mathrm{i}^{t-a_2+a_3-2a_4} \cdot (-1)^{a_2+a_3+a_4} = \mathrm{i}^t \cdot \mathrm{i}^{a_2} \cdot 1^{a_4} \cdot (-\mathrm{i})^{a_3}.$$

Now suppose that $p_1, p_2, \ldots, p_{35}$ (or $p_1, p_2, \ldots, p_{36}$) have b_2 numbers congruent to 2 modulo 5, b_3 numbers congruent to 3 modulo 5 and b_4 numbers congruent to 4 modulo 5, and then

$$Z = \mathrm{i}^t (1+\mathrm{i})^{b_2} (1+1)^{b_4} (1-\mathrm{i})^{b_3}.$$

From the above equation, it is known that Arg$Z$ is a multiple of $\frac{\pi}{4}$ and $|Z|$ is a positive integer power of $\sqrt{2}$ with the power no less than 35.

When $|Z| = \sqrt{2^{35}}$, $n = p_1^{\alpha_1} p_2^{\alpha_2} \cdots p_{35}^{\alpha_{35}}$ and $b_4 = 0$. At this point it is easy to find out that $ArgZ$ is an odd multiple of $\frac{\pi}{4}$, and thus $\frac{\pi}{4}$. Therefore,

$$\sum_{1 \le i < j \le 5} (c_i - c_j)^2 \ge \left(\frac{3}{2} + \frac{7}{2} - 1\right) 2^{34} = 4 \cdot 2^{34} = 2^{36}.$$

When $|Z| = \sqrt{2^{36}}$, there is

$$\begin{aligned} \sum_{1 \le i < j \le 5} (c_i - c_j)^2 &= \frac{3}{2}a^2 + \frac{7}{2}b^2 + ab \\ &= a^2 + 3b^2 + \frac{1}{2}(a+b)^2 \\ &\ge a^2 + b^2 = |Z|^2 \ge 2^{36}. \end{aligned}$$

To summarize, the proof is completed. □

Second Day
(8:00 – 12:30; November 25, 2020)

4 As shown in Fig. 4.1, let acute $\triangle ABC$ be inscribed to circle ω, $AB > AC$. M is the midpoint of minor arc $\overparen{BC}$ of circle ω and K is the antipodal point of point A on circle ω. Construct a parallel line of AM through the center of circle ω, intersecting segment AB at point D and the extension of CA at point E. Suppose that line BM intersects CK at point P and line CM intersects BK at point Q. Prove that $\angle OEB + \angle OPB = \angle ODC + \angle OQC$.

(Contributed by He Yijie)

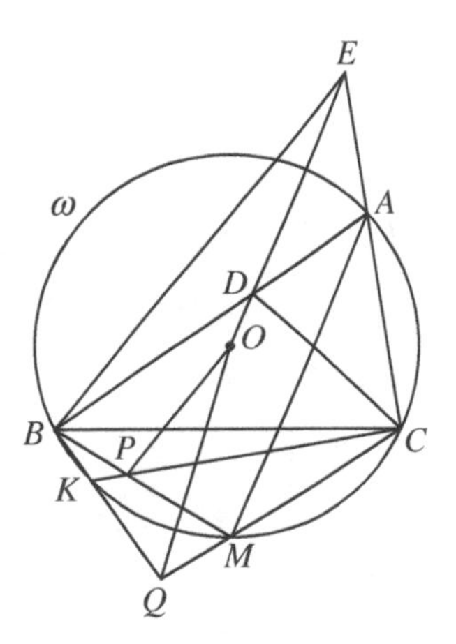

Fig. 4.1

Solution As shown in Fig. 4.2, connect OB and OC. By the given conditions, it follows that AM bisects $\angle BAC$ and $KB \perp AB$, $KC \perp AC$. Since O is the excentre of acute $\triangle ABC$, $\angle OBD = 90° - \angle ACB = \angle PCB$. And by $DO//AM$, we know that

$$\angle BDO = \angle BAM = \angle CAM = \angle MBC.$$

So $\triangle BOD \backsim \triangle CPB$. Then $\dfrac{BO}{CP} = \dfrac{BD}{BC}$, and hence $\dfrac{CO}{BD} = \dfrac{BO}{BD} = \dfrac{CP}{BC}$. And since

$$\begin{aligned}\angle OCP &= \angle OCB + \angle BCP \\ &= \angle OBC + \angle OBD = \angle DBC,\end{aligned}$$

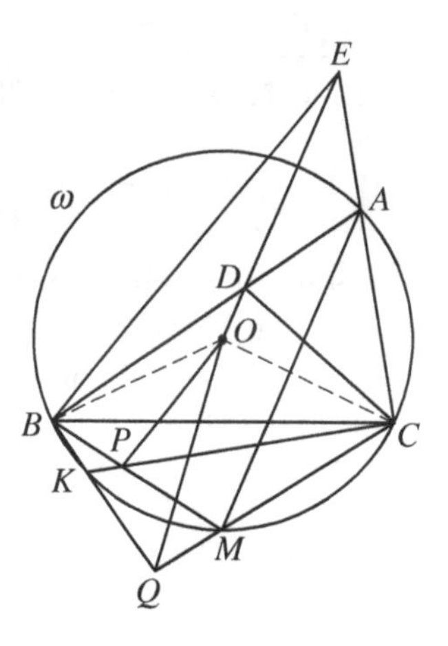

Fig. 4.2

$\triangle OCP \backsim \triangle DBC$. Therefore,

$$\begin{aligned}\angle OPB &= \angle BPC - \angle OPC \\ &= \angle BOD - \angle BCD \\ &= \angle OBC + \angle ODC. \qquad ①\end{aligned}$$

Similarly, we can obtain $\triangle COE \backsim \triangle BQC$ and $\triangle OBQ \backsim \triangle ECB$, which leads to

$$\begin{aligned}\angle OQC &= \angle BQC - \angle OQB \\ &= \angle COE - \angle EBC \\ &= \angle OCB + \angle OEB. \qquad ②\end{aligned}$$

By subtracting ① and ② and noting that $\angle OBC = \angle OCB$, we get

$$\angle OPB - \angle OQC = \angle ODC - \angle OEB,$$

namely,

$$\angle OEB + \angle OPB = \angle ODC + \angle OQC. \qquad \square$$

5 We consider a convex polyhedron P with the following properties: (i) each vertex of P belongs to exactly three faces; (ii) for any integer $k \geq 3$, the number of faces of k-sided polygon of P is even.

An ant starts from the midpoint of a certain edge and crawls on the surface of P along a closed path L composed by the edges of P. It passes through each point on L exactly once and finally returns to the starting point.

It is known that L divides the surface of P into two regions and that for any k, the number of faces of k-sided polygon in both regions is equal.

Prove that during the above process of crawling, the number of times the ant turns to the left at vertex P is the same as the number of times it turns to the right.

(Contributed by Yao Yijun)

Solution We consider graph G consisting of L and all the vertices and edges on its left side. It is a planar graph.

According to Euler's formula for planar graphs, we have

$$V - E + F = 1,$$

where V, E, F denote the number of vertices, edges and faces (without considering infinite faces), respectively.

In graph G, we note the following facts:

(1) Except for some vertices on L whose degree is 2, the degree of all other vertices in this graph is 3;
(2) the number of vertices with degree 2, say a, is exactly the number of left turns along L;
(3) the number of right turns along L is exactly the number of vertices with degree 3 on L in G, denoted as b.

Now we consider graph G' consisting of L and all the vertices and edges on it's the other side. It is also a planar graph. We have

$$V' - E' + F' = 1,$$

where V', E', F' denote the number of vertices, edges and faces, respectively.

Suppose that on convex polyhedron P, the k-sided polygon has N_k faces, $k = 3, 4, \ldots$. By the condition, it is know that the number of faces of k-sided polygon on each side of L is $\frac{N_k}{2}$. Let the number of vertices on L be ℓ. Then we have the following:

(0′) $F' - F = \frac{N_3}{2} + \frac{N_4}{2} + \cdots$;

(0″) $2E' - \ell = 2E - \ell = \frac{N_3}{2} \times 3 + \frac{N_4}{2} \times 4 + \cdots$;

(1′) Except for some vertices on L whose degree is 2, the degree of all other vertices in G' is 3;

(2′) the number of vertices with degree 2, say a', is exactly the number of right turns, namely b, along L;

(3′) the number of left turns along L is exactly the number of vertices with degree 3 on L in G', denoted as b'.

According to (1) and (1′), as well as the basic relationship between the number of edges and degrees in the graph, we get

$$2E = 2a + 3(V - a) = 3V - a;$$

$$2E' = 2a' + 3(V' - a') = 3V' - a' = 3V - b.$$

Then combining (0″), we get $E = E'$. Therefore, $a = b$ and the proposition holds. □

6 Let $\mathbb{N}_+$ be the set of all positive integers. Find all functions $f : \mathbb{N}_+ \to \mathbb{N}_+$ satisfying that for any $x, y \in \mathbb{N}_+$, $f(f(x) + y)$ divides $x + f(y)$.

(Contributed by Xiao Liang)

Solution All maps that satisfy the question are one of the following three categories:

(1) For any $x \in \mathbb{N}_+$, $f(x) = x$;

(2) for any $x > 1$, $f(x) = 1$ and $f(1)$ can be any positive integer;

(3) for any $x > 1$, $f(x) = \begin{cases} 1, & x \text{ is odd}, \\ 2, & x \text{ is even}, \end{cases}$ and $f(1)$ can be any odd positive integer.

It is easy to verify that all the three types of functions satisfy the conditions. In the following, we will prove that only the above three types of functions satisfy the conditions. The proof is done in three steps.

Step 1: We will prove that either $f(x) = x$ or f is not an injection.

For example, $f(1) > 1$, by taking $x = 1$ in the problem, we know that $f(f(1) + y)$ divides $1 + f(y)$. In particular,

$$f(f(1) + y_0) \leq 1 + f(y_0).$$

By induction it is easy to see that for fixed $y_0 \in \{1, \ldots, f(1)\}$ and any $t \in \mathbb{Z}_+$, there is

$$f(t \cdot f(1) + y) \leq t + f(y).$$

Therefore, for any y large enough, $f(y) < y$. So when y is large enough, $f : \{1, \ldots, y\} \to \{1, \ldots, y - 1\}$ cannot be an injection.

In the following, we will discuss the case $f(1) = 1$. By taking $x = 1$ in the problem, we know that $f(1 + y)$ divides $1 + f(y)$. In particular, $f(1 + y) \leq 1 + f(y)$. By induction it is easy to know that for any $x \in \mathbb{Z}_+$, there is $f(x) = x$ or $f(y) < y$ for y large enough. Similarly, we obtain that f cannot be an injection.

Step 2: We will show that if f is not an injection, then for x large enough, either $f(x) = 1$ or $f(x) = \begin{cases} 1, & x \text{ is odd,} \\ 2, & x \text{ is even.} \end{cases}$

Suppose f is not an injection and we denote A as the smallest positive integer such that there exists a positive integer $x_0 \in \mathbb{N}_+$ satisfying

$$f(x_0 + A) = f(x_0). \qquad ①$$

By substituting $x = x_0$ and $x = x_0 + A$ respectively into the problem, we get

$$f(f(x_0) + y) \mid x_0 + f(y)$$

and

$$f(f(x_0 + A) + y) \mid x_0 + A + f(y).$$

Subtracting the above two, we get

$$f(f(x_0) + y) \mid A. \qquad ②$$

This shows that $f(z)$ can only take values in the factors of A when z is greater than $f(x_0)$. Suppose that A has a total of D factors. According to the Pigeonhole principle, we investigate $D + 1$ consecutive positive integers $z, z + 1, \ldots, Z + D$ that are greater than $f(x_0)$, of which there must be two numbers that have equal images under f and the difference between these two numbers is less than or equal to D. This leads to $A \leq D$.

Such positive integer A can only be $A = 1$ or $A = 2$. If $A = 1$, according to ② we know that for x large enough, $f(x) = 1$. If $A = 2$, according to ② we see that for x large enough, $f(x) = 1$ or $f(x) = 2$. But $A = 2$ is the smallest A that makes ① valid, from which we know that

(a) Either for all x large enough, $f(x) = \begin{cases} 1, & x \text{ is odd,} \\ 2, & x \text{ is even,} \end{cases}$

(b) or for all x large enough, $f(x) = \begin{cases} 1, & x \text{ is even,} \\ 2, & x \text{ is odd.} \end{cases}$

However, (b) is impossible because taking two even numbers x and y large enough and substituting them into the problem gives

$$2 \mid x + 1,$$

a contradiction. This completes the proof of *Step 2*.

Step 3: We will prove that if f is not an injection, then for $x > 1$, then $f(x) = 1$ or $f(x) = \begin{cases} 1, & x \text{ is odd,} \\ 2, & x \text{ is even.} \end{cases}$

Let A be as set in *Step 2*. Taking x_0 large enough so that $f(x_0) = 1$ and substituting $x = x_0$ and $x = x_0 + A$ into the problem, we get

$$f(1 + y) \mid x_0 + f(y) \quad \text{and} \quad f(1 + y) \mid x_0 + A + f(y).$$

Subtracting the above two, we get

$$f(1 + y) \mid A.$$

When $A = 1$, this equation immediately leads to $f(y + 1) = 1$, and thus we obtain the function of (2) mentioned before.

When $A = 2$, this equation gives $f(y + 1) = 1$ or $f(y + 1) = 2$. However, the minimality of $A = 2$ shows that $f(x)$ must be $\ldots, 1, 2, 1, 2, \ldots$ alternatively in the case of $x > 1$. Combining the results of *Step 2* yields the function of the above mentioned (3). □

China Mathematical Olympiad

2021 (Fuzhou, Fujian)

2021 China Mathematical Olympiad (The Final) and the 37th National Mathematics Winter Camp for Middle School Students), was held at Affiliated High School of Fujian Normal University from December 19 to 25, 2021. The contest is the highest level and most advanced mathematics competition for middle school students in China. This winter camp has 33 domestic teams from all provinces, municipalities and autonomous regions in China's mainland, Hong Kong Special Administrative Region and Macao Special Administrative Region, and one foreign team from Singapore, with nearly 600 middle school students competing together, of which 60 students have qualified for the China national training team.

First Day
(8:00 – 12:30; December 21, 2021)

1 Given positive numbers a, b and segment AB of length a on a plane, suppose two moving points C, D in this plane satisfy that $ABCD$ is a non-degenerate convex quadrilateral with $BC = CD = b$, $DA = a$. It is easy to see that there exists circle I tangent to all the four sides of quadrilateral $ABCD$. Find the trajectory of center I.

(Contributed by Xiong Bin)

Solution Denote the center of the inscribed circle as I. Then BI is the intersection of the angle bisector of $\angle ABC$ and AC, so

$$\frac{AI}{AC} = \frac{a}{a+b}.$$

Hence, the trajectory of I is the homothetic image of the trajectory of C with respect to point A. (This can also be calculated by the coordinate method, basing on the two simple geometric facts that "I is on AC" and "I is at the same distance from AB and BC".)

And since the length of BC is a fixed value b, the position of point C is on the circle, denoted by ω_1, with center B and radius b.

Therefore, I must be on a circle, denoted by ω_2, with center O on segment AB, satisfying $\frac{AO}{AB} = \frac{a}{a+b}$ or $AO = \frac{a^2}{a+b}$, and radius $AO = \frac{ab}{a+b}$.

In order for $ABCD$ to be a convex quadrilateral, it is necessary and sufficient that both $\angle ACB$ and $\angle CAB$ are acute angles in $\triangle ABC$.

We discuss the following cases:

- If $a > b$, the range of point C can be two open arcs (both ends cannot be taken and being symmetric about the line on which AB lies) on ω_1. Each arc has one end at C' on the extension of AB satisfying $BC' = b$, and the other end at C'' on the circle satisfying

$$\cos \angle ABC'' = \frac{b}{a}.$$

This extreme case corresponds to an isosceles triangle with three sides of length a, a, $2b$. Therefore, the trajectory of the center of the inscribed circle is two arcs on ω_2, satisfying

$$\angle AOI \in \left(\arccos \frac{b}{a}, \pi\right);$$

- If $a = b$, then ω_2 is a circle with AB as its diameter on which only two points, A and B, cannot be taken;
- If $a < b$, the range of point C can be two open arcs (both ends cannot be taken and being symmetric about the line on which AB lies) on ω_1. Each arc has one end at C' on the extension of AB satisfying $BC' = b$, and the other end at C'' on the circle satisfying

$$\cos \angle ABC'' = \frac{a}{b}.$$

This extreme case corresponds to an isosceles triangle with three sides of length b, b, $2a$. Therefore, the trajectory of the center of the inscribed circle is two arcs on ω_2, satisfying

$$\angle AOI \in \left(\arccos \frac{a}{b}, \pi\right).$$

□

Remark This question is a very classical geometry question, slightly modified from the elementary mathematics group test question of the French National Competition (Concours Général) in 1879. (The original question considered circles tangent to the lines containing the four sides, so when $a \neq b$, there are two circles, and the convexity of the quadrilateral is not required.) In general, if the sum of the opposite sides of a quadrilateral is equal, then there must be an inscribed circle. Fixing the lengths of the four sides and the position of one side, allowing the other two vertices to move in the plane, then the trajectory of the incenter is still two arcs.[1]

The average score of this question is 18.6 points, and the average score of students who won silver medals or above is more than 19 points, but the average score of the 60 people who entered the China national training team did not reach 20 points. Most of the candidates can start smoothly, and the scoring criteria are relatively relaxed. Proving that the trajectory of I is part of a circle (specifying the position of the center and the radius of the circle) can get 15 points. However, there were also 9 students who entered the China national training team who got "point I is on a circle" but did not make any meaningful discussion on the range of values, and stopped there.

2 Find the maximum real number λ that satisfies the following conditions: for any positive real numbers p, q, r, s, there exists a complex number $z = a + bi$ $(a, b \in \mathbb{R})$ such that $|b| \geq \lambda|a|$ and

$$(pz^3 + 2qz^2 + 2rz + s)(qz^3 + 2pz^2 + 2sz + r) = 0.$$

(Contributed by He Yijie)

Solution We first prove that the maximum value of λ satisfying the conditions is $\sqrt{3}$.

[1]The original question and the extended version are both exercises in Jacques Hadamard's *Lessons in Geometry* (*Plane Geometry*). Hadamard himself won the second place in the mathematics group of the third grade (Troisième) of the Paris and Versailles school district in the 1879 competition. As the highest-ranking writer of plane geometry in the entire history of mathematics, apart from Euclid, he left an indelible mark on many fields of mathematics. In the early 1960s, this book (along with other elementary mathematics textbooks in the same series edited by G. Darboux) was translated into Chinese and published by Shanghai Scientific & Technical Publishers. In the early 1980s, the publishers republished this series, and Wu Xinmou, a researcher at the Institute of Mathematics of the Chinese Academy of Sciences, wrote a very good preface to the series, which is worth reading.

Assuming $\lambda > \sqrt{3}$ and taking $p = q = r = s = 1$, the original equation becomes $(z^3 + 2z^2 + 2z + 1)^2 = 0$, i.e.,

$$(z+1)^2\left(z+\frac{1}{2}+\frac{\sqrt{3}}{2}\mathrm{i}\right)\left(z+\frac{1}{2}-\frac{\sqrt{3}}{2}\mathrm{i}\right)=0,$$

and each complex root z satisfies $|\operatorname{Im} z| \leq \sqrt{3}|\operatorname{Re} z| < \lambda|\operatorname{Re} z|$, which is inconsistent with the question. Therefore, $\lambda \leq \sqrt{3}$.

In the following we prove that $\lambda = \sqrt{3}$ satisfies the conditions. The proposition group has prepared three methods, all of which all of which make use of proof by contradiction.

Method 1 (by He Yijie) Assuming otherwise, each complex root z of the following two equations

$$pz^3 + 2qz^2 + 2rz + s = 0 \qquad ①$$

$$qz^3 + 2pz^2 + 2sz + r = 0 \qquad ②$$

satisfies $|\operatorname{Im} z| < \sqrt{3}|\operatorname{Re} z|$.

Considering Equation ①, we now prove that $qr > ps$.

Note that Equation ① is a real coefficient cubic equation with one variable, and it must have real roots. In addition, by the fact that $p, q, r, s > 0$, it is clear that the real roots must be negative. There are two cases as follows.

(i) Equation ① has three negative roots. At this point we can set

$$pz^3 + 2qz^2 + 2rz + s = p(z+u_1)(z+u_2)(z+u_3),$$

where $u_1, u_2, u_3 > 0$. Comparing the coefficients, we know that

$$2q = p(u_1+u_2+u_3), \quad 2r = p(u_1u_2+u_2u_3+u_1u_3), \quad s = pu_1u_2u_3.$$

Therefore,

$$\begin{aligned} qr &= \frac{p^2}{4}(u_1+u_2+u_3)(u_1u_2+u_2u_3+u_1u_3) \\ &\geq \frac{p^2}{4}\cdot 3\sqrt[3]{u_1u_2u_3}\cdot 3\sqrt[3]{u_1u_2\cdot u_2u_3\cdot u_1u_3} \\ &= \frac{9}{4}p^2u_1u_2u_3 = \frac{9}{4}ps > ps. \end{aligned}$$

(ii) Equation ② has one negative root and a pair of conjugate imaginary roots. In this case, combining the hypothesis of proof by contradiction, we can set

$$\begin{aligned} pz^3 + 2qz^2 + 2rz + s &= p(z+u)(z-(v+kvi))(z-(v-kvi)) \\ &= p(z+u)(z^2 - 2vz + (1+k^2)v^2), \end{aligned} \qquad ③$$

where $u > 0$, $v, k \in \mathbb{R}$, $v \neq 0$, $0 < |k| < \sqrt{3}$.

Denote $m = 1 + k^2$, and then $0 < m < 4$. Comparing the coefficients of the terms in equation ③, we have

$$2q = p(u - 2v), \quad 2r = p(-2uv + mv^2), \quad s = pmuv^2. \qquad ④$$

Therefore,

$$0 < \frac{2r}{p} = -2uv + mv^2 < -2uv + 4v^2 - 2v(u - 2v) = -2v \cdot \frac{2q}{p},$$

which implies that $v < 0$. Furthermore, by using Equation ④, we know that

$$\frac{4}{p^2}(qr - ps) = \frac{2q}{p} \cdot \frac{2r}{p} - 4 \cdot \frac{s}{p} = (u - 2v)(-2uv + mv^2) - 4muv^2.$$

The right side of the above equation can be regarded as a linear function $g(m)$ of m, and combining $u > 0$, $v < 0$, we know that

$$g(0) = (u - 2v)(-2uv) > 0, \quad g(4) = -2v(u + 2v)^2 \geq 0.$$

Therefore, when $0 < m < 4$, there must be $\frac{4}{p^2}(qr - ps) > 0$, that is, $qr > ps$.

Combining the two cases (i) and (ii), there is always $qr > ps$.

Considering Equation ②, the same reasoning gives $ps > qr$, a contradiction! Therefore, the hypothesis is not valid, that is, $\lambda = \sqrt{3}$ satisfies the condition.

In conclusion, the desired maximum value of λ is $\sqrt{3}$.

Method 2 (by Xu Disheng) We can conversely assume that $\lambda = \sqrt{3}$ does not satisfy the condition, so that Equations ① and ② have no roots in region $A_1 := \left\{ z \in \mathbb{C} \,\middle|\, \frac{\pi}{3} \leq \arg z \leq \frac{2\pi}{3} \text{ or } \frac{4\pi}{3} \leq \arg z \leq \frac{5\pi}{3} \right\}$. We first prove that Equations ① and ② also have no roots in region $A_2 := \left\{ z \in \mathbb{C} \,\middle|\, 0 \leq \arg z \leq \frac{\pi}{3} \text{ or } \frac{5\pi}{3} \leq \arg z \leq 2\pi \right\}$.

If Equation ① has a root in region A_2, then since p, q, r, s are all positive, it is impossible to be a real root. From the fact that the complex roots of polynomials with real coefficients are conjugate to each other, we know that Equation ① has a root z_0 in $\left\{z \in \mathbb{C} \,\middle|\, 0 < \arg z \leq \frac{\pi}{3}\right\}$. However, as seen from the arguments, the imaginary parts of pz_0^3, $2qz_0^2$, $2rz_0$ are all greater than 0 at this point, and then $pz_0^3 + 2qz_0^2 + 2rz_0 + s \neq 0$, a contradiction! Similarly, Equation ② has no roots in A_2.

Therefore, the roots of Equations ① and ② are not in $A_1 \cup A_2$. In particular, if we denote the roots of ① and ② as (z_1, z_2, z_3), (w_1, w_2, w_3), respectively, then the real parts of z_i, ω_i are both negative, and we can set $z_1, w_1 \in \mathbb{R}$. By Vieta's formulas, we have

$$(z_1 + z_2 + z_3)\left(\frac{1}{z_1} + \frac{1}{z_2} + \frac{1}{z_3}\right) = \frac{4qr}{ps},$$

$$(w_1 + w_2 + w_3)\left(\frac{1}{w_1} + \frac{1}{w_2} + \frac{1}{w_3}\right) = \frac{4ps}{qr}.$$

It is sufficient to prove the following inequalities to derive the contradiction.

$$(z_1 + z_2 + z_3)\left(\frac{1}{z_1} + \frac{1}{z_2} + \frac{1}{z_3}\right) > 4, \qquad ⑤$$

$$(w_1 + w_2 + w_3)\left(\frac{1}{w_1} + \frac{1}{w_2} + \frac{1}{w_3}\right) > 4. \qquad ⑥$$

By symmetry we only prove ⑤.

(i) If $z_1, z_2, z_3 \in \mathbb{R}$, then from Cauchy's inequality, we have

$$(z_1 + z_2 + z_3)\left(\frac{1}{z_1} + \frac{1}{z_2} + \frac{1}{z_3}\right) \geq 9 > 4.$$

(ii) If z_1 is negative and the real part of imaginary number $z_2 = \overline{z_3}$ is negative, then from $z_i \notin A_1 \cup A_2$, we have

$$\frac{|\operatorname{Re} z_2|}{|z_2|} > \frac{1}{2}. \qquad ⑦$$

And from the fact that all the real parts of z_i have the same sign, and using Equation ⑦ and the inequality of arithmetic and geometric means, we get

$$\begin{aligned}(z_1+z_2+z_3)\left(\frac{1}{z_1}+\frac{1}{z_2}+\frac{1}{z_3}\right) &- (z_1+2\operatorname{Re} z_2)\left(\frac{1}{z_1}+\frac{2\operatorname{Re} z_2}{|z_2|^2}\right)\\ &= 1+\frac{4\,|\operatorname{Re} z_2|^2}{|z_2|^2}+2\left(\frac{\operatorname{Re} z_2}{z_1}+z_1\cdot\frac{\operatorname{Re} z_2}{|z_2|^2}\right)\\ &> 1+1+4\sqrt{\frac{(\operatorname{Re} z_2)^2}{|z_2|^2}} > 4.\end{aligned}$$

Method 3 (by Xiao Liang) Similar to the previous two methods, under the hypothesis of the proof by contradiction, equation

$$pz^3+2qz^2+2rz+s=0$$

has no positive roots. Let its negative real root be $-\alpha(\alpha>0)$. Using the relation between the roots and the coefficients and comparing the coefficients of the expansion gives

$$pz^3+2qz^2+2rz+s=(z+\alpha)\left(pz^2+(2q-\alpha p)z+\frac{s}{\alpha}\right).$$

Here,

$$2q-\alpha p=\frac{2r-\dfrac{s}{\alpha}}{\alpha}. \qquad ⑧$$

The solution of the above quadratic equation is $\dfrac{-(2q-\alpha p)\pm\sqrt{(2q-\alpha p)^2-4p\cdot\dfrac{s}{\alpha}}}{2p}$. By the hypothesis of the proof by contradiction, we get $(2q-\alpha p)^2>p\cdot\dfrac{s}{\alpha}$. Using ⑧, we obtain

$$(2q-\alpha p)\frac{2r-\dfrac{s}{\alpha}}{\alpha}>p\cdot\frac{s}{\alpha}\Rightarrow 4qr+ps-\left(2\alpha pr+\frac{2}{\alpha}qs\right)>ps.$$

Therefore,

$$4qr>2\alpha pr+\frac{2}{\alpha}qs\geq 4\sqrt{pqrs}\Rightarrow qr>ps.$$

Yet from the other equation, similarly we can get $ps>qr$, a contradiction.

□

Remark This is an algebra question within the scope of the high school mathematics competition.

In addition to the above methods, the following approach was used by some candidates in the examination to demonstrate that $\lambda = \sqrt{3}$ was satisfied.

Substituting $z = a + b\mathrm{i}$ into Equation ① and separating the real and imaginary parts gives

$$p(a^3 - 3ab^2) + 2q(a^2 - b^2) + 2ra + s = 0,$$

$$p(3a^2b - b^3) + 4qab + 2rb = 0.$$

(When $b \neq 0$, there is $p(3a^2 - b^2) + 4qa + 2r = 0$.)

To prove that when p, q, r, s satisfy certain conditions, there must be an imaginary root $z = a + b\mathrm{i}$ satisfying $b^2 \geq 3a^2$. This is equivalent to proving that the cubic polynomial of a

$$f(a) = 8pa^3 + 16qa^2 + \left(4r + \frac{8q^2}{p}\right)a + \frac{4qr}{p} - s, \qquad ⑨$$

obtained by substituting $b^2 = 3a^2$ into the above equation, has zeros satisfying $4qa + 2r \geq 0$, i.e., $a \geq -\dfrac{r}{2q}$.

Since the leading coefficient of $f(a)$ is a positive real number, a sufficient condition for the above conclusion to hold is that $f\left(-\dfrac{r}{2q}\right) \leq 0$, that is,

$$pr^3 - 2q^2r^2 + q^3s \geq 0. \qquad ⑩$$

Similarly, for Equation ②, a sufficient condition for the equation to have a root $z = a + b\mathrm{i}$ satisfying $b^2 \geq 3a^2$ is

$$qr^3 - 2p^2s^2 + p^3r \geq 0. \qquad ⑪$$

Thus, if ⑩ and ⑪ are not true, that is, there is

$$pr^3 - 2q^2r^2 + q^3s < 0, \quad qs^3 - 2p^2s^2 + p^3r < 0.$$

Then from these two inequalities we can get

$$pr^3 + q^3s < 2q^2r^2 \Rightarrow \sqrt{ps} < \sqrt{qr},$$

$$qs^3 + p^3r < 2p^2s^2 \Rightarrow \sqrt{qr} < \sqrt{ps},$$

respectively, an obvious contradiction.

Therefore, at least one of ⑩ and ⑪ are is true, and it may be appropriate to assume that ⑩ is true. At this point, $b = \pm\sqrt{3a^2 + \frac{4q}{p}a + \frac{2r}{p}}$ satisfies $b \neq 0$, $|b| \geq \sqrt{3}|a|$ and $z = a + bi$ satisfies Equation ①, so that the proposition holds. (We incidentally have proved that the roots of the sixth-order equation in the question cannot be real numbers.)

The average score of this question is 10.2 points, which is the only question among all the questions in this competition that has a large number of candidates in the score range of "almost all correct" and "almost all wrong". For the candidates who won the gold medal, the average score of this question was above 18 points, but for the candidates who won the silver medal, the average score of this question was less than 8 points.

From the marking process of this question, we noticed that although complex numbers play an irreplaceable role in modern mathematics, some of the excellent middle school students who participated in the competition still have a very limited understanding of complex numbers. (When encountering complex number problems, it is not a very good mathematical approach to solve the problem by transforming the problem into a real algebraic problem by "considering the separation of real and imaginary parts".)

3 Find all integers a such that there exists a set X of 6-tuple integers satisfying the following conditions: for each $k = 1, 2, \ldots, 36$, there exist $x, y \in X$ such that $ax + y - k$ is divisible by 37.

(Contributed by Wang Xinmao)

Solution 1 (Wang Xinmao) Let a and $X = \{x_1, x_2, \ldots, x_6\}$ satisfy the conditions in the question. Denote $p = 37$, $\overline{X} = \{\overline{x_1}, \overline{x_2}, \ldots, \overline{x_6}\}$, $Y = \{\overline{ax_i + x_j} \mid i, j = 1, 2, \ldots, 6\}$, where x represents the congruence class of x modulo p.

Since $|Y| = 36$, there is $\bar{0} \notin \overline{X}$. If $a \equiv 0 \pmod{p}$, then $|Y| \leq 6$. If $a \equiv 1 \pmod{p}$, then $|Y| \leq 21$. If $a \equiv -1 \pmod{p}$, then $\bar{0} \in Y$. Hence, $a \not\equiv 0, \pm 1 \pmod{p}$.

By $Y = \{\overline{a(ax_i + x_j)} \mid i, j = 1, 2, \ldots, 6\} = \{\overline{ax_j + a^2x_i)} \mid i, j = 1, 2, \ldots, 6\}$ and the lemma (see the following for its statement and proof), we have $\overline{X} = \left\{\overline{a^2x_1}, \overline{a^2x_2}, \ldots, \overline{a^2x_6}\right\}$. Hence, $a^{12}x_1x_2\cdots x_6 \equiv x_1x_2\cdots x_6 \pmod{p}$, and thus $a^{12} \equiv 1 \pmod{p}$. Let d be the exponent of a^2 modulo p. Then $d \mid 6$ and $d \neq 1$.

If $d=2$, then $a^2 \equiv -1 \pmod p, a \equiv \pm 6 \pmod p$. Suppose $a=6$, $X=\{\pm 1, \pm 3, \pm 5\}$, and then $\{ax_i + x_j \mid i,j = 1,2,\ldots,6\} = \{\pm 1, \pm 3, \ldots, \pm 35\}$, satisfying the conditions in the question. Similarly, any $a \equiv \pm 6 \pmod p$ and $X = \{\pm 1, \pm 3, \pm 5\}$ satisfy the conditions in the question.

If $d=3$, then $\overline{X}$ is of the form $\left\{\bar{x}, \overline{a^2x}, \overline{a^4x}, \bar{y}, \overline{a^2y}, \overline{a^4y}\right\}$. Also, $a^6 \equiv 1 \pmod p \Rightarrow a^3 \equiv \pm 1 \pmod p$. If $a^3 \equiv 1 \pmod p$, then $a\cdot x + a^2x \equiv a \cdot a^4x + a^4x \pmod p$, a contradiction. If $a^3 \equiv -1 \pmod p$, then $\overline{a\cdot x + a^4x} = \bar{0} \in Y$, which is also a contradiction.

If $d=6$, then

$$\overline{X} = \{\overline{a^{2i}x_1} \mid i = 0,1,\ldots,5\},$$

$$Y = \{\overline{(a^{2i+1}+a^{2j})x_1} \mid i,j = 0,1,\ldots,5\}.$$

From $a^6 \equiv -1 \pmod p$, and the polynomial equation

$$\prod_{j=0}^{5}(t+a^{2j}) \equiv \prod_{j=0}^{5}(t-a^{2j}) \equiv t^6 - 1 \pmod p,$$

we get $\prod\limits_{0\le i,j\le 5}(a^{2i+1}+a^{2j}) = \prod\limits_{0\le i\le 5}(a^{6(2i+1)}-1) \equiv 2^6 \not\equiv 1 \equiv 36! \pmod p$, a contradiction.

Therefore, the desired $a = 37k \pm 6$, $k \in \mathbb{Z}$.

Lemma *Let $a_1,\ldots,a_6$, $b_1,\ldots,b_6 \in \{0,1,2,\ldots,36\}$. If $\{a_i + b_j \mid 1 \le i, j \le 6\}$ is a reduced system modulo 37, then $\{b_1,\ldots,b_6\}$ is uniquely determined by $\{a_1,\ldots,a_6\}$.*

Proof of the lemma Let $f(x) = \sum\limits_{i=1}^{6} x^{a_i}$, $g(x) = \sum\limits_{i=1}^{6} x^{b_i}$. Then

$$f(x)g(x) \equiv \sum_{k=1}^{36} x^k \pmod{x^{37}-1},$$

and thus $f(x)g(x) \equiv -1 \pmod{q(x)}$, where $q(x) = 1 + x + \cdots + x^{36}$. Since there exists a unique $h(x) \in \mathbb{Q}[x]$ such that $f(x)h(x) \equiv -1 \pmod{q(x)}$ and $\deg h(x) < \deg q(x) = 36$, we have $g(x) = h(x)$ or $g(x) = h(x) + q(x)$. Since the number of monomials of $h(x)$ and $h(x)+q(x)$ cannot both be less than or equal to 6, $g(x)$ is uniquely determined by $f(x)$.

Solution 2 (Fu Yunhao) First, $a \not\equiv \pm 1 \pmod{37}$. Because when $a \equiv 1 \pmod{37}$,

$$ax_1 + x_2 \equiv ax_2 + x_1 \pmod{37},$$

which does not meet the requirements of the question; when $a \equiv -1 \pmod{37}$, $ax_1 + x_1 \equiv 0 \equiv ax_2 + x_2 \pmod{37}$, which also does not meet the requirements of the question.

(i) From $\sum_{i=1}^{6}\sum_{j=1}^{6}(ax_i + x_j) \equiv 1 + 2 + \cdots + 36 \equiv 0 \pmod{37}$, we get

$$6(a+1)\sum_{i=1}^{6} x_i \equiv 0 \pmod{37},$$

so $\sum_{i=1}^{6} x_i \equiv 0 \pmod{37}$. (All the following congruences are in the sense of modulo 37, and all unspecified single summations for i are from 1 to 6.)

(ii) From $\sum_{i=1}^{6}\sum_{j=1}^{6}(ax_i + x_j)^2 \equiv \sum_{i=1}^{36} k^2 \equiv 0$, we get

$$6(a^2+1)\sum x_i^2 + 2a\left(\sum x_i\right)^2 \equiv 0.$$

Hence,

$$(a^2+1)\sum_{i=1}^{6} x_i^2 \equiv 0.$$

Suppose $a \not\equiv \pm 6$, and then $a^2 + 1 \not\equiv 0$, so $\sum x_i^2 \equiv 0$.

(iii) From $\sum_{i=1}^{6}\sum_{j=1}^{6}(ax_i + x_j)^3 \equiv \sum_{i=1}^{36} k^3 \equiv 0$, we get

$$6(a^3+1)\sum x_i^3 + 3(a^2+a)\left(\sum x_i\right)\left(\sum x_i^2\right) \equiv 0.$$

Hence,

$$(a^3+1)\sum x_i^3 \equiv 0. \qquad ①$$

(iv) From $\sum_{i=1}^{6}\sum_{j=1}^{6}(ax_i + x_j)^4 \equiv \sum_{i=1}^{36} k^4 \equiv 0$, we get

$$6(a^4+1)\sum x_i^4 + 4(a^3+a)\left(\sum x_i\right)\left(\sum x_i^3\right) + 6a^2\left(\sum x_i^2\right)^2 \equiv 0.$$

Note that $a^4 + 1 \not\equiv 0$. (Otherwise, we have $(a^3+1)\sum x_i^3 \equiv 0$, $a^4 \not\equiv 1$, that is, the order of a modulo 37 is 8, which is impossible because 8 does not divide 36, a contradiction.) Therefore, $\sum x_i^4 = 0$.

(v) From $\sum_{i=1}^{6}\sum_{j=1}^{6}(ax_i+x_j)^5 \equiv \sum_{i=1}^{36} k^5 \equiv 0$, we get

$$6(a^5+1)\sum x_i^5 + 5(a^4+a)\left(\sum x_i\right)\left(\sum x_i^4\right)$$
$$+10(a^3+a^2)\left(\sum x_i^2\right)\left(\sum x_i^3\right) \equiv 0.$$

Hence $(a^5+1)\sum x_i^5 \equiv 0$. Therefore, $\sum x_i^5 \equiv 0$. (If $a^5+1\equiv 0$, then $a^{10}-1\equiv 0$, which leads to $a^2-1\equiv 0$, a contradiction.)

(vi) From $\sum_{i=1}^{6}\sum_{j=1}^{6}(ax_i+x_j)^6 \equiv \sum_{i=1}^{36} k^6 \equiv 0$, we get

$$6(a^6+1)\sum x_i^6 + 6(a^5+a)\left(\sum x_i\right)\left(\sum x_i^5\right)$$
$$+15(a^4+a^2)\left(\sum x_i^2\right)\left(\sum x_i^4\right) + 20a^3\left(\sum x_i^3\right)^2 \equiv 0.$$

Hence,

$$6(a^6+1)\sum x_i^6 + 20a^3\left(\sum x_i^3\right)^2 \equiv 0.$$

- If $37 \nmid a^3+1$, $37 \nmid a^6+1$, then from ① and ② we can get $\sum x_i^3 \equiv 0$, $\sum x_i^6 \equiv 0$. According to Newton's identity, we have $(z-x_1)(z-x_2)\cdots(z-x_6)\equiv z^6$, a contradiction.
- If $37 \mid a^3+1$, since $37 \nmid a+1$, then $37 \mid a^2-a+1$. By trying one by one, we can get $a\equiv 8$ or $a\equiv 28$. We get

$$\{ax_i+x_j\} \text{ is a reduced system modulo}$$
$$37 \Leftrightarrow \left\{\frac{1}{a}x_i+x_j\right\} \text{ is a reduced system modulo 37}$$

Therefore, we can set $x_2=8x_1$, $x_3=64x_1$, $x_5=8x_4$, $x_6=64x_4$. However, there is

$$ax_1+x_2 \equiv ax_3+x_2.$$

- If $37 \mid a^6+1$, since $37 \mid a^{12}-1$, then $37 \nmid a^6-1$. Also, we have $37 \nmid a^4-1$ (otherwise $37 \mid a^2+1, a\equiv \pm 6$), so the order of a modulo 37 is 12. From ①, we see $\sum x_i^3 \equiv 0$. By Newton's identity, we know that

$$(z-x_1)(z-x_2)\cdots(z-x_6)\equiv x^6+A.$$

We can set $x_2=a^2x_1, x_3=a^4x_1,\ldots,x_6=a^{10}x_1$.

Following the same method as Solution 1, a contradiction can be obtained. Therefore, $a \equiv \pm 6$. The example is also the same as Solution 1. $\square$

Remark The two main approaches to this problem are the generating function method and the power sum method.

The generating function method maps a set to a polynomial, and the addition of a set corresponds to the multiplication of a polynomial. If the numbers in the set are modulo n, then the corresponding polynomial is modulo $x^n - 1$. The generating function method is a relatively ideal way to handle information, and it can also be completely transformed in this problem.

In addition, the property that the polynomial

$$\frac{x^p - 1}{x - 1} = 1 + x + \cdots + x^{p-1}$$

is not reducible in $\mathbb{Z}[x]$ played a major role in solving the problem.

The power sum method is related to the reduced system. The reduced systems for modulo prime p are integer power sums

$$1^m + 2^m + \cdots + (p-1)^m = \begin{cases} 0 \pmod{p}, & \text{if } m = 1, 2, \ldots, p-2; \\ p-1 \pmod{p}, & \text{if } m = p-1. \end{cases}$$

Indeed, it is easy to prove, using Newton's power sum formula, that the above equation is also a sufficient condition for a set of $p-1$ numbers to form a reduced system modulo p.

This question can also be solved by discussing the power sum of $x_1, x_2, \ldots, x_6$. The power sum method may have a slightly larger computation, but the goal is clearer in the calculation process. In case no construction is found for $a \equiv \pm 6$, the power sum method can also concretely work out this set of solutions and prove that they are unique in the multiplication translation sense. This is also an advantage.

The background of this problem is a certain factorization problem, and we can consider a more general problem: For the 36 numbers in two integer sets $X = \{x_1, \ldots, x_6\}$, $Y = \{y_1, \ldots, y_6\}$ satisfying

$$X + Y = \{x_i + y_j : i, j = 1, 2, \ldots, 6\},$$

find the reduced system modulo 37. According to the results of the proposer's programming search, the new problem has a total of 1998 ordered set pairs (X, Y) satisfying the requirements in the sense of modulo 37.

If further standardized, that is, requiring $x_1+x_2+\cdots+x_6 \equiv 0$ (correspondingly, $y_1+y_2+\cdots+y_6 \equiv 0$), there are $\frac{1998}{37} = 54$ pairs of standardized (X, Y), all of which are related to the following three sets:

$$\mathbf{X}_0 = \{\pm 1, \pm 3, \pm 5\},$$
$$\mathbf{X}_1 = \{\pm 2, \pm 3, \pm 4\},$$
$$\mathbf{X}_2 = \{\pm 1, \pm 4, \pm 16\}.$$

These include a total of 18 for $(X, Y) \equiv (k\mathbf{X}_0, 6k\mathbf{X}_0)$ (i.e., solutions satisfying the proportionality of X and Y in the original problem); 18 for $(X, Y) \equiv (k\mathbf{X}_1, 8k\mathbf{X}_2)$; and 18 for $(X, Y) \equiv (8k\mathbf{X}_2, k\mathbf{X}_1)$. That is to say, after standardization, there is essentially only one set of solutions (i.e. solutions outside the original problem) for which X and Y are not proportional, i.e.,

$$(\mathbf{X}_1, 8\mathbf{X}_2) \equiv (\{\pm 2, \pm 3, \pm 4\}, \{\pm 5, \pm 8, \pm 17\}).$$

The rest are obtained by multiplying this set of solutions by some constant, or by exchanging X and Y.

The average score of this problem is 3.9 points, which is a difficult problem that can be done. For those who eventually entered the China national training team, the average score of this question reached over 16. However, for those who did not entered the national training team but won the gold medal, the average score of this question was less than 5.

If the 37 in the question is replaced by 17 or 101, then for the above solution there is essentially no change in the difficulty of the question and only some (not much) difference in the amount of calculation. However, for the discussion methods used by many candidates in the examination, the amount of calculation may vary considerably, thus affecting the score rate of this question.

Second Day
(8:00 – 12:30; December 22, 2021)

4 Given that $n (n \geq 3)$ scientists attend a conference, each scientist has some friends attending the conference (the friendship is mutual and no one is their own friend). It is known that no matter how these scientists are divided into non-empty two groups, there always exist two scientists in the same group who are friends, and there also exist two scientists in different groups who are friends. On the first day, a topic was proposed at the conference, and the degree of approval of each scientist for this topic can be represented by a non-negative integer. Starting from the second day, the degree of approval of each scientist becomes the integer part of the average of the degree of approval of all his friends on the previous day. Prove that after several days, the degree of approval of all scientists is the same.

(Contributed by Qu Zhenhua)

Solution Suppose n scientists form a set V. The friendship relation can be represented by a simple graph $G = (V, E)$. From the conditions of the question, it is known that G is a connected graph and not a bipartite graph. On the nth day, the degree of approval of each scientist for the topic is defined by function $f_n : V \to \mathbb{Z}$. Since

$$f_{n+1}(x) = \left[\frac{1}{\deg(x)}\left(\sum_{xy\in E} f_n(y)\right)\right] \geq \min_{xy\in E} f_n(y) \geq \min_{y\in V} f_n(y), \quad ①$$

it follows that $\min f_{n+1} \geq \min f_n$. Similarly, we can get $\max f_{n+1} \leq \max f_n$. Therefore, $\min f_n$ eventually becomes a constant, denoted by a; $\max f_n$ also eventually becomes a constant, denoted by b.

If $a = b$, then f_n eventually becomes a constant function, and the conclusion holds. In the following, we assume that $a < b$.

Since there are only a finite number of integer functions from V to $[a, b]$, there exists $i \neq j$ such that $f_i = f_j$, and thus we known that $\{f_n\}_{n\geq 1}$ is eventually periodic. Let the period be $t > 0$, and there exists a positive integer n_0 such that for all $n \geq n_0$, there is $f_n = f_{n+t}$. In the following we can derive a contradiction. For $n \geq n_0$, denote

$$M_n = \{x \in V \mid f_n(x) < b\}.$$

Let $n \geq n_0$, $y \in M_n$. From ①, if $xy \in E$, we have $f_{n+1}(x) < b$, that is, $x \in M_{n+1}$.

Three methods are given in the following to complete the proof.

Method 1 For any $x \in V$, we will show that there exists positive integer k (depending on x, which can be denoted as $k = k(x)$) such that such that a path from x can return to x after k edges, or after $k+1$ edges. Since G is not a bipartite graph, there exists an odd cycle C. Let the number of edges of C be $2p+1$. Take a point y on C, and due to the connectivity of G, there exists a path P from x to y. Let the number of edges of P be q. Thus, following P from x to y, then going around C once, and returning to x along P, it takes $2q+2p+1$ edges. Also, following P from x to y, going around C after p edges, and then returning to x along the same path, it takes $2q+2p$ edges. It is sufficient to take $k = 2q+2p$.

Let $x \in M_{n_0}$, $k = k(x)$. Then we have $x \in M_{n_0+k}$, $x \in M_{n_0+k+1}$, and hence $x \in M_{n_0+ku+(k+1)v}$ holds for any $u, v \in \mathbb{Z}_{\geq 0}$. Since k and $k+1$ are coprime,

$$ku + (k+1)v(u, v \in \mathbb{Z}_{\geq 0})$$

can take on all sufficiently large integers. Combing with the t-periodicity of M_n, it follows that $x \in M_n$ holds for all $n \geq 0$. Thus, for any neighbor y of x, we also have $y \in M_n$, $n \geq n_0$. Continuing this process, since M_{n_0} is nonempty and G is connected, it follows that for all $z \in V$, we have $z \in M_n$, $n \geq n_0$. Hence, $M_n = V$, $n \geq n_0$, which contradicts $\max f_n = b$. The conclusion is proved.

Method 2 For $x \in V$, denote $N(X) = \{y \in V \mid \exists x \in X, xy \in E\}$. Then the proof above shows that $N(M_n) \subset M_{n+1}$, $n \geq n_0$. Thus, $N(N(M_n)) \subset N(M_{n+1}) \subset M_{n+2}$. Note that $N(N(M_n)) \supset M_n$, so $M_n \subset M_{n+2}$. Hence,

$$M_{n_0} \subset M_{n_0+2} \subset M_{n_0+4} \subset \cdots \subset M_{n_0+2t} = M_{n_0}.$$

Therefore, $M_{n_0} = M_{n_0+2} = M_{n_0+4} = \cdots$ (denoted as A). Similarly, there is $M_{n_0+1} = M_{n_0+3} = M_{n_0+5} = \cdots$ (denoted as B). Then, $N(A) \subset B$, $N(B) \subset A$.

If $A \cap B = \varnothing$, then by the connectivity of G, $A \cup B = V$. In this case, G is a bipartite graph defined on (A, B), which contradicts the conditions of the question.

If $A \cap B \neq \varnothing$, let $x \in A \cap B$. Then, we have $N(x) \subset A$ and $N(x) \subset B$. Thus, for any neighbor y of x, we also have $y \in A \cap B$. Using this conclusion repeatedly, and combining with the connectivity of G, it is known that $A = B = V$, and thus $M_{n_0} = V$, which contradicts $\max f_{n_0} = b$. The conclusion is proved.

Method 3 Let $G = (V, E)$ be a connected non-bipartite graph and $f_i : V \to \mathbb{Z}_{\geq 0}$ be the degree of approval of each scientist on the ith day.

Considering $S_i = \sum\limits_{x \in V} \deg(x) f_i(x)^2$, then

$$\begin{aligned}
S_{i+1} &= \sum_{x \in V} \deg(x) f_{i+1}(x)^2 \\
&= \sum_{x \in V} \deg(x) \left[\frac{1}{\deg(x)} \sum_{xy \in E} f_i(y) \right]^2 \\
&\le \sum_{x \in V} \frac{1}{\deg(x)} \left(\sum_{xy \in E} f_i(y) \right)^2 \\
&\le \sum_{x \in V} \sum_{xy \in E} f_i(y)^2 = \sum_{y \in V} \deg(y) f_i(y)^2 = S_i.
\end{aligned}$$

The first inequality is obtained by removing the rounding symbol, and the second inequality is obtained by applying Cauchy's inequality. Thus, $S_1, S_2, \ldots$ is a monotonically nonincreasing sequence of nonnegative integers, and eventually converges to a constant, so there exists a positive integer m such that $S_m = S_{m+1}$.

Let us prove that f_m is a constant function. From the above derivation, it follows that when the equal sign holds, for each $x \in V$, f_m is a constant on the set $N(\{x\})$ composed of all the adjacent vertices of x. Thus, for any $y, z \in V$, if there is a path of length 2 between y and z (i.e., there is a vertex adjacent to both of them), then $f_m(y) = f_m(z)$. Furthermore, if there is a path of even length between y and z, then $f_m(y) = f_m(z)$. Since G is a connected nonbipartite graph, there is an odd cycle in G, and any two points have a trail of even length, so f_m takes a constant value on V. $\square$

Remark This question is related to Markov chains. Starting from an initial state, it will eventually converge to a periodic state. The solutions to this question all require considering the condition of equality after a certain semi-invariant is stabilized, and some basic techniques of combinatorics and graph theory are examined. The solutions of most candidates are similar to the first two methods or their variations, considering the maximum value of degree of approval. Some solutions also use the semi-invariant in Method 3. In addition, there were also methods by considering semi-invariant variables like $\sum\limits_{e \in E} g_m(e)$ (where $g_m(e)$ is the number of two endpoints of edge e belonging to $V \backslash M_m$), as well as methods of linear algebra.

The average score of this problem is 15.0 points, which is exactly in line with expectations. The average score of students who won the silver medal is more than 17 points, which is almost 3 points lower than the average score of students who won the gold medal, but much higher than the average score of students who won the bronze medal (4.0 points).

5 It is known that there are only two types of one-dimensional geometric objects, circles and lines, that appear in ruler-and-compass construction. Let there be two points marked on a paper with a distance of 1. Prove that a line and two points on this line with a distance of $\sqrt{2021}$ can be made on this paper with ruler and compass, such that the total number of different circles and lines that appear in the drawing process does not exceed 10.

Remark The solution should include clear steps of drawing, and number the circles and lines appearing in sequence. If the total number of circles and lines in your drawing process exceeds 10, then there may be partial scores according to the size of the total number.

(Contributed by Yao Yijun)

Solution The following are several methods of making $\sqrt{2021}$ with a total number of 9 and 10 of lines and circles.

Method 1 We number the objects of the one-dimensional set in the order in which they appear.

(1) Denote the two given points as O_1, A_1. Draw a circle with O_1 as the center and $O_1A_1 = 1$ as the radius.
(2) Draw line O_1A_1 and intersect the circle at another point B_1.
(3) Draw a circle with B_1 as the center and $A_1B_1(= 2)$ as the radius, and intersect the line at another point B_2.
(4) Draw a circle with A_1 as the center and $O_1B_2(= 3)$ as the radius, and intersect the line at point A_2 (on the same side of O_1 as A_1).
(5) Draw a circle with B_2 as the center and $A_2B_1(= 5)$ as the radius, and intersect the circle in (4) at points C_1, C_2, and intersect the line at point B_3 (on the same side of O_1 as B_2).
(6) Draw a circle with B_3 as the center and $A_2B_3(= 12)$ as the radius, and intersect the line at another point B_4.
(7) Draw a circle with B_4 as the center and $A_2B_4(= 24)$ as the radius, and intersect the line at another point B_5.

(8) Draw lines C_1C_2, $C_1B_2 = 5$, $A_1B_2 = 4$, $A_1C_1 = 3$, and thus $C_1A_1 \perp A_1B_1$.

(9) Draw a circle with B_1 as the center and $A_1B_5(= 45)$ as the radius, and intersect the line in (8) at two points C_3, C_4. Then $A_1C_3 = \sqrt{45^2 - A_1B_1^2} = \sqrt{2021}$ is the desired result.

Method 2 We number the objects of the one-dimensional set in the order in which they appear.

(1)–(3) are the same as the ones in Method 1.

(4) Draw a circle with A_1 as the center and $A_1B_2(= 4)$ as the radius, and intersect the line at another point A_2'.

(5) Draw a circle with B_2 as the center and $B_2B_1(= 2)$ as the radius, and intersect the circle in (3) at two points D_1, D_2.

(6) Draw line D_1D_2, which is the perpendicular bisector of segment B_1B_2.

(7) Take a point D far away from the midpoint M of segment B_1B_2 on line D_1D_2 as the center of the circle, and DA_2' as the radius (choose D so that this circle intersects the circle in (1) at two points E_1, E_2).

(8) Draw line E_1O_2 and intersect the circle in (7) at point F_1.

(9) Draw a circle with O_1 as the center and O_1F_1 $\left(= \dfrac{A_2'O_1 \cdot (O_1M + MA_2')}{1} = \dfrac{5 \times 9}{1} = 45\right)$ as the radius, and intersect the line in (7) at two points G_1, G_2, and then

$$G_1M = \sqrt{G_1O_1^2 - O_1M^2} = \sqrt{45^2 - 2^2} = \sqrt{2021}.$$

In the exam, there were not many different methods of drawing. Quite a few of the candidates' methods were similar to the following.

Method 3 We number the objects of the one-dimensional set in the order in which they appear.

(1) Denote the two given points as O_1, A_1. Draw a circle with O_1 as the center and $O_1A_1 = 1$ as the radius.

(2) Draw line O_1A_1 and intersect the circle at another point B_1.

(3) Draw a circle with B_1 as the center and O_1B_1 as the radius, intersect the line at another point B_2', and intersect the circle in (1) at two points C_1', C_2'.

(4) Draw a circle with B_2' as the center and $A_1B_2'(— 3)$ as the radius, and intersect the line at another point B_3'.

(5) Draw a circle with B_3' as the center and $A_1B_3'(=6)$ as the radius, and intersect the line at another point B_4'.
(6) Draw a circle with B_4' as the center and $A_1B_4'(=12)$ as the radius, and intersect the line at another point B_5'.
(7) Draw line $C_1'C_2'$, which is the perpendicular bisector of segment O_1B_1, passing through the midpoint M' of O_1B_1.
(8) Draw a circle with M' as the center and $M'B_5'\left(=22\frac{1}{2}\right)$ as the radius, and intersect the line in (2) at another point A'.
(9) Draw a circle with A' as the center and $A_1B_1(=2)$ as the radius, and intersect the circle in (8) at two points X_1, X_2.
(10) Connect X_1B_5', and it has a length of $\sqrt{\left(2\times 22\frac{1}{2}\right)^2-2^2}=\sqrt{2021}$.

□

Remark This question tests the students' mastery of the basic rules of ruler-and-compass construction (as described in the mathematics textbook of the compulsory education stage) and their imagination and understanding of basic geometric figures. Like the first and second questions, all the knowledge points covered in this question are also in the secondary school textbook. When writing the question, we deliberately avoided the expression "draw a segment of length $\sqrt{2021}$" to avoid ambiguity about whether "giving two endpoints of a segment" counts as giving the segment.

In addition to the above methods (there were very few candidates who completed the task with 9 one-dimensional geometric objects in the exam), some students used $2021 = 2048 - 27 = (32\sqrt{2})^2 - (3\sqrt{3})^2$ to give a method of drawing with 10 one-dimensional geometric objects. After the exam, there was a method of drawing $\sqrt{2021}$ with only 8 one-dimensional geometric objects on the Internet, and the basic idea was to use $2021 = 46^2 - (12^2 - 7^2)$.

On the other hand, since we can prove that using 6 one-dimensional geometric objects we can only get segments of length 32 at most, making $\sqrt{2021}$ requires at least 7 one-dimensional geometric objects.

Since this question is not common in recent mathematics competitions, the scoring criteria are relatively relaxed. Points were awarded according to the total number of one-dimensional geometric objects that actually appear in the end: for a total of 11 objects, 9 points; for a total of 12 objects, 6 points; for a total of 13–15 objects, 3 points.

The average score of this question is 15.8 points, but the average score for students who won gold and silver medals is slightly lower than that of the fourth question. However, the average score for students who won bronze medals is close to 10 points.

Overall, the performance on this question was better than expected, reflecting the fact that most candidates had a good grasp of the basic operations of ruler-and-compass construction in plane geometry. However, from the examination records and the answer sheets, there are still some candidates who have not mastered the basic operations written in the textbooks, such as "how to use a compass".

In history, French (non-professional) mathematician Émile Lemoine[2] published a book *Geometrography: or The Art of Geometric Constructions* (*La Géométrographie: Ou L'art Des Constructions Géometriques*) in 1893, in which he listed the number of times each basic operation needs to be performed after some geometric drawing methods are decomposed into several basic operations.

6 For integers $0 \leq a \leq n$, denote $f(n, a)$ as the number of coefficients in the expansion of $(x+1)^a(x+2)^{n-a}$ that are divisible by 3. For example, $(x+1)^3(x+2)^1 = x^4 + 5x^3 + 9x^2 + 7x + 2$, so $f(4, 3)$. For positive integer n, define $F(n)$ to be the minimum of $f(n, 0), f(n, 1), \ldots, f(n, n)$. Prove:

(1) There exist infinite numbers of positive integers n such that $F(n) \geq \dfrac{n-1}{3}$;

(2) for any positive integer n, there is $f(n) \leq \dfrac{n-1}{3}$.

(Contributed by Fu Yunhao)

Solution (1)

Method 1 (Fu Yunhao) We prove that for $n = 2(3^k - 1)$, $k \in \mathbb{N}_+$, there is

$$F(n) \geq \frac{n-1}{3}.$$

[2]Émile Lemoine (1840–1912), engineer, (amateur) mathematician (geometer), (amateur) musician. The Lemoine point and Lemoine circle in plane geometry are named after him.

We use the mathematical induction. When $k = 1$, there is $n = 4$, it is easy to know by enumeration that $F(n) = \frac{n-1}{3}$. Assume that the conclusion holds when $n = 2(3^k - 1)$, and we consider the case when $n = 2(3^{k+1} - 1)$. (All the following congruence relations are taken in the sense of modulo 3.)

For $0 \le a \le 2(3^{k+1} - 1)$, if $a \equiv 0$, then let $a = 3b$, and we have

$$\begin{aligned}(x+1)^a(x+2)^{n-a} &= (x+2)(x+1)^{3b}(x+2)^{3(2\cdot 3^k-1-b)} \\ &\equiv (x+2)(x^3+1)^b(x^3+2)^{2\cdot 3^k-1-b},\end{aligned}$$

where the congruence of polynomials refers to the congruence of coefficients in each term modulo 3. After the last equation is expanded, there exists no term of the form x^{3t+2}. Therefore, after expanding $(x+1)^a(x+2)^{n-a}$, the coefficients of terms $x^2, x^5, \ldots, x^{2\cdot 3^{k+1}-4}$ are all multiples of 3, so

$$f(n,a) \ge \frac{2\cdot 3^{k+1} - 4 - 2}{3} + 1 = 2\cdot 3^k - 1 = \frac{n-1}{3}.$$

If $a \equiv 1$, let $a = 3b+1$, we have

$$\begin{aligned}(x+1)^a(x+2)^{n-a} &= (x+1)(x+1)^{3b}(x+2)^{3(2\cdot 3^k-1-b)} \\ &\equiv (x+1)(x^3+1)^b(x^3+2)^{2\cdot 3^k-1-b}.\end{aligned}$$

Similarly, there is $f(n,a) \ge \frac{n-1}{3}$.

If $a \equiv 2$, let $a = 3b+2$, we have

$$\begin{aligned}(x+1)^a(x+2)^{n-a} &= (x+1)^2(x+2)^2(x+1)^{3b}(x+2)^{3(2\cdot 3^k-2-b)} \\ &\equiv (x^4+x^2+1)(x^3+1)^b(x^3+2)^{2\cdot 3^k-2-b}.\end{aligned}$$

By the induction hypothesis, $f(2\cdot 3^k - 2, b) \ge \frac{2\cdot 3^k - 2 - 1}{3} = 2\cdot 3^{k-1} - 1$, so the expansion of

$$(x^3+1)^b(x^3+2)^{2\cdot 3^k-2-b}$$

has at most $(2\cdot 3^k - 2) + 1 - (2\cdot 3^{k-1} - 1) = 4\cdot 3^{k-1}$ terms whose coefficients are not multiples of 3. Hence, the expansion of $(x^4+x^2+1)(x^2+1)^k(x^2+2)^{2\cdot 3^k-2-b}$ has at most $4\cdot 3^k$ terms whose coefficients are not multiples of 3, so

$$f(n,a) \ge n + 1 - 4\cdot 3^k = \frac{n-1}{3}.$$

Therefore, $F(n) \ge \frac{n-1}{3}$. The proof is complete.

Remark When $n = 2(3^k - 1)$, $a = \lambda \cdot 3^{k-1} - 1$, $\lambda = 1, 2, 4, 5$, there is always

$$f(n) = \frac{n - 1}{3}.$$

Method 2 (Xiao Liang) We prove that for $n = 3^k - 2$, $k \in \mathbb{N}_+$, there is $F(n) \geq \dfrac{n-1}{3}$. (The meanings of congruence and other symbols are the same as in Method 1.)

We use the mathematical induction. When $k = 1$, there is $n = 1$, the conclusion is obviously valid.

For general k, since $(n+1) - \dfrac{n-1}{3} = \dfrac{2n+4}{3} = 2 \cdot 3^{k-1}$ at this time, we only need to prove that for any $0 \leq a \leq 3^k - 2$, the number of terms in the expansion of $(x+1)^2(x+2)^{n-a}$ whose coefficients are not divisible by 3 does not exceed $2 \cdot 3^{k-1}$.

If $0 \leq a \leq 3^{k-1} - 2$, then $n - a \geq 2 \cdot 3^{k-1}$. By $(a+b)^3 \equiv a^3 + b^3$ and the induction, it is easy to see that

$$(x+2)^{3^{k-1}} \equiv x^{3^{k-1}} + 2^{3^{k-1}} \equiv x^{3^{k-1}} + 2.$$

Therefore,

$$(x+1)^a(x+2)^{n-a} \equiv (x+1)^a(x+2)^{n-a-2\cdot 3^{k-1}}(x^{2\cdot 3^k} + x^{3^k} + 1).$$

By the induction hypothesis, there are no more than $2 \cdot 3^{k-2}$ terms in the expansion of $(x+1)^a(x+2)^{n-a-2\cdot 3^{k-1}}$ whose coefficients are not divisible by 3. Therefore, the number of terms in the expansion of $(x+1)^a(x+2)^{n-a}$ whose coefficients are not divisible by 3 does not exceed $2 \cdot 3^{k-2}$.

If $a \geq 2 \cdot 3^{k-1}$, the same conclusion can be obtained by similar reasoning.

If $3^{k-1} \leq a \leq 2 \cdot 3^{k-1} - 2$, then $n - a \geq 3^{k-1}$. Thus,

$$\begin{aligned}(x+1)^a(x+2)^{n-a} &\equiv (x+1)^{a-3^{k-1}}(x+2)^{n-a-2\cdot 3^{k-1}}(x^2+2)^{3^{k-1}} \\ &\equiv (x+1)^{a-3^{k-1}}(x+2)^{n-a-2\cdot 3^{k-1}}(x^{2\cdot 3^{k-1}} + 2).\end{aligned}$$

By the induction hypothesis, the number of terms in the expansion of $(x+1)^a(x+2)^{n-a}$ whose coefficients are not divisible by 3 does not exceed $4 \cdot 3^{k-2}$.

Finally, if $a = 3^{k-1} - 1$, then

$$(x+1)^a(x+2)^{n-a} \equiv (x^2-1)^{3^{k-1}-1}(x^{3^{k-1}} - 1);$$

if $a = 2 \cdot 3^{k-1} - 1$, then

$$(x+1)^a(x+2)^{n-a} \equiv (x^2-1)^{3^{k-1}+1}(x^{3^{k-1}} - 1).$$

In both cases, the number of terms in the expansion of $(x+1)^a(x+2)^{n-a}$ whose coefficients are not divisible by 3 does not exceed $2 \cdot 3^{k-1}$.

Therefore, $F(n) \geq \frac{n-1}{3}$, which completes the proof.

(2) We prove a stronger proposition: for any positive integer n, there exists $a \in \{0, 1, \ldots, n\}$ such that

$$f(n,a) \leq \frac{n-1}{3}, \quad f(n+1, a+1) \leq \frac{n}{3}. \tag{$*$}$$

We proceed by induction on n.

For $n = 1$, take $a = 1$; for $n = 2$, take $a = 0$; for $n = 3$, take $a = 2$; for $n = 4$, take $a = 4$. It is easy to verify that they satisfy $(*)$. If $(*)$ holds for all positive integers less than n, we consider the case of n.

Case 1: Let $n = 3m+2, m \in \mathbb{N}_+$. By the induction hypothesis, there exists $a' \in \{0, 1, \ldots, m\}$ such that

$$f(m, a') \leq \frac{m-1}{3}, \quad f(m+1, a'+1) \leq \frac{m}{3}.$$

Let $a = 3a'$, and then

$$\begin{aligned}(x+1)^a(x+2)^{n-a} &= (x+2)^2((x+1)^{a'}(x+2)^{m-a'})^3 \\ &\equiv (x^2+x+1)(x^3+1)^{a'}(x^3+2)^{m-a'}\end{aligned}$$

By the induction hypothesis, there are at least $(m+1) - \frac{m-1}{3} = \frac{2m+4}{3}$ terms in the expansion of $(x^3+1)^{a'}(x^3+2)^{m-a'}$ whose coefficients are not multiples of 3. These terms multiplied by x^2+x+1 give at least $2m+4 > \frac{2n+4}{3}$ terms whose coefficients are not divisible by 3, so $f(n,a) \leq \frac{n-1}{3}$. On the other hand, we have

$$\begin{aligned}(x+1)^{a+1}(x+2)^{n-a} &= (x+2)^2(x+1)((x+1)^{a'}(x+2)^{m-a'})^3 \\ &\equiv (x^3+2x^2+2x+1)(x^3+1)^{a'}(x^3+2)^{m-a'} \\ &\equiv (2x^2+2x)(x^3+1)^{a'}(x^3+2)^{m-a'} \\ &\quad + (x^3+1)^{a'+1}(x^3+2)^{m-a'}.\end{aligned}$$

Similarly, there are at least $\frac{4m+8}{3}$ terms in the expansion of $(2x^2 + 2x)(x^3+1)^{a'}(x^3+2)^{m-a'}$ whose coefficients are not divisible by 3, and at least $\dfrac{2m+6}{3}$ terms in the expansion of $(x^3+1)^{a'+1}(x^3+2)^{m-a'}$ whose coefficients are not divisible by 3, so $f(n+1,a+1) < \dfrac{n}{3}$.

Case 2: Let $n = 3m+3, m \in \mathbb{N}_+$. Take the same a' as in *Case 1* and let $a = 3a'+1$. Then the expansion of $f(n,a)$ is the same as $f(n+1,a+1)$ in *Case 1*, which can be proved to satisfy the requirement. And

$$\begin{aligned}(x+1)^{a+1}(x+2)^{n-a} &= (x+1)^2(x+2)^2(x+1)^{3a'}(x+2)^{3(m-a')}\\ &\equiv (x^4+x^2+1)(x^3+1)^{a'}(x^3+2)^{m-a'}.\end{aligned}$$

By the induction hypothesis, there are at least $m+1-\dfrac{m-1}{3}=\dfrac{2m+4}{3}$ terms in the expansion of $(x^3+1)^{a'}(x^3+2)^{m-a'}$ whose coefficients are not divisible by 3. These terms multiplied by x^4+x^2+1 give $2m+4$ terms whose coefficients are not divisible by 3, so there are at most $(n+2)-(2m+4) = m+1 = \dfrac{n}{3}$ terms whose coefficients are multiples of 3.

Case 3: Let $n = 3m+4$, $m \in \mathbb{N}_+$. Take the same a' as in *Case 1* and *Case 2*. Let $a = 3a'+2$. Then the expansion of $f(n,a)$ is the same as $f(n+1,a+1)$ in *Case 1* and *Case 2*, which can be proved to satisfy the requirement. And

$$\begin{aligned}(x+1)^{a+1}(x+2)^{n-a} &= (x+2)^2(x+1)^{3(a'+1)}(x+2)^{3(m-a')}\\ &\equiv (x^2+x+1)(x^3+1)^{a'}(x^3+2)^{m-a'}.\end{aligned}$$

By the induction hypothesis, there are at least $m+2-\dfrac{m}{3}=\dfrac{2m+6}{3}$ terms in the expansion of $(x^3+1)^{a'}(x^3+2)^{m-a'}$ whose coefficients are not multiples of 3. These terms multiplied by x^2+x+1 give $2m+6$ terms whose coefficients are not multiples of 3, so there are at most $(n+2)-(2m+6) = m < \dfrac{n}{3}$ terms whose coefficients are multiples of 3.

In conclusion, $(*)$ holds, so the original proposition is proved. □

Remark This question is a binomial problem on $\mathbf{F}_3$, and the main idea is to use

$$(x+y)^3 \equiv x^3+y^3 \pmod 3$$

to simplify each polynomial. In this process, if the same digit code of a and $n-a$ in trinary is 1 and 2, respectively, the formula will affect the other digits, while the other cases will not affect each other. Therefore, the first question is to choose n so that this situation does not occur, and the second question is to choose appropriate a so that even if this more chaotic situation occurs, it will not affect the processing. The method of strengthening induction is derived from this idea.

In addition, the n in the first question can be guessed by the following method: combining the conclusion of the second question, we know that $F(n) = \frac{n-1}{3}$, so there should be at most $\frac{2n+4}{3}$ terms in the expansion whose coefficients are not multiples of 3. Therefore, in the expression of trinary, $\frac{2n+4}{3}$ must be a relatively regular number. Combined with $n \equiv 1 \pmod 3$, it can be guessed that $n+2$ should be a regular number, such as 3^n, $2 \cdot 3^n$, etc.

The average score of this question is 2.9 points. More than 100 students solved the first question, most of them basically followed Method 1, and a few students' answers were close to Method 2; few students solved the second question. The average score of the students who entered the Chinese national training team for this question was close to 11 points, while the average score of the students who did not enter the national training team but won the gold medal was less than 5 points.

During the marking process of this question, we noticed that when answering the first question, many students found that the case of $3n+4$ can be reduced to the case of n, but they did not consider that the inequality can only be established when n must be congruent to 1 modulo 3. Thus, they mistakenly thought that all numbers n congruent to 1 modulo 3 satisfied the conclusion of the question. In fact, it is enough to verify up to $n = 10$ to find out that this is not the case. Similarly, in the second question, some students tried to first fix a and use mathematical induction to prove it, but they did not verify the induction basis when inducting, resulting in serious mistakes. For example, it was thought that $f(4,2) \leq \frac{4-1}{3} = 1$, but in fact $f(4,2) = 2$.

From the scores of Questions 4 and 5, it can be inferred that the candidates who made substantial progress on Question 6 had plenty of time, but still had such problems with basic knowledge. This indicates that some candidates did not have a good grasp of mathematical induction.

China National Team Selection Test

2021

The main task of the Chinese National Training Team for the 62nd International Mathematical Olympiad (IMO) in 2021 was to select members of the Chinese national team for China to participate in the 62nd International Mathematical Olympiad held online by Russia in 2021.

From March 14th to March 23rd, 2021, the first round training and selection were hosted by Hangzhou Xuejun High School in Zhejiang Province. A total of 60 contestants participated in the examination. After two tests (with equal weights), 15 contestants were selected to enter the second round.

The second round was held from April 5th to April 13th, 2022 at Suzhou High School of Jiangsu Province. During this period, two tests were conducted, and the total scores of the four tests were compared (with equal weights). The top six scorers were identified to be the Chinese national team members for the 62nd IMO. The six members were Wang Yichuan (No. 2 High School of East China Normal University, 12th grade), Feng Chenxu (Shenzhen Middle School, 10th grade), Chen Ruitao (the High School Affiliated to Renmin University of China, 11th grade), Peng Yebo (Shenzhen Middle School, 11th grade), Wei Chen (Beijing National Day School, 12th grade), and Xia Yuxing (No. 1 Middle School Affiliated to Central China Normal University, 11th grade).

The coaches of the national training team are (in alphabet order): Ai Yinghua (Tsinghua University), Fu Yunhao (Southern University of Science and Technology), He Yijie (East China Normal University), Lin Tianqi (East China Normal University), Qu Zhenhua (East China Normal University), Wang Bin (Institute of Mathematics and Systems Science,

Chinese Academy of Sciences), Wang Xinmao (University of Science and Technology of China), Xiao Liang (Peking University), Xiong Bin (East China Normal University), Xu Disheng (Peking University), Yao Yijun (Fudan University), Yu Hongbing (Soochow University).

Test I, First Day
(8:00 – 12:30 pm; March 16, 2021)

1 Let m, n be positive integers, a_{ij} ($1 \leq i \leq m$, $1 \leq j \leq n$) be nonnegative real numbers such that for any i, j, the inequalities

$$a_{i,1} \geq a_{i,2} \geq \cdots \geq a_{i,n}, \quad a_{1,j} \geq a_{2,j} \geq \cdots \geq a_{m,j}$$

hold. For $i = 1, 2, \ldots, m$ and $j = 1, 2, \ldots, n$, define

$$X_{i,j} = a_{1,j} + \cdots + a_{i-1,j} + a_{i,j} + a_{i,j-1} + \cdots + a_{i,1},$$

$$Y_{i,j} = a_{m,j} + \cdots + a_{i+1,j} + a_{i,j} + a_{i,j+1} + \cdots + a_{i,n}.$$

Prove:

$$\prod_{i=1}^{m}\prod_{j=1}^{n} X_{i,j} \geq \prod_{i=1}^{m}\prod_{j=1}^{n} Y_{i,j}.$$

(Contributed by Ai Yinghua)

Solution 1 The problem conditions imply that

$$X_{i,j} \geq (i+j-1) \cdot a_{i,j}, \quad Y_{i,j} \leq (m+n-i-j+1) \cdot a_{i,j}.$$

Hence, we have

$$\begin{aligned}\prod_{i=1}^{m}\prod_{j=1}^{n} X_{i,j} &\geq \prod_{i=1}^{m}\prod_{j=1}^{n}(i+j-1) \cdot a_{i,j} \\ &= \prod_{i=1}^{m}\prod_{j=1}^{n}(m+n-i-j+1) \cdot a_{i,j} \\ &\geq \prod_{i=1}^{m}\prod_{j=1}^{n} Y_{i,j},\end{aligned}$$

where the equality on the second line is due to the one-to-one correspondence $(i, j) \leftrightarrow (m+1-i,\ n+1-j)$, and $i+j-1$ corresponds to $m+n-i-j+1$.

Solution 2 We shall use the following inequality:

For $0 \le c_1 \le \cdots \le c_t$, $0 \le d_1 \le \cdots \le d_t$,

$$(c_1 + d_1)(c_2 + d_2) \cdots (c_t + d_t) \le (c_1 + d_t)(c_2 + d_{t-1}) \cdots (c_t + d_1). \quad ①$$

This is because, for each i,

$$(c_i + d_i)(c_{t+1-i} + d_{t+1-i}) \le (c_i + d_{t+1-i})(c_{t+1-i} + d_i).$$

Multiplying the above inequalities for $i = 1, \ldots, t$ gives

$$\prod_{i=1}^{t}((c_i + d_i)(c_{t+1-i} + d_{t+1-i})) \le \prod_{i=1}^{t}((c_i + d_{t+1-i})(c_{t+1-i} + d_i)),$$

and taking the square root on both sides of the above inequality gives ①.

For the original problem, let $L_{i,j} = \sum\limits_{q=1}^{j} a_{i,q}$, $R_{i,j} = \sum\limits_{q=j}^{n} a_{i,q}$,

$$U_{i,j} = \sum_{p=1}^{i} a_{p,j}, \quad D_{i,j} = \sum_{p=i}^{m} a_{p,j}.$$

Then

$$X_{i,j} = L_{i,j} + U_{i,j} - a_{i,j}, \quad Y_{i,j} = D_{i,j} + R_{i,j} - a_{i,j}.$$

Apply ① to derive

$$\begin{aligned}
\prod_{i=1}^{m}\prod_{j=1}^{n} X_{i,j} &= \prod_{i=1}^{m}\prod_{j=1}^{n}(L_{i,j} + (U_{i,j} - a_{i,j})) \\
&\ge \prod_{i=1}^{m}\prod_{j=1}^{n}(L_{i,n+1-j} + (U_{i,j} - a_{i,j})) \\
&\quad \text{(since } \{L_{i,j}\}_{j=1}^{n} \text{ increases while } \{U_{i,j} - a_{i,j}\}_{j=1}^{n} \text{ decreases)} \\
&\ge \prod_{i=1}^{m}\prod_{j=1}^{n}(R_{i,j} + (U_{i,j} - a_{i,j})) \text{ (since } L_{i,n+1-j} \ge R_{i,j}) \\
&= \prod_{j=1}^{n}\prod_{i=1}^{m}((R_{i,j} - a_{i,j}) + U_{i,j}) \\
&\ge \prod_{j=1}^{n}\prod_{i=1}^{m}((R_{i,j} - a_{i,j}) + U_{m+1-i,j}) \\
&\quad \text{(since} \{R_{i,j} - a_{i,j}\}_{i=1}^{m} \text{ decreases while } \{U_{i,j}\}_{i=1}^{m} \text{ increases)}
\end{aligned}$$

$$\geq \prod_{j=1}^{n}\prod_{i=1}^{m}((R_{i,j}-a_{i,j})+D_{i,j}) \quad (\text{since } U_{m+1-i,j} \geq D_{i,j})$$

$$= \prod_{i=1}^{m}\prod_{j=1}^{n} Y_{i,j}.$$

□

2 Given positive integers n and k, $n > k^2 > 4$. In an $n \times n$ grid, any k squares in distinct rows and distinct columns are called a k-set. Find the largest positive integer N satisfying that: one can choose N squares of the $n \times n$ grid and colour them in a certain way, such that for any coloured k-set, some two squares have the same colour, and some two squares have different colours.

(Contributed by Qu Zhenhua)

Solution $N = (k-1)^2 n$.

Choose $(k-1)^2$ rows of the grid: colour the first $k-1$ rows in colour c_1; the second $k-1$ rows in colour c_2; ...; the last $k-1$ rows in colour c_{k-1}. Altogether, $(k-1)^2 n$ squares are coloured. For any coloured k-set, as there are only $k-1$ colours, some two squares must have the same colour. On the other hand, if all the squares in this k-set have the same colour, then by definition they are in k different rows, yet there are only $k-1$ rows in that colour, a contradiction. This implies that some two squares have different colours. Therefore, $N \geq (k-1)^2 n$.

In a k-set, if all the squares have the same colour, call it a mono k-set; if the squares have distinct colours, call it a poly k-set. We assert that for any colouring of $(k-1)^2 n + 1$ squares, there must exist a mono k-set or a poly k-set. This will give $N \leq (k-1)^2 n$ and the conclusion. First, we need a lemma.

Lemma *In an $n \times n$ grid, among any $(m-1)n+1$ squares, $1 \leq m \leq n$, there exists an m-set.*

Proof of lemma Divide the squares of the $n \times n$ grid into n groups, such that the square in the ith row and jth column is in group a $(1 \leq i, j, a \leq n)$ if and only if $i - j \equiv a \pmod{n}$. Notice that for each group, the squares are in distinct rows and distinct columns. By the pigeonhole principle, among any $(m-1)n+1$ squares, m of them are in a group, and they form an m-set.

For the original problem, assume that $(k-1)^2 n + 1$ squares are coloured in a certain way. According to the lemma, there exists a $((k-1)^2+1)$-set, call it A. If there are $k-1$ or fewer colours in A, then by the pigeonhole principle some k squares are in the same colour, and they form a mono k-set; if there are k or more colours in A, then choose k squares of distinct colours, and they form a poly k-set.

Therefore, the largest N is $(k-1)^2 n$.

Alternative proof of lemma Choose any $(m-1)n+1$ squares and colour them black. Take m rows with the most black squares: say they are row $1, 2, \ldots, m$, with $x_1, x_2, \ldots, x_m$ black squares, respectively, and the other rows have $x_{m+1}, \ldots, x_n$ black squares, respectively. If there exists l, $1 \leq l \leq m$, such that in the first m rows, black squares of some l rows are distributed in $l-1$ (or fewer) columns, say, black squares of row $1, \ldots, l$ are all in the first $l-1$ columns. Then, $x_1, \ldots, x_l \leq l-1$; for $i > m$, $x_i \leq l-1$ as well (since there are fewer black squares in those rows). Hence,

$$
\begin{aligned}
(m-1)n+1 = \sum_{i=1}^{n} x_i &\leq l(l-1) + (m-l)n + (n-m)(l-1) \\
&= mn - n + m - l(m+1-l) \\
&\leq mn - n + m - m = (m-1)n,
\end{aligned}
$$

which is a contradiction. It follows that for any l of the first m rows, the black squares are located in at least l columns. By Hall's marriage theorem, there are m black squares in distinct rows and distinct columns, which is an m-set. The lemma is proved. □

3 Fix a positive integer $n \geq 2$. Prove: for any n distinct integers $a_1, a_2, \ldots, a_n$, the set

$$\left\{1, 2, \ldots, \frac{n(n-1)}{2}\right\}$$

contains at least $\left\lceil \dfrac{n(n-6)}{19} \right\rceil$ elements that cannot be expressed as the difference of some a_i and a_j. Here, $\lceil x \rceil$ is the least integer greater than or equal to x.

(Contributed by Wang Bin)

Solution Let $m = \dfrac{n(n-1)}{2}$. Without loss of generality, assume $a_1 > a_2 > \cdots > a_n$. Denote the multiset of $a_i - a_j$, $1 \leq i < j \leq n$ by $D = \{d_1, d_2, \ldots, d_m\}$ (elements can be repeated).

Suppose that $m-t$ elements of $M=\{1,2,\ldots,m\}$ appear in D and they form the subset A, while the other t elements do not appear in D and they form $B=\{b_1,\ldots,b_t\}$. Let $C=D\backslash A=\{c_1,\ldots,c_t\}$ (elements can be repeated).

Consider the generating function $F(z)=z^{a_1}+z^{a_2}+\cdots+z^{a_n}$ and

$$\begin{aligned}
G(z) &= F(z)F(z^{-1}) \\
&= (z^{a_1}+z^{a_2}+\cdots+z^{a_n})(z^{-a_1}+z^{-a_2}+\cdots+z^{-a_n}) \\
&= n+\sum_{i<j}(z^{a_i-a_j}+z^{a_j-a_i}) \\
&= n+\sum_{i<j}(z^{|a_i-a_j|}+z^{-|a_i-a_j|}) \\
&= n+\sum_{k=1}^{m}(z^{d_k}+z^{-d_k}) \\
&= n+\sum_{k=1}^{m}(z^k+z^{-k})-\sum_{k=1}^{t}(z^{b_k}+z^{-b_k})+\sum_{k=1}^{t}(z^{c_k}+z^{-c_k}).
\end{aligned}$$

If z is a unit complex number,

$$F(z^{-1})=F(\underline{z})=\overline{F(z)},$$

$$G(z)=F(z)F(z^{-1})=|F(z)|^2$$

is a nonnegative real. We want to choose an appropriate ϑ and

$$z=\mathrm{e}^{2\vartheta\mathrm{i}}=\cos 2\vartheta+\mathrm{i}\sin 2\vartheta,$$

such that

$$\begin{aligned}
H(z) &= 1+\sum_{k=1}^{m}(z^k+z^{-k}) \\
&= \sum_{k=-m}^{m}\mathrm{e}^{2k\vartheta\mathrm{i}} \\
&= \frac{\mathrm{e}^{(2m+1)\vartheta\mathrm{i}}-\mathrm{e}^{-(2m+1)\vartheta\mathrm{i}}}{\mathrm{e}^{\vartheta\mathrm{i}}-\mathrm{e}^{-\vartheta\mathrm{i}}} \\
&= \frac{\sin(2m+1)\vartheta}{\sin\vartheta}
\end{aligned}$$

is a nonnegative real and its value is as large as possible. To this end, take $\vartheta = \dfrac{3\pi}{2(2m+1)}$, $z_0 = \mathrm{e}^{\frac{3\pi \mathrm{i}}{2m+1}}$ to get

$$H(z_0) = H(\mathrm{e}^{\frac{3\pi \mathrm{i}}{2m+1}}) = \frac{-1}{\sin\vartheta} < \frac{-1}{\vartheta}$$
$$= -\frac{2(2m+1)}{3\pi} = -\frac{2(n^2-n+1)}{3\pi}.$$

Furthermore, as

$$0 \le G(z_0) = (n-1) + H(z_0) + \sum_{k=1}^{t}(z_0^{c_k} + z_0^{-c_k}) - \sum_{k=1}^{t}(z_0^{b_k} + z_0^{-b_k})$$
$$\le (n-1) + H(z_0) + 4t,$$

it follows that

$$t \ge \frac{1}{4}[-H(z_0) - (n-1)]$$
$$> \frac{(n^2-n+1)}{6\pi} - \frac{(n-1)}{4}$$
$$= \frac{n^2 - (1.5\pi+1)(n-1)}{6\pi}$$
$$> \frac{n^2-6n}{19}.$$

□

Test I, Second Day
(8 – 12:30 pm; March 17, 2021)

4 Let $f(x)$ and $g(x)$ be integral polynomials. Suppose that for infinitely many prime numbers p, there exists an integer m_p such that $f(a) \equiv g(a+m_p) \pmod p$ for every integer a. Prove that $f(x) = g(x+r)$ for some rational number r.

(Contributed by Qu Zhenhua)

Solution Let P be the set of all prime numbers p satisfying the problem condition. First, notice that $\deg f = \deg g$, and their leading coefficients are equal; otherwise, the leading term of $f(x) - g(x+m_p)$ is independent of p and m_p. Take $p \in P$, $p > \max\{\deg f, \deg g\}$. As $f(a) - g(a+m_p)$ is always divisible by p and the degree of $f(x) - g(x+m_p)$ is less than p, it follows that every term of $f(x) - g(x + m_p)$ has coefficient divisible by p. Since p can be arbitrarily large, while the leading coefficient of $f(x) - g(x + m_p)$ is fixed, we arrive at an absurdity.

Let $f(x) = a_n x^n + a_{n-1}x^{n-1} + \cdots + a_0$,

$$g(x) = b_n x^n + b_{n-1}x^{n-1} + \cdots + b_0,$$

where $a_n = b_n \neq 0$. Take $r = \dfrac{a_{n-1} - b_{n-1}}{nb_n}$ rational. We show $f(x) = g(x + r)$.

Take arbitrary $p \in P$, $p > \max\{n, b_n\}$. Since

$$f(a) \equiv g(a + m_p) \pmod p$$

for any integer a and the degrees are both less than p, the two polynomials in x must satisfy

$$f(x) \equiv g(x + m_p) \pmod p \qquad ①$$

Compare the coefficients of x^{n-1} in ① to obtain

$$nb_n \cdot m_p \equiv a_{n-1} - b_{n-1} \pmod p,$$

in particular, $nb_n r \equiv nb_n \cdot m_p \pmod p$. Now ① yields

$$\begin{aligned}(nb_n)^n \cdot g(x + r) &\equiv (nb_n)^n g(x + m_p) \\ &\equiv (nb_n)^n f(x) \pmod p.\end{aligned}$$

Since the above equation holds for infinitely many primes p, it must be

$$(nb_n)^n g(x + r) = (nb_n)^n f(x),$$

or $g(x + r) = f(x)$. □

5 As illustrated in Fig. 5.1, circle Ω is tangent to AB, AC at B and C, respectively; D is the midpoint of AC; O is the circumcentre of $\triangle ABC$. A circle Γ through A and C meets the minor arc $\overset{\frown}{BC}$ of Ω at P, and meets AB at Q other than A. It is known that the midpoint R of the minor arc $\overset{\frown}{PQ}$ satisfies $CR \perp AB$. Let L be the intersection of the rays PQ and CA; M be the midpoint of AL; N be the midpoint of DR; $MX \perp ON$ with foot X. Show that the circumcircle of $\triangle DNX$ passes through the centre of circle Γ.

(Contributed by Lin Tianqi)

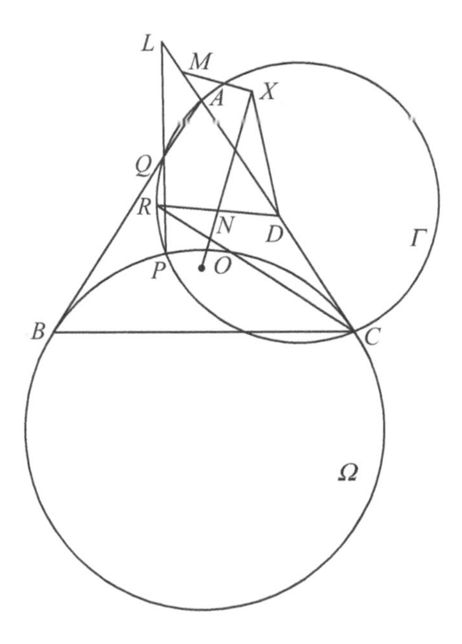

Fig. 5.1

Solution As shown in Fig. 5.2, let K and S be the centres of Γ and Ω, respectively. Clearly, O is the midpoint of AS. Let RR' be a diameter of Γ. Draw circle ω passing through A, Q and tangent to AC; let T be its centre.

First, we show that R, P, S are collinear; R, Q, T are collinear. Let the inscribed angles in Γ that subtend $\overparen{AQ}$, $\overparen{PQ}$, and $\overparen{CP}$ be respectively α, 2ϑ, and β. Since $CR \perp AB$,

$$\begin{aligned}90^\circ &= \angle BAC + \angle ACR \\ &= (2\vartheta + \beta) + (\alpha + \vartheta) \\ &= \alpha + \beta + 3\vartheta.\end{aligned}$$

Since $\angle CPR = 180^\circ - (\vartheta + \beta)$,

$$\angle CPS = \angle SCP = 90^\circ - \angle ACP = 90^\circ - (\alpha + 2\vartheta),$$

and hence

$$\begin{aligned}&\angle CPR + \angle CPS = [180^\circ - (\vartheta + \beta)] + [90^\circ - (\alpha + 2\vartheta)] \\ &= 180^\circ + 90^\circ - (\alpha + \beta + 3\vartheta) = 180^\circ,\end{aligned}$$

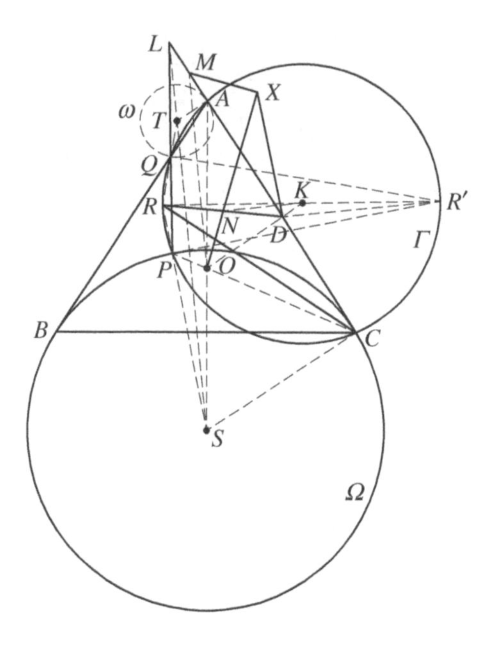

Fig. 5.2

which implies that R, P, S are collinear. Moreover, $\angle AQR = 180° - (\alpha + \vartheta)$,

$$\begin{aligned}\angle AQT &= \angle TAQ \\ &= 90° - \angle CAQ \\ &= 90° - (\beta + 2\vartheta),\end{aligned}$$

and

$$\begin{aligned}&\angle AQR + \angle AQT = [180° - (\alpha + \vartheta)] + [90° - (\beta + 2\vartheta)] \\ &= 180° + 90° - (\alpha + \beta + 3\vartheta) = 180°,\end{aligned}$$

hence R, Q, T are collinear.

Note that RR' is a diameter of Γ. So, $PR' \perp RS$, $QR' \perp RT$, indicating that PR', QR' are tangent to circles Ω and ω, respectively. As R' is the midpoint of $\overset{\frown}{PAQ}$, $PR' = QR'$, and R' has identical power with respect to circles Ω and ω. Moreover, AC is a common external tangent of Ω and ω, D is the midpoint of AC. Hence, D lies on the radical axis of Ω and ω as well, the line DR' is the radical axis, $DR' \perp ST$.

Next, we show that S, T, L are collinear. Let the radii of circles Ω, ω be r_1, r_2, respectively. Then

$$r_1 = \frac{PC}{2\cos\angle PCS} = \frac{PC}{2\sin\angle ACP};$$

similarly, $r_2 = \dfrac{AQ}{2\sin\angle CAQ}$. It follows that

$$\frac{r_1}{r_2} = \frac{PC\cdot\sin\angle CAQ}{AQ\cdot\sin\angle ACP} = \frac{PC\cdot CQ}{AQ\cdot AP} = \frac{S_{\triangle PCQ}}{S_{\triangle PAQ}} = \frac{LC}{LA},$$

and S, T, L are collinear.

Since $DR'\perp ST$, $DR'\perp SL$. Since N, K, O, M are the midpoints of RD, RR', AS, AL, respectively, we have $DR' \,//\, NK$, $SL \,//\, OM$, and $NK\perp OM$. Moreover, $DK\perp AC$, and hence $\angle NKD = \angle OMD$. Meanwhile, $\angle OXM = \angle ODM = 90°$ implies that O, D, X, M all lie on a circle, $\angle OMD = \angle OXD = \angle NXD$. Together, $\angle NKD = \angle NXD$, so D, N, X, K are concyclic, and we conclude that the circumcircle of $\triangle DNX$ passes through K, the centre of Γ. □

6 Given positive integers n, r and distinct prime numbers $p_1, p_2, \ldots, p_r$. Initially, there are $(n+1)^r$ numbers on the blackboard: $p_1^{i_1}p_2^{i_2}\cdots p_r^{i_r}$ $(0\le i_1, i_2, \ldots, i_r\le n)$. Alice and Bob take turns (Alice goes first) to make the following moves, until only one number is left on the blackboard:

- Every time, Alice erases two numbers (can be identical) and writes their greatest common divisor on the blackboard.
- Every time, Bob erases two numbers (can be identical) and writes their least common multiple on the blackboard.

Find the least integer M, such that Alice can guarantee the remaining number does not exceed M.

(Contributed by Xiao Liang)

Solution The least M is $(p_1\cdots p_r)^{\left\lfloor\frac{n}{2}\right\rfloor}$.

Denote $N = (p_1\cdots p_r)^n$, $M = (p_1\cdots p_r)^{\left\lfloor\frac{n}{2}\right\rfloor}$. For each divisor a of N, call $\left(a, \dfrac{N}{a}\right)$ a "pair". Evidently, $\gcd\left(a, \dfrac{N}{a}\right)$ is a divisor of M, while $\operatorname{lcm}\left(a, \dfrac{N}{a}\right)$ is a multiple of M.

First, Alice can guarantee that the remaining number on the board is a divisor of M.

Suppose n is odd, $M = (p_1 \cdots p_r)^{(n-1)/2}$. Alice adopts the following strategy so that her move always leaves some pairs and a divisor C of M on the board. First, she erases 1, N, and writes 1, which leaves $\frac{1}{2}(n+1)^r - 1$ pairs and 1 on the board.

(1) If Bob erases a, b from two pairs $\left(a, \frac{N}{a}\right)$, $\left(b, \frac{N}{b}\right)$, and writes $\operatorname{lcm}(a, b)$, then Alice erases $\frac{N}{a}$, $\frac{N}{b}$, and writes $\gcd\left(\frac{N}{a}, \frac{N}{b}\right) = \frac{N}{\operatorname{lcm}(a, b)}$ which forms a pair with $\gcd(a, b)$.

(2) If Bob erases a pair $\left(a, \frac{N}{a}\right)$ and writes $\operatorname{lcm}\left(a, \frac{N}{a}\right)$, then Alice erases $\operatorname{lcm}\left(a, \frac{N}{a}\right)$ and C, and writes their greatest common divisor C', which is a divisor of M, leaving some pairs and C' on the board.

(3) If Bob erases C and a from $\left(a, \frac{N}{a}\right)$ and writes $\operatorname{lcm}(a, C)$, then Alice erases $\frac{N}{a}$, $\operatorname{lcm}(a, C)$, and writes $C' = \gcd\left(\frac{N}{a}, \operatorname{lcm}(a, C)\right)$

which is a divisor of $\operatorname{lcm}\left(C, \gcd\left(a, \frac{N}{a}\right)\right)$. Since $\gcd\left(a, \frac{N}{a}\right)$ is a divisor of M, C' is likewise. Still, it leaves some pairs and a divisor C' of M on the board.

Based on the above argument, Alice is certain that her move leaves some pairs and a divisor of M on the board. Eventually, there is only one number left which is a divisor of M.

Now suppose n is even, $M = (p_1 \cdots p_r)^{n/2}$. Alice adopts the following strategy so that her move always leaves some pairs and two divisors C_1 and C_2 of M on the board. First, she erases 1, N, and writes 1, leaving pairs and 1, M. In the subsequent moves, if Bob erases one or two numbers from pair(s), Alice follows (1)-(3) in the previous discussion; if Bob erases C_1, C_2 and writes $\operatorname{lcm}(C_1, C_2) = C_1'$ which is also a divisor of M, then Alice erases any pair $\left(a, \frac{N}{a}\right)$ and writes $\gcd\left(a, \frac{N}{a}\right) = C_2'$ which is again a divisor of M. In any case, every time after Alice's move, some pairs and two divisors of M are left on the board. Eventually, Bob must be confronted with two divisors of M, and his move will leave a divisor of M on the board.

Next, Bob can guarantee that the remaining number on the board is a multiple of M.

Indeed, Bob adopts the following strategy so that his move always leaves some pairs and a multiple C_1 of M (when n is even), or two multiples C_1 and C_2 of M (when n is odd). Initially, take $C_1 = (p_1 \cdots p_r)^{n/2}$ (when n is even), or $C_1 = (p_1 \cdots p_r)^{(n-1)/2}$, $C_2 = (p_1 \cdots p_r)^{(n+1)/2}$ (when n is odd). Bob just reverses Alice's moves as follows:

(1)′ If Alice erases a, b from two pairs $\left(a, \frac{N}{a}\right)$, $\left(b, \frac{N}{b}\right)$ and writes gcd(a, b), then Bob erases $\frac{N}{a}$, $\frac{N}{b}$, and writes $\operatorname{lcm}\left(\frac{N}{a}, \frac{N}{b}\right) = N/\gcd(a, b)$.

(2)′ If Alice erases a pair $\left(a, \frac{N}{a}\right)$ and writes $\gcd\left(a, \frac{N}{a}\right)$, then Bob erases this number and C_1.

(3)′ If Alice erases C_i and a from $\left(a, \frac{N}{a}\right)$ and writes $\gcd(a, C_i)$, then Bob erases $\frac{N}{a}$ and $\gcd(a, C_i)$.

(4)′ If Alice erases C_1, C_2 and writes $\gcd(C_1, C_2)$, then Bob erases any pair $\left(a, \frac{N}{a}\right)$.

Similar to the discussion of Alice's strategy, we just interchange the terms "gcd" and "lcm" to arrive at the conclusion that Bob guarantees the remaining number is a multiple of M.

Therefore, the least M is $(p_1 \cdots p_r)^{\left\lfloor \frac{n}{2} \right\rfloor}$. □

Test II, First Day
(8 – 12:30 pm; March 21, 2021)

1 As shown in Fig. 1.1, the quadrilateral $ABCD$ is inscribed in the circle Γ, $AB + BC = AD + DC$. Let E be the midpoint of the arc $\overparen{BCD}$; F be the antipode of A ($F \neq C$); I be the incentre of $\triangle ABC$; J be the excentre of $\triangle ABC$ against $\angle BAC$; K is the incentre of $\triangle BCD$. Let P satisfy that $\triangle BIC$ and $\triangle KPJ$ are similar triangles with their corresponding vertices in order. Prove that the lines EK, PF intersect on Γ.

(Contributed by Lin Tianqi)

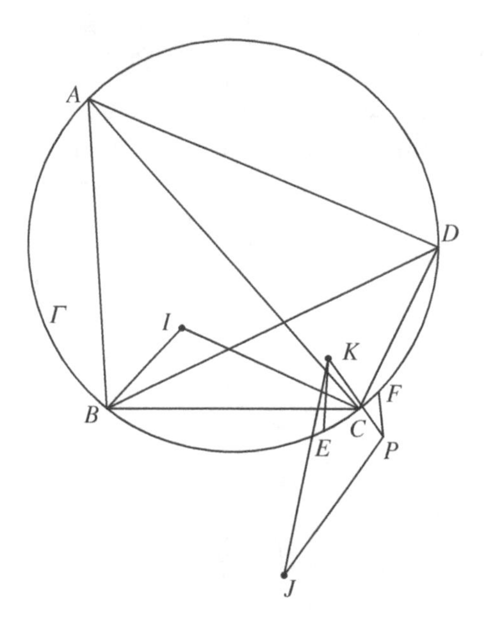

Fig. 1.1

Solution As illustrated in Fig. 1.2, let L be the excentre of $\triangle ACD$ relative to A. Assume that the escribed circles $\odot J$ of $\triangle ABC$, $\odot L$ of $\triangle ACD$ meet the extension of AC at M_1, M_2, respectively (not shown in the figure). Then $CM_1 = \dfrac{AB+BC-AC}{2}$, $CM_2 = \dfrac{AD+DC-AC}{2}$. Since $AB + BC = CD + DA$, it follows that $CM_1 = CM_2$, $M_1 = M_2$, and $JL \perp AC$.

First, we show that P is the orthocentre of $\triangle AJL$, and hence A, C, P are collinear. It is easy to see that B, I, K, C, J are concyclic. Since $\angle KJP = \angle BCI = \dfrac{1}{2}\angle ACB$, it follows that

$$\begin{aligned}\angle IJK &= \angle IBK = \angle IBC - \angle KBC \\ &= \frac{1}{2}\angle ABC - \frac{1}{2}\angle DBC = \frac{1}{2}\angle ABD.\end{aligned}$$

Thus,

$$\begin{aligned}\angle AJP &= \angle IJK + \angle KJP \\ &= \frac{1}{2}\angle ABD + \frac{1}{2}\angle ACB \\ &= \frac{1}{2}\angle BCD,\end{aligned}$$

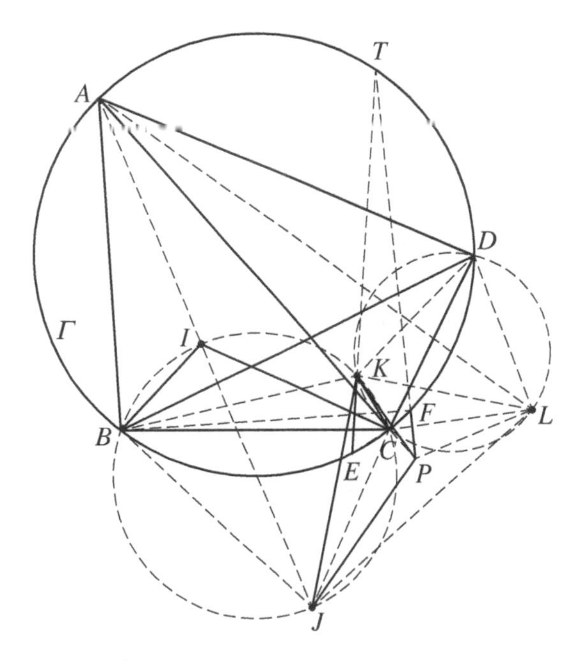

Fig. 1.2

while $\angle JAL = \dfrac{1}{2}\angle BAD$, it follows that $\angle AJP + \angle JAL = 90°$, and $JP \perp AL$.

Meanwhile, as $CI \perp CJ$, $AC \perp JL$, hence $\angle CJL = \angle ACI = \angle BCI = \angle KJP$, and JC, JP are isogonal lines of $\angle KJL$. As C, K, D, L lie on a circle, we have

$$\begin{aligned}\angle LKC = \angle LDC &= 90° - \frac{1}{2}\angle ADC \\ &= \frac{1}{2}\angle ABC = \angle CBI = \angle JKP,\end{aligned}$$

and KC, KP are isogonal lines of $\angle JKL$. Therefore, C and P are isogonal conjugate points in $\triangle KJL$, implying that

$$\begin{aligned}\angle PLJ &= \angle KLC = \angle KDC \\ &= \frac{1}{2}\angle BDC = \frac{1}{2}\angle BAC \\ &= \angle JAC.\end{aligned}$$

Since $AC \perp JL$, it follows that $LP \perp AJ$, P is the orthocentre of $\triangle AJL$, and hence A, C, P are collinear.

Next, we show $\triangle ECF \backsim \triangle KCP$ and $\triangle ECK \backsim \triangle FCP$.

Since AF is a diameter of Γ, $AC \perp CF$. It is shown that A, C, P are collinear, hence $CP \perp CF$; in addition, as CK, CE are the interior and exterior angle bisectors of $\angle BCD$, hence $CE \perp CK$. Together, they lead to $\angle ECF = \angle KCP$. Finally,

$$\begin{aligned}\measuredangle CKP &= \measuredangle JKP - \measuredangle CKJ\\ &= \measuredangle CBI - \measuredangle CBJ\\ &= 2\measuredangle CBI - 90^\circ\\ &= \measuredangle CBA - \measuredangle FBA\\ &= \measuredangle FBC = \measuredangle FEC.\end{aligned}$$

We deduce that $\triangle ECF \backsim \triangle KCP$, and $\triangle ECK \backsim \triangle FCP$.

Let the lines EK and FP meet at T. From $\triangle ECK \backsim \triangle FCP$ (the corresponding vertices are in order), it follows that

$$\angle ETF = 180^\circ - \measuredangle(EK, FP) = 180^\circ - \measuredangle ECF,$$

and hence T, E, C, F are concyclic, that is, the intersection T of EK, FP lies on Γ. □

2 Given positive integers k and n ($n \geq 2$), find the minimum constant c satisfying this assertion: if G is a simple kn-regular graph (the degree of each vertex is kn) with m vertices, then each vertex can be coloured one of n colours, such that the number of "mono edges" is at most cm. Here, a mono edge is an edge incident to two vertices of the same colour.

(Contributed by Fu Yunhao)

Solution $c = \dfrac{k(kn-n+2)}{2(kn+1)}$.

First, we prove $c \geq \dfrac{k(kn-n+2)}{2(kn+1)}$. Consider a complete graph K_{kn+1} with any n-colouring of the vertices. Suppose that the numbers of vertices coloured by $1, 2, \ldots, n$ are $a_1, a_2, \ldots, a_n$, respectively. Then

$$\begin{aligned} a_1^2 + a_2^2 + \cdots + a_n^2 &\geq \frac{(a_1 + a_2 + \cdots + a_n)^2}{n} \\ &= \frac{(kn+1)^2}{n} \\ &= k^2 n + 2k + \frac{1}{n}. \end{aligned}$$

As $a_1, a_2, \ldots, a_n$ are integers,

$$a_1^2 + a_2^2 + \cdots + a_n^2 \geq k^2 n + 2k + 1.$$

Therefore, the number of mono edges is

$$\begin{aligned} \sum_{i=1}^{n} \mathrm{C}_{a_i}^2 &= \frac{1}{2}\left(\sum_{i=1}^{n} a_i^2 - \sum_{i=1}^{n} a_i\right) \\ &\geq \frac{1}{2}(k^2 n + 2k + 1 - (kn+1)) \\ &= \frac{1}{2}k(kn - n + 2), \end{aligned}$$

which implies $c \geq \dfrac{k(kn-n+2)}{2(kn+1)}$.

Next, we give three proofs that $c = \dfrac{k(kn-n+2)}{2(kn+1)}$ satisfies the problem assertion.

Proof 1 For any kn-regular graph G, it is well known that G has a $(kn+1)$-colouring such that adjacent vertices are of different colours. Divide the $kn+1$ colours randomly into n groups such that one group contains $k+1$ colours while each of the others contains k colours, and for all these divisions, they have equal probabilities. Now change all colours of a group into one colour. For any edge of G, initially, it is incident to vertices of different colours; after changing the colours, it is incident to vertices of the same colour, namely, it becomes a mono edge with a probability

$$\frac{\mathrm{C}_{k+1}^2 + (n-1)\mathrm{C}_k^2}{\mathrm{C}_{kn+1}^2} = \frac{k(k+1) + k(k-1)(n-1)}{(kn+1)(kn)} = \frac{kn-n+2}{n(kn+1)}.$$

Therefore, after changing the colours, the expected number of mono edges is

$$\frac{kn-n+2}{n(kn+1)}\cdot\frac{knm}{2}=m\cdot\frac{k(kn-n+2)}{2(kn+1)}=cm.$$

There must exist an n-colouring of G such that the number of mono edges is less than or equal to cm. This verifies the problem assertion.

Proof 2 Suppose that among all n-colourings of G, the minimum number of mono edges is $T(G)$. Let V consist of vertices of G. For a permutation $\pi=(v_1,v_2,\ldots,v_m)$ of V, colour $v_1,v_2,\ldots,v_m$ in order (by one of n colours) as follows. For a vertex v, suppose that in π, $A_\pi(v)$ of kn neighbouring vertices of v appear before v. We colour v with the colour that appears least frequently among those $A_\pi(v)$ neighbours (it appears at most $\left\lfloor\frac{A_\pi(v)}{n}\right\rfloor$ times). According to this greedy algorithm, it is guaranteed that

$$T(G)\le F(\pi):=\sum_{v\in V}\left\lfloor\frac{A_\pi(v)}{n}\right\rfloor.$$

The above inequality holds for any permutation π. It suffices to prove, for all $m!$ permutations, the average of $F(\pi)$ is $\frac{k(kn-n+2)}{2(kn+1)}\cdot m$, so that some permutation π gives a desired n-colouring. For a vertex v, let its neighbours be $\{u_1,u_2,\ldots,u_{kn}\}$. Consider $B=\{v,u_1,u_2,\ldots,u_{kn}\}$: among the $m!$ relative orders of elements of B given by all $m!$ permutations, v appears at position $1,2,\ldots,kn+1$ with equal probability $\frac{1}{kn+1}$. Hence, as π traverses all $m!$ permutations, the average of $\left\lfloor\frac{A_\pi(v)}{n}\right\rfloor$ is

$$\begin{aligned}\frac{1}{kn+1}\sum_{a=0}^{kn}\left\lfloor\frac{a}{n}\right\rfloor&=\frac{1}{kn+1}(0\times n+1\times n+\cdots+(k-1)\times n+k)\\&=\frac{k(kn-n+2)}{2(kn+1)}.\end{aligned}$$

As a result, the average of $F(\pi)$ is $\frac{k(kn-n+2)}{2(kn+1)}\cdot m=cm$.

Proof 3 For a kn-regular graph G, let $A=|V(G)|$ be the number of vertices of G. Since there are only finitely many n-colourings of G, there must be one with the minimum number of mono edges, say C. We have the following claims on C.

Claim 1 For every vertex v, v has at most k neighbours that are in the same colour as v; if this does occur, then v has exactly k neighbours in each of the n colours (we call this v a "balanced vertex").

Proof of claim 1 Let v be of colour P. If there is another colour Q such that among v's neighbours, the number of vertices in colour P is larger than that in colour Q, then we change the colour of v from P to Q. This will reduce the number of mono edges, which is a contradiction. Hence, there are fewest neighbours of v in colour P (or one of the fewest); as v has kn neighbours, the claim is verified.

Claim 2 Adjacent balanced vertices must be monochromatic.

Proof of claim 2 Let v, w be adjacent balanced vertices, v in colour P while w in colour Q. Changing v to Q and w to P, vw will not be a mono edge, but the numbers of mono edges incident to v or w will each decrease by 1, a contradiction.

Claim 3 If v is not a balanced vertex, v is adjacent to r balanced vertices which have different colours from v, then there are at most $k - \dfrac{r}{k+1}$ mono edges incident to v.

Proof of claim 3 Consider the edges between these r balanced vertices. As each balanced vertex is adjacent to at most k balanced vertices of its colour, in the subgraph of these vertices, the degree of each vertex does not exceed k. Now we can choose at least $\dfrac{r}{k+1}$ of these balanced vertices such that any two of them are not adjacent (assume that at most t vertices can be chosen, then they have tk or fewer neighbours, $tk+t \geq r$, or $t \geq \dfrac{r}{k+1}$). Change them to the colour of v. From the properties of balanced vertices, we see that the number of mono edges does not change; however, if the number of mono edges emanating from v exceeds k, then changing v's colour will reduce the number of mono edges, a contradiction. Therefore, the number of mono edges emanating from v does not exceed k, and hence there are at most $k - \dfrac{r}{k+1}$ mono edges incident to v.

Let a be the number of balanced vertices in G. There are two situations.

(i) If $a \leq \dfrac{k+1}{kn+1}A$. By calculating the number of mono edges emanating from each vertex in two ways, we find that the total number of mono edges does not exceed

$$\begin{aligned}\frac{1}{2}(ak+(A-a)(k-1)) &= \frac{1}{2}((k-1)A+a)\\ &\le \frac{1}{2}\left((k-1)A+\frac{k+1}{kn+1}A\right)\\ &= \frac{k(kn-n+2)}{2(kn+1)}A.\end{aligned}$$

(ii) If $a > \dfrac{k+1}{kn+1}A$. From Claim 2, it follows that the heterochromatic neighbours of a balanced vertex are never balanced vertices. Thus, the number of edges connecting a balanced vertex and a heterochromatic neighbour is $ak(n-1) > \dfrac{k(k+1)(n-1)}{kn+1}A$. Now for each non-balanced vertex, consider its heterochromatic balanced (vertex) neighbours: let the numbers be $r_1, r_2, \ldots, r_{A-a}$. We have

$$r_1+r_2+\cdots+r_{A-a} > \frac{k(k+1)(n-1)}{kn+1}A.$$

From Claim 3, the number of mono edges does not exceed

$$\begin{aligned}\frac{1}{2}\left(ka+\sum_{i=1}^{A-a}\left(k-\frac{r_i}{k+1}\right)\right) &= \frac{1}{2}\left(kA-\frac{1}{k+1}\sum_{i=1}^{n-a}r_i\right)\\ &< \frac{1}{2}\left(kA-\frac{k(k+1)(n-1)}{(k+1)(kn+1)}A\right)\\ &= \frac{k(kn-n+2)}{2(kn+1)}A.\end{aligned}$$

Therefore, $c = \dfrac{k(kn-n+2)}{2(kn+1)}$ as desired. □

3 Given positive integers a, b, c which are pairwise coprime. Let $f(n)$ represent the number of nonnegative integer solutions (x, y, z) of the equation $ax+by+cz=n$. Prove: there exist real constants α, β, γ, such that for every nonnegative real number n,

$$|f(n)-(\alpha n^2+\beta n+\gamma)| < \frac{a+b+c}{12}.$$

(Contributed by Wang Bin)

Solution 1 Consider the generating function of $\{f(n)\}$:

$$\begin{aligned}
G(t) &= \sum_{n=0}^{\infty} g(n)t^n \\
&= (1 + t^a + t^{2a} + \cdots)(1 + t^b + t^{2b} + \cdots)(1 + t^c + t^{2c} + \cdots) \\
&= \frac{1}{(1-t^a)(1-t^b)(1-t^c)}.
\end{aligned}$$

Since a, b, c are pairwise coprime, the denominator $(1-t^a)(1-t^b)(1-t^c)$ contains no repeated factors other than $(1-t)^3$. Let $\omega = \mathrm{e}^{2\pi\mathrm{i}/a}$, $\tau = \mathrm{e}^{2\pi\mathrm{i}/b}$, $\rho = e^{2\pi\mathrm{i}/c}$ be the primitive a, b, cth roots of unity, respectively. Consider the partial fraction decomposition

$$\begin{aligned}
\frac{1}{(1-t^a)(1-t^b)(1-t^c)} &= \frac{h_0}{(1-t)} + \frac{h_1}{(1-t)^2} + \frac{h_2}{(1-t)^3} \\
&\quad + \sum_{k=1}^{a-1} \frac{d_k}{1-\omega^k t} + \sum_{k=1}^{b-1} \frac{e_k}{1-\tau^k t} + \sum_{k=1}^{c-1} \frac{f_k}{1-\rho^k t},
\end{aligned}$$

in which the numerators $h_0, h_1, h_2, d_1, \ldots, d_{a-1}, e_1, \ldots, e_{b-1}, f_1, \ldots, f_{c-1}$ are all complex numbers. To find d_k in $\dfrac{d_k}{1-\omega^k t}$, multiply $(1-\omega^k t)$ on both sides and let $t \to \omega^{-k}$ (or just let $1-\omega^k t = 0$). As all the other terms vanish, we arrive at

$$\begin{aligned}
d_k &= \lim_{t\to\omega^{-k}} \frac{1-\omega^k t}{(1-(\omega^k t)^a)(1-t^b)(1-t^c)} \\
&= \frac{1}{a} \times \frac{1}{(1-\omega^{-kb})(1-\omega^{-kc})}.
\end{aligned}$$

In the expansion of $G(t)$, the coefficient of t^n is

$$\begin{aligned}
f(n) &= h_0 + h_1 \times (n+1) + h_2 \times \frac{(n+1)(n+2)}{2} \\
&\quad + \sum_{k=1}^{a-1} d_k \omega^{kn} + \sum_{k=1}^{b-1} e_k \tau^{kn} + \sum_{k=1}^{c-1} f_k \rho^{kn}.
\end{aligned}$$

Observe that

$$\begin{aligned}\left|\sum_{k=1}^{a-1} d_k\omega^{kn}\right| &\leq \sum_{k=1}^{a-1}|d_k| \\ &= \frac{1}{a}\times\sum_{k=1}^{a-1}\frac{1}{|1-\omega^{-kb}||1-\omega^{-kc}|} \\ &\leq \frac{1}{a}\times\sum_{k=1}^{a-1}\frac{1}{|1-\omega^{k}|^2} \\ &= \frac{a^2-1}{12a} < \frac{a}{12}.\end{aligned}$$

Here, the equality in the last row requires a lemma which will be proved. Likewise, we have in the following $\left|\sum_{k=1}^{b-1} e_k\tau^{kn}\right| < \frac{b}{12}$ and $\left|\sum_{k=1}^{c-1} f_k\rho^{kn}\right| < \frac{c}{12}$.

Taking $\beta_2 = \frac{h_2}{2}$, $\beta_1 = h_1 + \frac{3}{2}h_2$, $\beta_0 = h_0 + h_1 + h_2$, it follows that

$$\begin{aligned}|\beta_2 n^2 + \beta_1 n + \beta_0 - f(n)| &\leq \left|\sum_{k=1}^{a-1} d_k\omega^{kn}\right| + \left|\sum_{k=1}^{b-1} e_k\tau^{kn}\right| + \left|\sum_{k=1}^{c-1} f_k\rho^{kn}\right| \\ &< \frac{a+b+c}{12}.\end{aligned}$$

Lemma (for the equality in the last row) *For any positive integer m, $\sum_{k=1}^{m-1}\frac{1}{\sin^2\frac{k\pi}{m}} = \frac{m^2-1}{3}$ holds; equivalently, $\sum_{k=1}^{m-1}\cot^2\frac{k\pi}{m} = \frac{m^2-3m+2}{3}$.*

Proof of lemma For $\vartheta \in A = \left\{\frac{\pi}{m}, \frac{2\pi}{m}, \ldots, \frac{(m-1)\pi}{m}\right\}$,

$$(\cos\vartheta + \mathrm{i}\sin\vartheta)^m = \cos(m\vartheta) + \mathrm{i}\sin(m\vartheta) = \pm 1,$$

where the imaginary part

$$\mathrm{C}_m^1 \sin\vartheta\cos^{m-1}\vartheta - \mathrm{C}_m^3\sin^3\vartheta\cos^{m-3}\vartheta + \mathrm{C}_m^5\sin^5\vartheta\cos^{m-5}\vartheta + \cdots = 0,$$

or $\mathrm{C}_m^1(\cot\vartheta)^{m-1} - \mathrm{C}_m^3(\cot\vartheta)^{m-3} + \mathrm{C}_m^5(\cot\vartheta)^{m-5} + \cdots = 0$. This implies that $\{\cot\frac{k\pi}{m} : k = 1, 2, \ldots, m-1\}$ satisfy the $(m-1)$th degree polynomial in $\cot\vartheta$. According to Viète's formulas, the sum of the $m-1$ roots is $\sigma_1 = 0$ while the sum of their products multiplied in pairs is

$$\sigma_2 = -\frac{C_m^3}{C_m^1} = -\frac{(m-1)(m-2)}{6}.$$

Consequently, the sum of the squares of the $m-1$ roots is

$$\sum_{k=1}^{m-1} \cot^2\frac{k\pi}{m} = \sigma_1^2 - 2\sigma_2 = \frac{m^2-3m+2}{3}. \qquad \square$$

Solution 2 By symmetry, assume $a \le b \le c$. If $a = b = c = 1$, $f(n)$ is the number of ways of writing n as three nonnegative integers, which is C_{n+2}^2, and by taking $\alpha = \frac{1}{2}$, $\beta = \frac{3}{2}$, $\gamma = 1$, the error is constantly 0. In the following, assume $c > 1$. For nonnegative integer m, let $g(m)$ represent the number of nonnegative integer solutions (x, y) of $ax + by = m$.

First, we try to find $g(m)$. Let $x \equiv ka \pmod b$, $m \equiv \ell b \pmod a$, where

$$k \in \{0, 1, \dots, b-1\}, \quad \ell \in \{0, 1, \dots, a-1\}.$$

Evidently, $x \equiv k \pmod b$, $y \equiv \ell \pmod a$. Suppose $x = k + ub$, $y = \ell + va$. Then

$$m = ka + \ell b + (u+v)ab,$$

giving $u + v = \dfrac{m-(ka+\ell b)}{ab}$, and the number of pairs (u, v) is $\dfrac{m-(ka+\ell b)}{ab} + 1$ (note that $\dfrac{m-(ka+\ell b)}{ab}$ is an integer greater than -2; when $\dfrac{m-(ka+\ell b)}{ab} = -1$ the conclusion is still valid). Therefore, $g(m) = \dfrac{m-(ka+\ell b)}{ab} + 1$.

Next, consider $f(n)$. In the Diophantine equation $ax + by + cz = n$, if z is given $(cz \le n)$, then the number of nonnegative solutions of $ax + by = n - cz$ is $g(n - cz)$. Hence,

$$f(n) = \sum_{i=0}^{\lfloor n/c \rfloor} g(n - ic).$$

For each integer i, define $k_i \in \{0, 1, \ldots, b-1\}$ and $\ell_i \in \{0, 1, \ldots, a-1\}$ such that

$$n - ic = k_i a \pmod{b}, \quad n - ic = \ell_i b \pmod{a}.$$

We have

$$\begin{aligned} f(n) &= \sum_{i=0}^{\lfloor \frac{n}{c} \rfloor} g(n - ic) \\ &= \sum_{i=0}^{\lfloor \frac{n}{c} \rfloor} \left(\frac{(n-ic) - k_i a + \ell_i b)}{ab} + 1 \right) \\ &= \left\lfloor \frac{n}{c} \right\rfloor \cdot \left(\frac{n}{ab} + 1 \right) - \frac{c}{2ab} \left\lfloor \frac{n}{c} \right\rfloor \left(\left\lfloor \frac{n}{c} \right\rfloor + 1 \right) - \left(\frac{1}{b} \sum_{i=0}^{\lfloor \frac{n}{c} \rfloor} k_i + \frac{1}{a} \sum_{i=0}^{\lfloor \frac{n}{c} \rfloor} \ell_i \right). \end{aligned}$$

Note that a, b, c are pairwise coprime. Hence, for $i = 0, 1, \ldots, b-1$, the integers $n - ic$ cover a complete system of residues modulo b; so do the integers $k_i a$ and the integers k_i. In addition, when i shifts by a multiple of b, k_i does not change.

Now we turn to $\sum\limits_{i=0}^{\lfloor n/c \rfloor} \left(k_i - \frac{b-1}{2} \right)$: notice that the summands add up to 0 as i goes over a complete system of residues modulo b. Therefore, $\sum\limits_{i=0}^{\lfloor n/c \rfloor} \left(k_i - \frac{b-1}{2} \right) = \sum\limits_{i=0}^{s} \left(k_i - \frac{b-1}{2} \right)$, where s is the least nonnegative residue of $\left\lfloor \frac{n}{c} \right\rfloor$ modulo b. In $\sum\limits_{i=0}^{s} \left(k_i - \frac{b-1}{2} \right)$, the summands are distinct and are all from $\left\{ \frac{1-b}{2}, \frac{3-b}{2}, \ldots, \frac{b-3}{2}, \frac{b-1}{2} \right\}$ (the positive and negative elements are symmetric). This indicates that $\left| \sum\limits_{i=0}^{s} \left(k_i - \frac{b-1}{2} \right) \right|$ cannot exceed the sum of the positive elements $\frac{2}{2} + \frac{4}{2} \cdots + \frac{b-3}{2} + \frac{b-1}{2} = \frac{b^2-1}{8}$ when b is odd, or $\frac{1}{2} + \frac{3}{2} \cdots + \frac{b-3}{2} + \frac{b-1}{2} = \frac{b^2}{8}$ when b is even. Hence,

$$\left| \sum_{i=0}^{\lfloor n/c \rfloor} \left(k_i - \frac{b-1}{2} \right) \right| \leq \frac{b^2}{8}. \qquad ①$$

Likewise,

$$\left|\sum_{i=0}^{\lfloor n/c\rfloor}\left(\ell_i-\frac{a-1}{2}\right)\right|\le\frac{a^2}{8}. \qquad ②$$

From ① and ②, it follows that

$$\begin{aligned}&\left|\left(\frac{1}{b}\sum_{i=0}^{\lfloor\frac{n}{c}\rfloor}k_i+\frac{1}{b}\sum_{i=0}^{\lfloor\frac{n}{c}\rfloor}l_i\right)-\left(\left\lfloor\frac{n}{c}\right\rfloor+1\right)\left(\frac{b-1}{2b}+\frac{a-1}{2a}\right)\right|\\&\le\frac{1}{b}\left|\sum_{i=0}^{\lfloor\frac{n}{c}\rfloor}\left(k_i-\frac{b-1}{2}\right)\right|+\frac{1}{a}\left|\sum_{i=0}^{\lfloor\frac{n}{c}\rfloor}\left(\ell_i-\frac{a-1}{2}\right)\right|\\&\le\frac{b}{8}+\frac{a}{8}. \qquad ③\end{aligned}$$

Let r be the least nonnegative residue of n modulo c. Then

$$\begin{aligned}&\left\lfloor\frac{n}{c}\right\rfloor\cdot\left(\frac{n}{ab}+1\right)-\frac{c}{2ab}\left\lfloor\frac{n}{c}\right\rfloor\cdot\left(\left\lfloor\frac{n}{c}\right\rfloor+1\right)-\left(\left\lfloor\frac{n}{c}\right\rfloor+1\right)\left(\frac{b-1}{2b}+\frac{a-1}{2a}\right)\\&=\frac{(n-r)(n+ab)}{abc}-\frac{(n-r)(n-r+c)}{2abc}-\frac{(2ab-a-b)(n-r+c)}{2abc}\\&=\frac{n^2-(a+b+c)n-(2abc+ac+bc)-r(r+a+b-c)}{2abc}. \qquad ④\end{aligned}$$

As r takes values $0,1,\ldots,c-1$, denote the maximum and the minimum of $r(r+a+b-c)$ as $A+B$ and $A-B$, respectively. By ③ and ④,

$$\left|f(n)-\frac{n^2-(a+b+c)n-(2abc+ac+bc)-A}{2abc}\right|\le\frac{a+b}{8}+\frac{B}{2abc}.$$

Let $\alpha=\dfrac{1}{2abc}$, $\beta=-\dfrac{a+b+c}{2abc}$,

$$\gamma=-\frac{2abc+ac+bc+A}{2abc}.$$

We will justify the problem statement

$$|f(n)-(\alpha n^2+\beta n+\gamma)|<\frac{a+b+c}{12} \qquad ⑤$$

for different a,b,c values. □

Case 1: $a = b = 1$ and $c > 1$. An investigation on the convexity and the axis of symmetry yields

$$A + B = c - 1, \quad A - B \geq -\frac{(c-2)^2}{4},$$

and hence $B \leq \frac{c^2}{8}$. In addition, the left-hand sides of ① and ② are constantly 0, thereby making no contribution to the error. It follows that

$$|f(n) - (\alpha n^2 + \beta n + \gamma)| \leq \frac{c^2}{8 \cdot 2abc} = \frac{c}{16} < \frac{a+b+c}{12}$$

which is ⑤.

Case 2: $a = 1$, $b > 1$, and $c \geq a + b$. Again, by considering the convexity and the axis of symmetry, we find

$$A + B = b(c-1), \quad A - B \geq -\frac{(c-1-b)^2}{4},$$

and $B \leq \frac{(b+c-1)^2}{8}$. The left-hand side of ① is constantly 0, contributing zero error. Therefore,

$$\begin{aligned}
|f(n) - (\alpha n^2 + \beta n + \gamma)| &\leq \frac{b}{8} + \frac{(b+c-1)^2}{8 \cdot 2abc} \\
&= \frac{b}{8} + \frac{b^2 + 2bc + c^2 - 2b - 2c + 1}{16bc} \\
&= \frac{b}{8} + \frac{3bc + c^2 - 2b - 2c + 1 - b(c-b)}{16bc} \\
&\leq \frac{b}{8} + \frac{3bc + c^2 - 2b - 2c + 1 - (c-1)}{16bc} \\
&< \frac{b}{8} + \frac{2bc + c^2 - 3c}{16bc} \\
&= \frac{b}{8} + \frac{3}{16} + \frac{c-3}{16b} \\
&\leq \frac{b}{8} + \frac{3}{16} + \frac{c-3}{32} \\
&= \frac{a+b+c}{12} - \frac{5a - 4b - 1}{96} \\
&< \frac{a+b+c}{12},
\end{aligned}$$

and ⑤ is verified.

Case 3: $a > 1$ and $c \geq a + b - 2$. The convexity and the axis of symmetry indicate

$$A + B = (a + b - 1)(c - 1), \quad A - B = -\frac{(c - a - b)^2}{4},$$

and thus $B \leq \dfrac{(a + b + c - 2)^2}{7} < \dfrac{c^2}{2}$. The error estimate

$$\begin{aligned}|f(n) - (\alpha n^2 + \beta n + \gamma)| &< \frac{a + b}{8} + \frac{c^2}{4abc} = \frac{a + b}{8} + \frac{c}{4ab} \\ &\leq \frac{a + b}{8} + \frac{c}{24} = \frac{a + b + c}{12} - \frac{c - a - b}{24} \\ &\leq \frac{a + b + c}{12},\end{aligned}$$

verifies ⑤.

Case 4: $a > 1$ and $c \leq a + b - 3$. As $r(r + a + b - c)$ is strictly increasing when $r \geq 0$, we have $2B = (c - 1)(a + b - 1)$ and

$$B = \frac{(c - 1)(a + b - 1)}{2} < \frac{(a + b)c}{2}.$$

From $c \leq a+b-3 \leq a+c-4$, it follows that $a \geq 4$. The error is bounded by

$$\begin{aligned}|f(n) - (\alpha n^2 + \beta n + \gamma)| &< \frac{a + b}{8} + \frac{(a + b)c}{4abc} \\ &= \frac{a + b}{8} + \frac{1}{4a} + \frac{1}{4b} \\ &< \frac{a + b}{8} + \frac{1}{16} + \frac{1}{20} \\ &< \frac{a + b + 1}{8} \\ &= \frac{3a + 3b + 3}{24} \\ &< \frac{2a + 2b + (a + b + 3)}{24} \\ &\leq \frac{2a + 2b + 2c}{24} \\ &= \frac{a + b + c}{12},\end{aligned}$$

and ⑤ holds.

Based on the above argument, the problem statement is verified.

Test II, Second Day
(8:00 – 12:30 pm; March 22, 2021)

4 For a positive integer n, let $\varphi(n)$ represent the number of positive integers not exceeding n and relatively prime to n. Find all functions $f:\mathbb{N}_+ \to \mathbb{N}_+$ satisfying that for any positive integers m, n with $m \geq n$,

$$f(m\varphi(n^3)) = f(m)\varphi(n^3).$$

(contributed by He Yijie)

Solution The desired functions have the form

$$f(n) = \begin{cases} c, & n = 1, \\ dn, & n \geq 2, \end{cases}$$

where c, d are positive integers. It is straightforward to check that the problem conditions are met.

In the governing equation, taking $n = 2$ leads to $f(4m) = 4f(m)$ for $m \geq 2$. By iterations, it follows that

$$f(4^k m) = 4^k f(m) \quad ①$$

for all $m \geq 2$.

Claim For positive integers m, p with $m \geq 2$, $f(pm) = pf(m)$.

Proof of claim Induct on p: when $p = 1$, the conclusion is trivial; assume $p \geq 2$ and the conclusion is valid for all positive integers less than p. Now, if p is composite, the conclusion is validated by the induction hypothesis. In the following, assume p is a prime. Taking $n = p$ in the governing equation, we obtain

$$f(p^2(p-1)m) = p^2(p-1)f(m)$$

for all $m \geq p$. By the induction hypothesis,

$$\begin{aligned} p^2(p-1)f(m) &= f(p^2(p-1)m) \\ &= (p-1)f(p^2m), \end{aligned}$$

and thus

$$f(p^2m) = p^2 f(m) \quad ②$$

for all $m \geq p$.

Next, take $n = p^2$ in the governing equation to get

$$f(p^5(p-1)m) = p^5(p-1)f(m)$$

for $m \geq p^2$. In a similar manner, it follows by induction that

$$f(p^5(p-1)m) = (p-1)f(p^5m).$$

Hence,

$$f(p^5m) = p^5f(m) \qquad \text{③}$$

for $m \geq p^2$.

Based on the above argument, we take k with $4^k \geq p^2$: for any $m \geq 2$,

$$4^kp^4f(pm) \overset{②}{=} 4^kf(p^5m) \overset{①}{=} f(4^kp^5m) \overset{③}{=} p^5f(4^km) \overset{①}{=} p^54^kf(m).$$

Therefore, $f(pm) = pf(m)$, and the induction is completed.

According to the claim, particularly for $m \geq 2$,

$$2f(m) = f(2m) = mf(2).$$

Taking $m = 3$, we infer that $f(2)$ is even. So for $m \geq 2$, $f(m) = dm$, where $d = \dfrac{f(2)}{2}$ is an integer. □

5 Let n be a positive integer, $a_1, a_2, \ldots, a_{2n+1}$ be $2n+1$ positive real numbers. For $k = 1, 2, \ldots, 2n+1$, define

$$b_k = \max_{0 \leq m \leq n} \left(\frac{1}{2m+1} \sum_{i=k-m}^{k+m} a_i \right),$$

where the subscript of a_i is taken modulo $2n+1$. Prove: the number of subscripts k satisfying that $b_k \geq 1$ does not exceed $2\sum_{i=1}^{2n+1} a_i$.

(Contributed by Ai Yinghua)

Solution Define $I = \{k|b_k \geq 1\}$. For every $k \in I$, assume that the maximum value b_k of $\dfrac{1}{2m+1}\sum_{i=k-m}^{k+m} a_i$ is attained at $m = m_k$, and call

$$[k-m_k, k+m_k] := \{k-m_k, k-m_k+1, \ldots, k+m_k\}$$

a "nice segment", where the subscripts are taken modulo $2n+1$. Obviously, the union of all nice segments contains I.

Claim There exists a collection of nice segments whose union contains I, and moreover, each $i \in \{1, 2, \ldots, 2n+1\}$ is contained in at most two segments.

Proof of claim If $[1,\ 2n+1]$ is a nice segment, the conclusion is trivial. In the following, assume this is not the case. If i is contained in r nice segments

$$[i-u_1, i+v_1], \ldots, [i-u_r, i+v_r]$$

where $r \geq 3$ and $0 \leq u_j,\ v_j < 2n$, let $u_j = \max\{u_1, \ldots, u_r\}$ and $v_k = \max\{v_1, \ldots, v_r\}$. We can keep the nice segments $[i-u_j, i+v_j]$, $[i-u_k, i+v_k]$ and drop the other $r-2$ segments. Now the segments still cover $1, \ldots, 2n+1$, and at most two of them cover i. For each $i \in \{1, 2, \ldots, 2n+1\}$, perform the above operation. Eventually, we find a collection of nice segments with the desired properties.

For the original problem, let $[i_1-m_1, i_1+m_1], \ldots, [i_r-m_r,\ i_r+m_r]$ be a collection of nice segments chosen in the claim. We have

$$2\sum_{i=1}^{2n+1} a_i \geq \sum_{\alpha=1}^{r} \sum_{k=i_\alpha-m_\alpha}^{i_\alpha+m_\alpha} a_k = \sum_{\alpha=1}^{r} (2m_\alpha+1) b_{i_\alpha} \geq \sum_{\alpha=1}^{r} (2m_\alpha+1) \geq |I|,$$

the first inequality is due to each i being contained in at most two nice segments; the equality and the next inequality are due to the definition of nice segments; the last inequality is due to the union of the nice segments containing I. □

6 Find the least positive real a satisfying this condition: for any three points A, B, C on the unit circle, there exists an equilateral triangle PQR with side length a, such that A, B, C are all inside or on the boundary of triangle PQR.

(Contributed by Wang Xinmao)

Solution $a = \dfrac{(2\sin 80^\circ)^2}{\sqrt{3}}$.

First, we prove $a = \dfrac{(2\sin 80^\circ)^2}{\sqrt{3}}$ is sufficient. For any three points A, B, C on the unit circle, let $\angle BAC = \alpha$, $\angle ABC = \beta$, $\angle ACB = \gamma$, $\alpha \leq \beta \leq \gamma$.

If $\beta \leq 60^\circ$, since $AB \leq 2 < a$ (here, $a > \dfrac{(2\sin 75^\circ)^2}{\sqrt{3}} = \dfrac{(\sqrt{6}+\sqrt{2})^2}{4\sqrt{3}} = \dfrac{8+4\sqrt{3}}{4\sqrt{3}} > 2$), one can draw a segment PQ of length a that contains A

and B, then find R such that $\triangle PQR$ is equilateral, C, R on the same side of PQ. It follows from $\alpha \le \beta \le 60°$ that C is inside $\triangle PQR$ or on the boundary.

If $\beta > 60°$, we first prove the following:

At least one of $\sin\beta\sin\gamma$ and $\sin\beta\sin(\alpha+60°)$ is less than or equal to $\sin^2 80°$. $(*)$

In fact, as $2(\beta+\alpha+60°)+(\beta+\gamma) = 480° - (\gamma-\beta) \le 480°$, one of $\beta+\alpha+60°$ and $\beta+\gamma$ does not exceed $160°$. From the simple property of sine function

$$\begin{aligned}\sin x\sin y &= \frac{1}{2}(\cos(x-y)-\cos(x+y))\\ &\le \frac{1}{2}(1-\cos(x+y))\\ &= \sin^2\frac{x+y}{2},\end{aligned}$$

$(*)$ follows.

(i) If $\sin\beta\sin\gamma \le \sin^2 80°$, let $AD \perp BC$ with foot D; take P, Q on the line BC such that $PD = DQ = \frac{a}{2}$; take R on the ray DA such that

$$DR = \frac{\sqrt{3}a}{2} = 2\sin^2 80°.$$

Notice that $AD = 2\sin\beta\sin\gamma \le DR$, and thus A lies inside $\triangle PQR$ or on the boundary; from $60° \le \beta, \gamma \le 120°$, it is easy to see that B, C lie on the segment PQ.

(ii) If $\sin\beta\sin(\alpha+60°) \le \sin^2 80°$, take $P = A$, extend AB to Q, $AQ = a$, and find R such that $\triangle PQR$ is equilateral and C, R lie on the same side of PQ. Let M be the intersection of the ray AC and the line QR. Notice that $AB \le 2 < a$, hence B is on the boundary of $\triangle PQR$; since

$$AM = a\cdot\frac{\sin 60°}{\sin(\alpha+60°)} = \frac{2\sin^2 80°}{\sin(\alpha+60°)} > 2\sin\beta = AC,$$

C is inside $\triangle PQR$ or on the boundary.

In conclusion, $a = \dfrac{(2\sin 80°)^2}{\sqrt{3}}$ satisfies the problem condition.

Next, we prove $a \ge \dfrac{(2\sin 80°)^2}{\sqrt{3}} = \dfrac{2\sin^2 80°}{\sin 60°}$ is necessary. To this end, choose A, B, C on the unit circle with $\angle BAC = 20°$, $\angle ABC = \angle ACB = 80°$.

Suppose the equilateral triangle PQR with side length a covers A, B, C (inside or on the boundary). By a translation if necessary, we can make one of A, B, C, say A on the boundary of $\triangle PQR$, and then by a rotation about A we can make B or C on the boundary. If neither of them is a vertex of $\triangle PQR$, assume they all lie on PQ or PR. Take P as the homothetic centre and rescale $\triangle PQR$ smaller such that all of A, B, C fall on the boundary or become a vertex of $\triangle PQR$. Since $\angle ABC$, $\angle ACB > 60°$, B, C cannot be vertices; in addition, there is symmetry between B and C. Thus, it suffices to consider four situations as follows.

(1) A is a vertex, say $A = P$. Let $PH \perp QR$ with foot H. Then the angle between PH and one of PB, PC is less than or equal to $10°$. It follows that $PH \geq 2\sin^2 80°$, and

$$a = PQ \geq \frac{2\sin^2 80°}{\sin 60°}.$$

(2) A and B lie on the same side, say PQ, with $PA \leq PB$. Let $PH \perp QR$ at H. Then the angle between AC and PH is $10°$. Similar to (1), we have

$$a \geq \frac{2\sin^2 80°}{\sin 60°}.$$

(3) B and C lie on the same side, say QR. Let $PH \perp QR$ at H. Then the lines AC and PH cross at $10°$. Similar to (1), we have

$$a \geq \frac{2\sin^2 80°}{\sin 60°}.$$

(4) A, B, C are all on different sides. Let $A \in PQ$, $B \in QR$, $C \in RP$, and $\vartheta = \angle RBC \leq \angle RCB$, $\vartheta \leq 60°$.

(i) If $\vartheta \leq 20°$, then the angle between AB and the altitude on QR does not exceed $10°$. Similar to (1), we get

$$a \geq \frac{2\sin^2 80°}{\sin 60°}.$$

(ii) If $40° \leq \vartheta \leq 60°$, then the angle between AC and the altitude on PQ does not exceed $10°$. Similar to (1), $a \geq \dfrac{2\sin^2 80°}{\sin 60°}$.

(iii) If $20^\circ \le \vartheta \le 40^\circ$, then

$$\begin{aligned} a = QR &= QB + BR \\ &= \frac{2\sin 80^\circ \sin(20^\circ + \vartheta)}{\sin 60^\circ} + \frac{2\sin 20^\circ \sin(60^\circ + \vartheta)}{\sin 60^\circ} \\ &\ge \frac{2\sin 80^\circ \sin 40^\circ + 2\sin 20^\circ \sin 80^\circ}{\sin 60^\circ} \\ &= \frac{2\sin^2 80^\circ}{\sin 60^\circ}. \end{aligned}$$

In summary, the minimum is $a = \dfrac{2\sin^2 80^\circ}{\sin 60^\circ}$. □

Test III, First Day
(8:00 am – 12:30 pm; April 7, 2021)

1 Given a convex n-gon $(n \ge 5)$ $\Omega : P_1P_2\dots P_n$, of which no three diagonals are concurrent inside Ω. Prove that one can choose a point inside every quadrilateral $P_iP_jP_kP_\ell (1 \le i < j < k < \ell \le n)$ and not on any diagonal of Ω, such that the C_n^4 points obtained are distinct and the segment connecting any two of them intersects with at least one diagonal of Ω.

(Contributed by Leng Fusheng)

We give three proofs as follows.

Solution 1 To begin, notice that the diagonals of Ω divide the polygon into $\mathrm{C}_n^4 + \mathrm{C}_{n-1}^2$ small regions. Every quadrilateral $P_iP_jP_kP_\ell (1 \le i < j < k < \ell \le n)$ has a 1-1 correspondence with the intersection of the diagonals P_iP_k and P_jP_ℓ, and additionally, every intersection is adjacent to four small regions. Therefore, it is enough to assign every intersection to one of the four adjacent small regions, such that the C_n^4 assigned small regions are all distinct.

Place Ω in the Cartesian coordinate plane such that no diagonal is parallel to the x axis. If two diagonals P_iP_k, P_jP_ℓ meet at Q, then we assign Q to the unique small region incident to Q and whose interior points all have y coordinates larger than that of Q (intuitively, above Q). Obviously, different intersections correspond to different small regions, as for each small region R, there is a unique vertex with the minimum y coordinate, and this vertex is assigned to R. This completes the proof. □

Solution 2 As in the first proof, we are required to assign each intersection (of two diagonals) to an adjacent small region, such that the assigned small regions are all distinct. In the plane of Ω, pick any point Q not concurrent with any diagonal. Suppose that two diagonals ℓ_1 and ℓ_2 intersect at R. Then Q lies inside one of the four zones of the plane divided by ℓ_1 and ℓ_2: in the anticlockwise direction, let T be the next zone, and assign R to the unique small region adjacent to R and contained in T. It remains to prove that each small region is assigned at most one intersection.

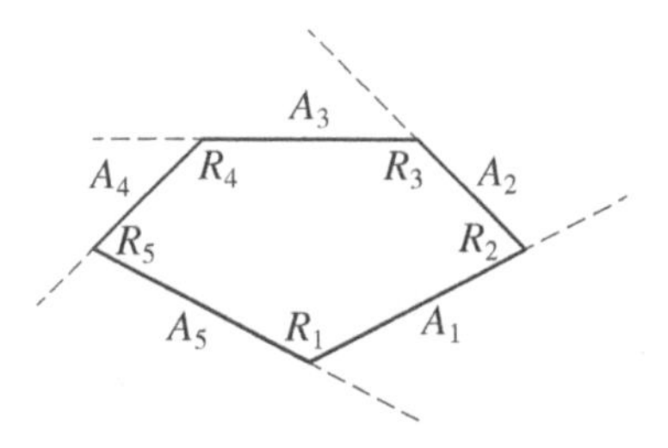

Fig. 1.1

For a small region illustrated in Fig. 1.1, denote its vertices anticlockwise as $R_1, R_2, \ldots, R_m (R_0 = R_m)$. The rays R_1R_2, $R_2R_3, \ldots, R_{m-1}R_m$, R_mR_1 divide the plane excluding the small region into $A_1, A_2, \ldots, A_m$, where A_i is the zone whose boundary consists of rays R_iR_{i+1} and $R_{i-1}R_i$. According to the correspondence between small regions and intersections, this small region is assigned R_i if and only if Q lies inside zone A_i, which could happen for at most one subscript i. Since this is true for any small region, we infer that each small region is assigned at most one intersection, and the proof is completed. □

Solution 3 Let all the diagonals of Ω be $\ell_1, \ldots, \ell_m$. We put these m diagonals in Ω in the order of the subscripts and define S_k as the set of intersections created by adding the diagonal ℓ_k (S_k could be empty), T_k as the set of all small regions divided by $I_1, \ldots, I_k$. We use induction to prove, for each k, there exists a one-to-one mapping

$$f_k : S_1 \cup \cdots \cup S_k \to T_k,$$

such that every $Q \in S_1 \cup \cdots \cup S_k$ is a vertex of the polygon $f_k(Q)$. After that, a discussion as in the beginning of the first proof will lead to the conclusion.

For $k = 1$, the assertion is obvious. Assume for $k - 1$, the one-to-one mapping f_{k-1} satisfies the condition. Now we add the diagonal I_k to Ω.

If S_k is empty, the assertion is clearly true for k. Assume S_k is nonempty. Then the intersections in S_k all lie on ℓ_k, say they are $Q_1, \ldots, Q_r$, and they traverse the small regions $A_1, \ldots, A_{r+1}$, respectively, dividing each A_i into two new small regions B_i and C_i.

Notice that $f_{k-1}^{-1}(A_i)$ can be either empty or determined, say a vertex R_i of B_i or C_i. Now we define f_k as follows: for each $i \in \{1, \ldots, r\}$,

(1) If $f_{k-1}^{-1}(A_i)$ is empty, then let $f(Q_i) = B_i$;
(2) If R_i is a vertex of B_i, then let $f_k(R_i) = B_i$, $f_k(Q_i) = C_i$;
(3) If R_i is a vertex of C_i, then let $f_k(R_i) = C_i$, $f_k(Q_i) = B_i$.

For all other intersections Q, let $f_k(Q) = f_{k-1}(Q)$. It is straightforward to check that f_k has the desired properties. Particularly, when $k = m$, the mapping $f = f_m$ gives a one-to-one correspondence between all interior intersections and small regions of Ω, satisfying that each intersection Q is a vertex of the small region $f(Q)$. □

2 Given 2021 distinct positive integers $a_1, a_2, \ldots, a_{2021}$. Define the sequence $\{a_n\}$ inductively as follows: for each integer $n \geq 2022$, a_n is the smallest positive integer different from $a_1, a_2, \ldots, a_{n-1}$ and not dividing the product $a_{n-1}a_{n-2} \ldots a_{n-2021}$. Prove: there exists a positive integer M, such that all integers greater than or equal to M appear in $\{a_n\}$.

(Contributed by Xu Disheng)

Solution Let $k = 2021$. In fact, we will prove the statement for any integer $k > 0$.

Lemma 1 *There exists $C > 0$ independent of k, such that $\tau(m) \leq Cm^{\frac{1}{k+1}}$ holds for all positive integer m, where $\tau(m)$ is the number of positive factors of m.*

Proof of lemma 1 Let $m = p_1^{\alpha_1} p_2^{\alpha_2} \ldots p_r^{\alpha_r}$ be the prime factorization of m. Evidently,

$$\tau(m) = (\alpha_1 + 1)(\alpha_2 + 1) \cdots (\alpha_r + 1).$$

Equivalently, there exists $C > 0$ such that

$$\prod_{i=1}^{r} \frac{(\alpha_i + 1)^{k+1}}{p_i^{\alpha_i}} \leq C^{k+1}$$

holds for all prime p_i and positive integers α_i. Note that polynomials grow slower than exponential functions, and hence for each prime $p \le 2^{k+1}$, the fraction $\frac{(\alpha+1)^{k+1}}{p^\alpha}$ has an upper bound when $\alpha \ge 0$. Since the number of primes $p \le 2^{k+1}$ is finite, there exists C' such that for any prime $p \le 2^{k+1}$,

$$\frac{(\alpha+1)^{k+1}}{p^\alpha} < C'.$$

On the other hand, for $p > 2^{k+1}$,

$$\frac{(\alpha+1)^{k+1}}{p^\alpha} < \left(\frac{\alpha+1}{2^\alpha}\right)^{k+1} < 1.$$

It follows that $\prod_{i=1}^{r} \frac{(\alpha_i+1)^{k+1}}{p_i^{\alpha_i}} \le C'^s$, where s is the number of distinct prime factors of m less than 2^{k+1}. The existence of $C > 0$ is now verified.

Lemma 2 *For any fixed integer $L > 0$, there exists $M > 0$ such that for any positive integer $m \ge M$, one can find a prime power $p^\alpha | m$, and $p^\alpha > L$.*

Proof of lemma 2 Consider all prime powers less than or equal to L: there are only finitely many of them. Define M equals the product of them plus 1. If $m \ge M$ and $p|m$, then its highest power in m satisfies $p^\alpha | m$, and $p^\alpha > L$.

Lemma 3 *There exists a positive integer D such that $a_n < 2n + D$ for every positive integer n.*

Proof of Lemma 3 We take D satisfying $D > \max_{1\le s\le k}\{a_s\}$ and

$$D > 2^k C^{k+1} + C D^{\frac{k}{k+1}},$$

where $C > 0$ is defined in Lemma 1, and use induction to prove $a_n < 2n + D$ for every n. By the choice of D, $a_n < 2n + D$ is true for $n = 1, 2, \ldots, k$. Assume $a_n < 2n + D$ holds for $1, 2, \ldots, n + k - 1$, where n is a positive integer. Now for a_{n+k}, it is known from the problem that a_{n+k} is the smallest positive integer not dividing $a_n a_{n+1} \cdots a_{n+k-1}$, and different from $a_1, \ldots, a_{n+k-1}$. Hence,

$$a_{n+k} \le \tau(a_n a_{n+1} \cdots a_{n+k-1}) + n + k.$$

From Lemma 1 and induction hypothesis, we have

$$\begin{aligned}a_{n+k} &\leq C(a_n a_{n+1}\dots a_{n+k-1})^{\frac{1}{k+1}} + n + k\\ &< C(2(n+k-1)+D)^{\frac{k}{k+1}} + n + k\\ &< C(2^{\frac{k}{k+1}}(n+k)^{\frac{k}{k+1}} + D^{\frac{k}{k+1}}) + n + k.\end{aligned}$$

In the last step, we use the fact that $x^{\frac{k}{k+1}}$ is concave down for $x > 0$. Now, if $n + k \leq 2^k C^{k+1}$, then

$$\begin{aligned}a_{n+k} &< C2^{\frac{k}{k+1}}(n+k)^{\frac{k}{k+1}} + CD^{\frac{k}{k+1}} + n + k\\ &\leq 2^k C^{k+1} + CD^{\frac{k}{k+1}} + n + k\\ &< D + 2(n+k);\end{aligned}$$

if $n + k > 2^k C^{k+1}$, then

$$C2^{\frac{k}{k+1}}(n+k)^{\frac{k}{k+1}} < n + k, CD^{\frac{k}{k+1}} < D,$$

and

$$a_{n+k} \leq n + k + D + n + k = 2(n+k) + D.$$

Therefore, $a_n < 2n + D$ is true for $n + k$.

Return to the original problem. For integer $m > 0$, if m does not appear in the sequence, then, as the elements in $\{a_n\}$ are distinct, there must exist a positive integer N_1, such that $a_n > m$ for $n > N_1$. By definition of $\{a_n\}$, $m|a_n a_{n+1}\dots a_{n+k-1}$ for all $n > N_1$.

According to Lemma 2, there exists $M > 0$, such that for any integer $m \geq M$, we can find a prime power $p^\alpha|m$, and $p^\alpha > (3k)^k$. We claim that such m must appear in $\{a_n\}$. Suppose to the contrary that for some $m > M$, m does not appear in $\{a_n\}$. Fix a prime power $p^\alpha|m$, $p^\alpha > (3k)^k$. By the previous argument, there exists a positive integer N_1, such that for $n > N_1$,

$$m|a_n a_{n+1}\dots a_{n+k-1}.$$

For this n, there exists i, $n \leq i \leq n + k - 1$ and $p^{\lceil \alpha/k \rceil}|a_i$. Let

$$N_2 > 3N_1 + D + 3k,$$

and consider the set $A = \{n : 1 \le n \le N_2, p^{\lceil \alpha/k \rceil} | a_n\}$. On one hand, from Lemma 3, we have $a_n < 2n + D$, and thus

$$|A| \le \frac{2N_2 + D}{p^{\lceil \frac{\alpha}{k} \rceil}} < \frac{2N_2 + D}{3k}.$$

On the other hand, since m does not appear in $\{a_n\}$, $m | a_n a_{n+1} \cdots a_{n+k-1}$ for $n > N_1$. Consequently, for any k consecutive terms in $\{N_1 + 1, N_1 + 2, \ldots, N_2\}$, at least one term belongs to A, implying that $|A| \ge \frac{N_2 - N_1}{k} - 1$. However, by the choice of N_2, we have

$$|A| \ge \frac{N_2 - N_1}{k} - 1 > \frac{2N_2 + D}{3k},$$

which contradicts the previous estimate. This means all integers $m \ge M$ must appear in $\{a_n\}$. □

3 Find the largest constant $C > 0$, such that for any integer $n \ge 2$, one can find real numbers $x_1, x_2, \ldots, x_n \in [-1, 1]$ satisfying

$$\prod_{1 \le i < j \le n} (x_i - x_j) \ge C^{\frac{n(n-1)}{2}}.$$

(Contributed by Wang Bin)

Solution 1 For sufficiently large n, we guess the optimal choice of $(x_1, \ldots, x_n)$ on $[-1, 1]$ is close to the projection of points uniformly distributed on the unit circle onto the x axis. To this end, we introduce the reference choice

$$a_k = \cos \vartheta_k = \cos \frac{2k-1}{2n}\pi, \quad k = 1, 2, \ldots, n.$$

For $A = \{a_1, a_2, \ldots, a_n\}$, the product

$$P_A = \prod_{1 \le i < j \le n} (a_i - a_j)$$

and

$$\sum_{i \ne k} \frac{1}{a_k - a_i}, \quad k = 1, \ldots, n$$

will be used later.

Consider the nth degree Chebyshev polynomial of the first kind $T(X) = T_n(X)$, with leading coefficient 2^{n-1} and $T(\cos\vartheta) = \cos n\vartheta$.

Substitute $X = \cos\vartheta$, and differentiate to find $T'(\cos\vartheta)$, $T''(\cos\vartheta)$:

$$T'(\cos\vartheta) = \frac{\mathrm{d}\cos n\vartheta}{\mathrm{d}\cos\vartheta} = \frac{-n\sin n\vartheta \times \mathrm{d}\vartheta}{-\sin\vartheta \times \mathrm{d}\vartheta} = n \times \frac{\sin n\vartheta}{\sin\vartheta};$$

$$\begin{aligned} T''(\cos\vartheta) &= \frac{\mathrm{d}T'(\cos\vartheta)}{\mathrm{d}\cos\vartheta} = n \times \frac{\mathrm{d}\dfrac{\sin n\vartheta}{\sin\vartheta}}{\mathrm{d}\cos\vartheta} \\ &= n \times \frac{\dfrac{n\cos n\vartheta}{\sin\vartheta} \times \mathrm{d}\vartheta - \dfrac{\cos\vartheta\sin n\vartheta}{\sin^2\vartheta \times \mathrm{d}\vartheta}}{-\sin\vartheta \times d\vartheta} \\ &= n \times \frac{\cos\vartheta\sin n\vartheta - n \times \cos n\vartheta\sin\vartheta}{\sin^3\vartheta}. \end{aligned}$$

For $k = 1, 2, \ldots, n$, denote $\vartheta_k = \dfrac{2k-1}{2n}\pi$. We have

$$T(a_k) = T(\cos\vartheta_k) = \cos(n\vartheta_k) = 0,$$

and hence all the roots of $T(X)$ are $A = \{a_1, a_2, \ldots, a_n\}$. Define

$$H(X) = \frac{1}{2^{n-1}}T(X) = (X - a_1)(X - a_2)\cdots(X - a_n).$$

From another perspective, the first derivative $H'(X) = \dfrac{\mathrm{d}H(X)}{\mathrm{d}X}$ is the sum of all products of $n-1$ terms in

$$B = \{X - a_1, X - a_2, \ldots, X - a_n\}$$

(there are $\dbinom{n}{n-1}$ products), denoted as $H'(X) = \sigma_{n-1}(B)$. Furthermore, the second derivative $H''(X) = \dfrac{\mathrm{d}H'(X)}{\mathrm{d}X}$ is twice the sum of all products of $n-2$ terms in B (there are $\dbinom{n}{n-2}$ products), denoted as $H''(X) = 2\sigma_{n-2}(B)$.

When $X = a_k \in A$, we have $X - a_k = 0$ in B, and

$$H'(a_k) = \sigma_{n-1}(B) = \prod_{i \neq k}(a_k - a_i),$$

$$H''(a_k) = 2\sigma_{n-2}(B) = 2\prod_{i \neq k}(a_k - a_i)\sum_{i \neq k}\frac{1}{a_{k-a_i}}.$$

Substitute $X = a_k = \cos\vartheta_k \in A$ in $T'(X)$ and $T''(X)$ to obtain

$$T'(a_k) = T'(\cos\vartheta_k) = n \times \frac{(-1)^{k-1}}{\sin\vartheta_k},$$

$$T''(a_k) = T''(\cos\vartheta_k) = n \times \frac{(-1)^{k-1}\cos\vartheta_k}{\sin^3\vartheta_k}.$$

Therefore,

$$\sum_{i \neq k}\frac{1}{a_k - a_i} = \frac{H''(a_k)}{2H'(a_k)} = \frac{T''(a_k)}{2T'(a_k)} = \frac{\cos\vartheta_k}{2\sin^2\vartheta_k}, \quad k = 1, 2, \ldots, n.$$

Moreover,

$$\begin{aligned} P_A^2 &= (-1)^{\frac{n(n-1)}{2}}\prod_{k=1}^{n}\prod_{i \neq k}(a_k - a_i) \\ &= (-1)^{\frac{n(n-1)}{2}}\prod_{k=1}^{n}H'(\cos\vartheta_k) \\ &= \frac{n^n}{2^{n(n-1)}\prod\limits_{k=1}^{n}\sin\vartheta_k}. \end{aligned}$$

Notice that the complex roots of $(z+1)^n + 1 = 0$ are $\{e^{\frac{2k-1}{n}\pi i} - 1,$ $k = 1, \ldots, n\}$ with magnitudes $2\sin\dfrac{2k-1}{2n}\pi = 2\sin\vartheta_k$. The product of the magnitudes is $\prod\limits_{k=1}^{n} 2\sin\vartheta_k = 2$, implying

$$P_A^2 = \frac{n^n}{2^{n(n-1)}\prod\limits_{k=1}^{n}\sin\vartheta_k} = \frac{n^n}{2^{n(n-1)}2^{-(n-1)}} \Rightarrow P_A = \left(\frac{1}{2}\right)^{\frac{n(n-1)}{2}} n^{\frac{n}{2}}2^{\frac{n-1}{2}},$$

and hence $C = \frac{1}{2}$ satisfies the problem statement. In the following, we prove $C \leq \frac{1}{2}$.

For any positive integer n, assume $1 \geq x_1 > x_2 > \cdots > x_n \geq -1$ and

$$P^* = \prod_{1 \leq i < j \leq n} (x_i - x_j) \geq C^{\frac{n(n-1)}{2}}.$$

We make a comparison with the reference choice:

$$\begin{aligned} F &= \sum_{1 \leq i < j \leq n} \frac{x_i - x_j}{a_i - a_j} \\ &= \sum_{k=1}^{n} \left(x_k \times \sum_{i \neq k} \frac{1}{a_k - a_j} \right) \\ &= \sum_{k=1}^{n} \left(x_k \times \frac{\cos \vartheta_k}{2 \sin^2 \vartheta_k} \right) \\ &\leq \frac{1}{2} \sum_{k=1}^{n} \frac{|\cos \vartheta_k|}{\sin^2 \vartheta_k}. \end{aligned}$$

Evaluating

$$\begin{aligned} G = \sum_{k=1}^{n} \cot^2 \vartheta_k &= \sum_{k=1}^{n} \cot^2 \frac{2k-1}{2n}\pi \\ &= \sum_{l=1}^{2n-1} \cot^2 \frac{l\pi}{2n} - \sum_{l=1}^{n-1} \cot^2 \frac{l\pi}{n}, \end{aligned}$$

using the well-known identity

$$\sum_{k=1}^{m-1} \cot^2 \frac{k\pi}{m} = \frac{(m-1)(m-2)}{3},$$

it follows that

$$G = \frac{(2n-1)(2n-2)}{3} - \frac{(n-1)(n-2)}{3} = n(n-1).$$

Now, by Cauchy's inequality,

$$\begin{aligned}(2F)^2 &\leq \left(\sum_{k=1}^{n} \frac{|\cos\vartheta_k|}{\sin^2\vartheta_k}\right)^2 \\ &\leq \sum_{k=1}^{n} \frac{\cos^2\vartheta_k}{\sin^2\vartheta_k} \times \sum_{k=1}^{n} \frac{1}{\sin^2\vartheta_k} \\ &= G \times (G+n) \\ &= (n-1)n^3.\end{aligned}$$

Finally, by AM-GM inequality, we arrive at

$$\begin{aligned}\frac{P^*}{P_A} &= \prod_{1\leq i<j\leq n} \frac{x_i - x_j}{a_i - a_j} \\ &\leq \left(\frac{2}{n(n-1)} \sum_{1\leq i<j\leq n} \frac{x_i - x_j}{a_i - a_j}\right)^{\frac{n(n-1)}{2}} \\ &< \left(\frac{n}{n-1}\right)^{\frac{n(n-1)}{4}} < \mathrm{e}^{\frac{n}{4}}.\end{aligned}$$

$$C^{\frac{n(n-1)}{2}} \leq P^* < \mathrm{e}^{\frac{n}{4}} \times P_A = \left(\frac{1}{2}\right)^{\frac{n(n-1)}{2}} \times n^{\frac{n}{2}} 2^{\frac{n-1}{2}} \mathrm{e}^{\frac{n}{4}}.$$

The above inequality holds for all positive integers n. By taking n sufficiently large, we deduce that $C \leq \dfrac{1}{2}$. □

Solution 2 (modified from Wang Yichuan's solution). First prove $C \leq \dfrac{1}{2}$. Without loss of generality, assume

$$1 \geq x_1 \geq \cdots \geq x_n \geq -1,$$

where $x_i = \cos\vartheta_i (1 \leq i \leq n)$, $0 \leq \vartheta_1 \leq \cdots \leq \vartheta_n \leq \pi$. Let $p_j = \mathrm{e}^{\mathrm{i}\vartheta_j}$, $q_j = \mathrm{e}^{-\mathrm{i}\vartheta_j}$ be complex numbers. Clearly, on the complex plane, $p_1, \ldots, p_n$,

$q_n, \ldots, q_1$ correspond to $2n$ points anticlockwise on the unit circle, which we denote in order as $A_1, \ldots, A_{2n}$. It follows that

$$\begin{aligned}\prod_{1\le j<k\le n}(x_j - x_k) &= \prod_{1\le j<k\le n}(\cos\vartheta_j - \cos\vartheta_k)\\ &= \prod_{1\le j<k\le n} 2\sin\frac{\vartheta_j+\vartheta_k}{2}\sin\frac{\vartheta_j-\vartheta_k}{2}. \qquad ①\end{aligned}$$

By the law of sine, for $1 \le j < k \le n$,

$$2\sin\frac{\vartheta_j+\vartheta_k}{2} = |p_j - q_k| = |p_k - q_j|,$$

$$2\sin\frac{\vartheta_j-\vartheta_k}{2} = |p_j - p_k| = |q_j - q_k|.$$

Hence,

$$\begin{aligned}&2\sin\frac{\vartheta_j+\vartheta_k}{2}\sin\frac{\vartheta_j-\vartheta_k}{2}\\ &= \frac{1}{2}\sqrt{|p_j - q_k||p_k - q_j||p_j - p_k||q_j - q_k|}\\ &= \frac{1}{2}\sqrt{|A_jA_{2n+1-k}||A_kA_{2n+1-j}||A_jA_k||A_{2n+1-j}A_{2n+1-k}|}.\end{aligned}$$

Plug the above identity into ①, to obtain

$$\begin{aligned}&\prod_{1\le j<k\le n}(x_j - x_k)\\ &= \prod_{1\le j<k\le n}\frac{1}{2}\sqrt{|A_jA_{2n+1-k}||A_kA_{2n+1-j}||A_jA_k||A_{2n+1-j}A_{2n+1-k}|}\\ &= \left(\frac{1}{2}\right)^{\frac{n(n-1)}{2}}\sqrt{\prod_{1\le j<k\le 2n j+k\ne 2n+1}|A_jA_k|}. \qquad ②\end{aligned}$$

Now we estimate $\prod\limits_{1\le j<k\le 2n j+k\neq 2n+1} |A_jA_k|$. Divide A_jA_k $(1 \le j < k \le 2n)$ into n groups

$$\Omega_1 : A_1A_2, A_2A_3, \ldots, A_{2n}A_1;$$

$$\Omega_2 : A_1A_3, A_2A_4, \ldots, A_{2n}A_2;$$

$$\cdots\ \cdots$$

$$\Omega_{n-1} : A_1A_2, A_2A_3, \ldots, A_{2n}A_{n-1};$$

$$\Omega_n : A_1A_{n+1}, A_2A_{n+2}, \ldots, A_nA_{2n}.$$

Notice that for each $1 \le j \le n-1$, the sum of the inscribed angles subtended by all chords in Ω_j is $j\pi$; for each Ω_j, let λ_j be the number of chords A_sA_t such that $s+t \neq 2n+1$, then it must be $\lambda_j = 2n$ or $\lambda_j = 2n-2$. From the law of sine and Jensen's inequality applied to the convex function $\ln(\sin x)$ $\left(x \in \left(0, \frac{\pi}{2}\right)\right)$, it follows that

$$\prod\{|A_sA_t| : A_sA_t \in \Omega_j, s+t \neq 2n+1\} \le \left(2\sin\frac{j\pi}{\lambda_j}\right)^{\lambda_j}.$$

On one hand, due to $\sin\frac{j\pi}{\lambda_j} \le 1$,

$$\left(2\sin\frac{j\pi}{\lambda_j}\right)^{\lambda_j} \le 2^{2n}\left(\sin\frac{j\pi}{\lambda_j}\right)^{2n-2} \le 2^{2n}\left(\sin\frac{j\pi}{2n-2}\right)^{2n-2}.$$

On the other hand, we have

$$\prod\{|A_sA_t| : A_sA_t \in \Omega_n\} \le 2^n.$$

Hence,

$$\begin{aligned}
\prod_{1\le j<k\le 2n j+k\neq 2n+1} |A_jA_k| &\le 2^n \cdot \prod_{j=1}^{n-1}\left(2^{2n}\left(\sin\frac{j\pi}{2n-2}\right)^{2n-2}\right)\\
&= 2^{3n-2}\left(\prod_{j=1}^{n-1}\left(2\sin\frac{j\pi}{2n-2}\right)\right)^{2n-2}\\
&\le 2^{3n-2}\cdot(2\sqrt{n})^{2n-2},
\end{aligned}$$

and in the last step we used the trigonometric identity

$$\prod_{j=1}^{m}\left(2\sin\frac{j\pi}{2m}\right)=2\sqrt{m}.$$

Now plug the inequality into ②, yielding

$$\begin{aligned} C^{\frac{n(n-1)}{2}} &\leq \prod_{1\leq j<k\leq n}(x_j-x_k) \\ &\leq \left(\frac{1}{2}\right)^{\frac{n(n-1)}{2}}\cdot 2^{\frac{3n-2}{2}}\cdot(2\sqrt{n})^{n-1}. \end{aligned}$$

Letting $n\to\infty$, $C\leq\frac{1}{2}$.

It remains to prove that $C=\frac{1}{2}$ can be attained. Let $(\vartheta_1,\ldots,\vartheta_n)=\left(\frac{\pi}{2n},\frac{3\pi}{2n},\ldots,\frac{(2n-1)\pi}{2n}\right)$. Evidently, the complex numbers $p_j=\mathrm{e}^{\mathrm{i}\vartheta_j}$, $q_j=\mathrm{e}^{-\mathrm{i}\vartheta_j}$ correspond to a regular $2n$-gon $A_1\ldots A_{2n}$ in the complex plane. From ②, we get

$$\prod_{1\leq j<k\leq n}(x_j-x_k)=\left(\frac{1}{2}\right)^{\frac{n(n-1)}{2}}\sqrt{\prod_{\substack{1\leq j<k\leq 2n\\ j+k\neq 2n+1}}|A_jA_k|}.$$

Since $A_1A_{n+1},\ldots,A_nA_{2n}$ are the longest diagonals (chords) in the regular $2n$-gon $A_1\cdots A_{2n}$,

$$\prod_{\substack{1\leq j<k\leq 2n\\ j+k\neq 2n+1}}|A_jA_k|\geq\prod_{\substack{1\leq j<k\leq 2n\\ k-j\neq n}}|A_jA_k|=\left(\prod_{j=1}^{n-1}2\sin\frac{j\pi}{2n}\right)^{2n}=(\sqrt{n})^{2n}\geq 1,$$

which implies

$$\prod_{1\leq j<k\leq n}(x_j-x_k)\geq\left(\frac{1}{2}\right)^{\frac{n(n-1)}{2}}.$$

In conclusion, $C=\frac{1}{2}$. □

Test III, Second Day
(8 am – 12:30 pm; April 8, 2021)

4 For each positive integer N, let $\tau(N)$ be the number of positive factors of N; $\omega(N)$ be the number of distinct prime factors of N; $\Omega(N)$ be the number of prime factors (counts multiplicities) of N. Prove: for each positive integer n,

$$\sum_{m=1}^{n} 5^{\omega(m)} \leq \sum_{k=1}^{n} \left\lfloor \frac{n}{k} \right\rfloor \tau(k)^2 \leq \sum_{m=1}^{n} 5^{\Omega(m)}.$$

Here, $\lfloor x \rfloor$ is the largest integer not exceeding x.

(Contributed by He Yijie)

Solution 1 First, note that $\left\lfloor \frac{n}{k} \right\rfloor$ represents the number of multiples of k among $1, 2, \ldots, n$. Hence,

$$\sum_{k=1}^{n} \left\lfloor \frac{n}{k} \right\rfloor \tau(k)^2 = \sum_{k=1}^{n} \sum_{1 \leq m \leq n,\, k|m} \tau(k)^2 = \sum_{m=1}^{n} \sum_{k|m} \tau(k)^2.$$

To prove the problem statement, it suffices to justify, for $m = 1, \ldots, n$,

$$5^{\omega(m)} \leq \sum_{k|m} \tau(k)^2 \leq 5^{\Omega(m)}.$$

When $m = 1$, it is obvious. When $m > 1$, let $m = p_1^{\alpha_1} p_2^{\alpha_2} \cdots p_r^{\alpha_r}$ be the prime factorization ($p_1, p_2, \ldots, p_r$ are distinct prime numbers and $\alpha_1, \alpha_2, \ldots, \alpha_r$ are positive integers). For $k = p_1^{\beta_1} p_2^{\beta_2} \cdots p_r^{\beta_r}$, $\tau(k) = (\beta_1 + 1)(\beta_2 + 1) \cdots (\beta_r + 1)$. Thus,

$$\begin{aligned} \sum_{k|m} \tau(k)^2 &= \sum_{\substack{0 \leq \beta_1 \leq \alpha_1 \\ \cdots \\ 0 \leq \beta_r \leq \alpha_r}} (\beta_1 + 1)^2 (\beta_2 + 1)^2 \cdots (\beta_r + 1)^2 \\ &= \prod_{i=1}^{r} (1^2 + 2^2 + \cdots + (\alpha_i + 1)^2). \end{aligned}$$

Now it suffices to show for each $1 \leq i \leq r$,

$$5 \leq 1^2 + 2^2 + \cdots + (\alpha_i + 1)^2 \leq 5^{\alpha_i}.$$

For $j \in \mathbb{N}_+$, define $T(j) = 1^2+2^2+\cdots+(j+1)^2 = \frac{1}{6}(j+1)(j+2)(2j+3)$. Then

$$\frac{T(j+1)}{T(j)} = \frac{j+2}{j+1} \cdot \frac{j+3}{j+2} \cdot \frac{2j+5}{2j+3} \in \left[1, \frac{2}{1} \cdot \frac{3}{2} \cdot \frac{5}{3}\right] = [1,5].$$

Since $T(1) = 5$, it is straightforward to check $T(j) \in [5, 5^j]$ $(\forall_j \in \mathbb{N}_+)$ by induction. This finishes the proof. □

5 Find all functions $f : \mathbb{R} \to \mathbb{R}$ such that for any $x, y \in \mathbb{R}$,

$$f(xf(y) + y^{2021}) = yf(x) + (f(y))^{2021}.$$

(Contributed by Fu Yunhao)

Solution $f(x) = 0$ or $f(x) = x$.

Denote $P(x,y)$ as the governing equation of the problem. From $P(x,0)$, we get $f(xf(0)) = (f(0))^{2021}$. If $f(0) \neq 0$, as x is arbitrary, f must be a constant function, say $f(x) \equiv c$. Then $c = yc + c^{2021}$ for $y \in \mathbb{R}$, $c = 0$, a contradiction. Hence, $f(0) = 0$.

If $f(a) = 0$ for some $a \neq 0$, then $P(x,a)$ leads to $f(x) = 0$ for all $x \in \mathbb{R}$. Clearly, this function satisfies the governing equation. In the following, assume $f(a) \neq 0$ whenever $a \neq 0$.

From $P(0,1)$, we get $f(1) = (f(1))^{2021}$, and $f(1) = \pm 1$. If $f(1) = -1$, $P(1,1)$ gives $f(0) = -2$, a contradiction. Thus, $f(1) = 1$.

Now $P(x,1)$ gives

$$f(x+1) = f(x) + 1. \quad ①$$

Compare $P(x,y)$ and $P(x+1,y)$, to obtain

$$f(xf(y) + f(y) + y^{2021}) \xlongequal{P(x+1,y)} yf(x+1)$$

$$+(y)^{2021} \xlongequal{①} yf(x) + y + f(y)^{2021} \xlongequal{P(x,y)} y + f(xf(y) + y^{2021}).$$

As $f(y) \neq 0$, $xf(y) + y^{2021}$ can take any real value z, and thus

$$f(z + f(y)) = y + f(z). \quad ②$$

In ②, letting $z = 0$, we get $f(f(y)) = y$. Then, letting $y = f(w)$ in ②, we arrive at the following equation

$$f(z + f(f(w))) = f(z+w) = f(z) + f(w)$$

for any $z, w \in \mathbb{R}$.

Note that $P(0,y)$ gives $f(y^{2021}) = (f(y))^{2021}$, and so $P(x,y)$ becomes

$$\begin{aligned} yf(x) + (f(y))^{2021} &= f(xf(y) + y^{2021}) \\ &= f(xf(y)) + f(y^{2021}) \\ &= f(xf(y)) + (f(y))^{2021}, \end{aligned}$$

which implies $yf(x) = f(xf(y))$. Let $y = f(z)$ and use $f(f(z)) = z$ to derive

$$f(xz) = f(x)f(z).$$

In particular, f maps positive reals to positive reals. By the well-known properties of Cauchy's multiplicative functional equation, $f(x) = x$, which clearly satisfies the governing equation.

We conclude that there are only two functions: $f(x) = 0$ and $f(x) = x$. □

6 Prove that there exists a constant $\lambda > 0$, such that: for any positive integer m, in the rectangular coordinates system, if all vertices of $\triangle ABC$ are integral points, and there is a unique interior (not on the side) integral point whose x and y coordinates are multiples of m, then the area of $\triangle ABC$ is less than λm^3.

(Contributed by Wang Bin and Fu Yunhao)

Solution We call (x, y) an "m-integral point" if x, y are integers and $m|x$, $m|y$. By a proper translation, we may assume that the unique interior m-integral point is at (0, 0). Extend AO, BO, and CO to meet the opposite sides at points D, E, and F, respectively. Denote

$$p = \frac{OD}{AD} = \frac{S_{\triangle OBC}}{S_{\triangle ABC}}, \quad q = \frac{OE}{BE} = \frac{S_{\triangle OCA}}{S_{\triangle ABC}}, \quad r = \frac{OF}{CF} = \frac{S_{\triangle OAB}}{S_{\triangle ABC}}.$$

Evidently, $p + q + r = 1$; assume $p \geq q \geq r > 0$ $\left(\text{so that } p \geq \frac{1}{3} \geq \frac{1}{m+2}\right)$. Let A', B', and D' be the respective symmetric points of A, B, and D about O. On the line AD, let UV be the intersection of the segments AD and $D'A'$, namely, U is A or D', V is D or A', whichever is closer to O. Since $\triangle BUV \subseteq \triangle ABC$, except for O, there is no m-integral point in the interior of $\triangle BUV$ or in the interior of UV, and neither in the symmetric $\triangle B'UV$. Together, they imply that the parallelogram $BUB'V$

(centered at O) does not have interior m-integral points other than O. By Minkowski's theorem,

$$4m^2 \geq S_{BUB'V} = 4S_{\triangle OBV} = 4S_{\triangle ABC} \times \min\{p, 1-p\} \times \frac{q}{q+r};$$

$$S_{\triangle ABC} \leq \frac{4m^2}{4\min\{p, 1-p\}} \times \frac{q+r}{q} \leq \frac{2m^2}{\min\{p, 1-p\}}.$$

If $p > \dfrac{m}{m+1}$, then

$$\frac{OD}{OA'} = \frac{OD}{AO} = \frac{p}{1-p} > m,$$

taking $OL = m \cdot OA'$ on segment OD, L must be an m-integral point inside $\triangle ABC$, which is contradictory.

Therefore, $p \leq \dfrac{m}{m+1}$, $\min\{p, 1-p\} \geq \dfrac{1}{m+2}$, and we deduce

$$S_{\triangle ABC} \leq 2m^2(m+2).$$

□

Test IV, First Day
(8:00 am – 12:30 pm; April 11, 2021)

1 Given integer $n \geq 2$. Find the least positive integer m, such that there are n^2 distinct positive real numbers $x_{i,j}$ $(1 \leq i, j \leq n)$ satisfying the following conditions:

(1) For every i, j,

$$x_{i,j} = \max\{x_{i,1}, x_{i,2}, \ldots, x_{i,j}\} \quad \text{or}$$

$$x_{i,j} = \max\{x_{1,j}, x_{2,j}, \ldots, x_{i,j}\};$$

(2) For every i, there are at most m indices k, with

$$x_{i,k} = \max\{x_{i,1}, x_{i,2}, \ldots, x_{i,k}\};$$

(3) For every j, there are at most m indices k, with

$$x_{k,j} = \max\{x_{1,j}, x_{2,j}, \ldots, x_{k,j}\}.$$

(Contributed by Xiao Liang and Fu Yunhao)

Solution $m = \left\lfloor \frac{n+3}{2} \right\rfloor$. Put these n^2 numbers $x_{i,j}$ in an $n \times n$ table: call $x_{i,j}$ a "row pivot" if $x_{i,j} = \max\{x_{i,1}, x_{i,2}, \ldots, x_{i,j}\}$; call $x_{i,j}$ a "column pivot", if $x_{i,j} = \max\{x_{1,j}, x_{2,j}, \ldots, x_{i,j}\}$.

First, prove $m \geq \left\lfloor \frac{n+3}{2} \right\rfloor$. Consider all $(n-1)^2$ numbers $x_{i,j}$ $(2 \leq i, j \leq n)$: every column (from 2 to n) contains at most $m-1$ column pivots (by condition (3), and notice that $x_{1,j}$ $(2 \leq j \leq n)$ is a pivot); every row (from 2 to n) contains at most $m-1$ row pivots (by condition (2), and notice that $x_{i,1}$ $(2 \leq i \leq n)$ is a pivot). Due to condition (1), every number is a pivot (row or column or both), and hence

$$(n-1)(m-1) + (n-1)(m-1) \geq (n-1)^2. \qquad ①$$

We claim the equality in ① cannot be attained. Let $x_{i_0,j_0} = \max x_{i,j}$ $(2 \leq i, j \leq n)$. If x_{i_0,j_0} is both row and column pivots, then x_{i_0,j_0} is counted twice on the left-hand side of ①, and the equality cannot hold. Suppose x_{i_0,j_0} is not a row pivot. Then $x_{i_0,1} > x_{i_0,j_0}$ and there are no other row pivots in this row, the equality in ① cannot hold. Similarly, if x_{i_0,j_0} is not a column pivot, then there are no other column pivots in this column, and the equality in ① cannot hold. Therefore, $2(m-1) > n-1$, $m \geq \left\lfloor \frac{n+3}{2} \right\rfloor$.

Next, construct tables for which $m = \left\lfloor \frac{n+3}{2} \right\rfloor$. When $n = 2t$, divide the $n \times n$ table into four $t \times t$ tables as follows:

Ω_1	Ω_2
Ω_3	Ω_4

Take ε_1, $\varepsilon_2 > 0$ sufficiently small, such that $\varepsilon_1 < \frac{1}{2n}\varepsilon_2 < \frac{1}{4n^2}$. For k =1, 4, let the number in row i and column j of Ω_k be $k - i\varepsilon_1 + j\varepsilon_2$. Note that in Ω_1 and Ω_4, the numbers decrease from the top right to the bottom left. For $k = 2, 3$, let the number in row i and column j of Ω_k be $k + i\varepsilon_1 - j\varepsilon_2$. Note that in Ω_2 and Ω_3 the numbers decrease from the bottom left to the top right.

It is easy to check that all numbers in Ω_1 and Ω_4 are row pivots, and all numbers in Ω_2 and Ω_3 are column pivots. In addition, except for the first and the $t+1$ rows, there are no other column pivots in Ω_1 and Ω_4; except for the first and the $t+1$ columns, there are no other row pivots in Ω_2 and Ω_3. Hence, conditions (2) and (3) are fulfilled, and $m=t+1$.

When $n=2t-1$, take the construction for $n=2t$ as above and remove the rightmost column and the bottom row. □

2 As illustrated in Fig. 2.1, in the acute $\triangle ABC$, $AB < AC$, I is the incentre, and $\odot O$ is the circumcentre. Let M, N be the midpoints of $\overparen{BAC}$ and $\overparen{BC}$, respectively. Let D be a point on $\overparen{AC}$ such that $AD \,//\, BC$. The escribed circle of $\triangle ABC$ against $\angle BAC$ touches BC at E. Let F be a point inside $\triangle ABC$, satisfying $IF \,//\, BC$ and $\angle BAF = \angle CAE$. The line NF meets $\odot O$ at R other than N, the lines AF and DI meet at K, and the lines AR and IF meet at L.

Prove: $NK \perp ML$.

(Contributed by Lin Tianqi)

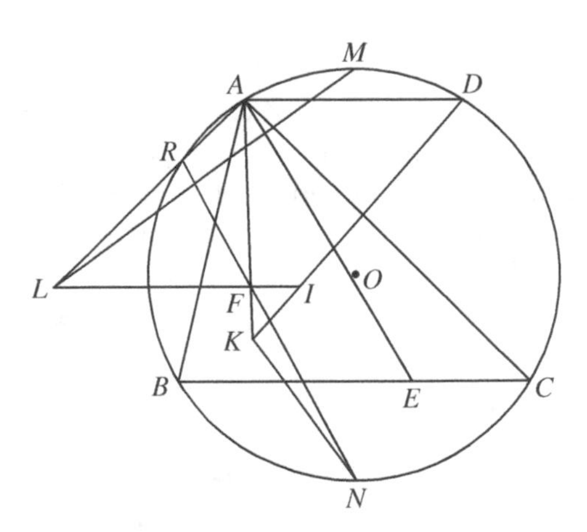

Fig. 2.1

Solution First, we need a lemma.

Lemma *Let R' be the midpoint of BC. Then $\angle AMI = \angle IR'B$, $IR' \,//\, AE$.*

Proof of lemma As illustrated in Fig. 2.2, let I_b and I_c be the escentres of $\triangle ABC$ relative to the vertices B and C, respectively. Then A, M, I_b, and I_c are collinear. As $\angle I_bBI_c = \angle I_bCI_c = 90°$, the points B, C, I_b, and I_c all lie on the circle with diameter I_bI_c; since $MB = MC$, M is the centre of this circle, $MI_b = MI_c$. Since $\triangle II_bI_c \sim \triangle ICB$, M, R' are midpoints of I_bI_c and BC, respectively, it follows that $\angle AMI = \angle IR'B$.

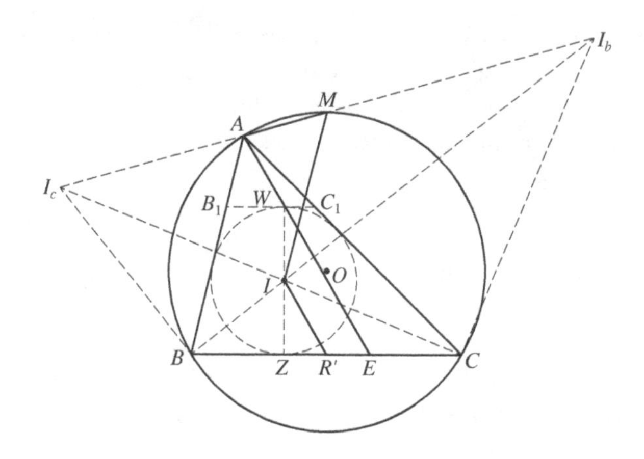

Fig. 2.2

Let the incircle $\odot I$ of $\triangle ABC$ touch BC at Z and ZW be a diameter. Through W, draw a line parallel to BC that crosses AB, AC at B_1, C_1, respectively. Then $B_1C_1 // BC$. Since $\triangle A_1B_1C_1$ and $\triangle ABC$ are homothetic with centre A, W and E are correspondent points, it follows that A, W, E are collinear. Finally, from properties of escribed circles, we infer that $BZ = CE$, R' is the midpoint of ZE, as I is the midpoint of ZW, $IR' // AE$. The lemma is verified.

Return to the original problem. Let S be the intersection of AI and BC. Then

$$\begin{aligned}\angle AIF &= \angle ASB = \angle SAC + \angle ACS \\ &= \angle BCN + \angle ACS \\ &= \angle ACN = 180^\circ - \angle ARN,\end{aligned}$$

indicating that A, I, F, and R lie on a circle, say ω, as shown in Fig. 2.3.

Suppose the lines AF and MI intersect at X. We show that X is on $\odot O$. By the lemma,

$$\angle AMI = \angle IRB = \angle AEB,$$

$$\text{also } \angle AMN = \angle ACN = \angle\ ASB, \text{ thus}$$

$$\begin{aligned}\angle IMN &= \angle AMN - \angle AMI \\ &= \angle ASB - \angle AEB \\ &= \angle IAE.\end{aligned}$$

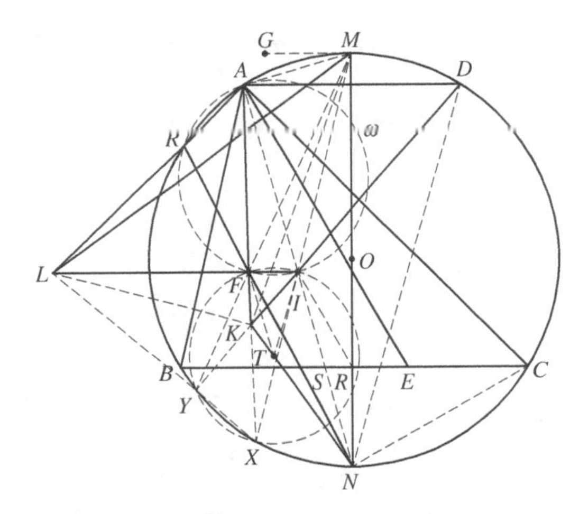

Fig. 2.3

It is given that $\angle IAE = \angle FAI$, hence $\angle IMN = \angle FAI$, A, M, N, and X are concyclic, that is, X lies on $\odot O$.

Let Y be the other intersection of the line DI and $\odot O$. Clearly,

$$\angle IYX = \angle DYX = \angle DAX = \angle IFX,$$

and I, X, Y, F lie on a circle, say $\odot T$.

Through M, draw the tangent line MG of $\odot O$ (G and Y are on the same side of MN), $MG // IF$. Then

$$\angle YMG = \angle MXY = \angle IXY = \angle YFL,$$

yielding the collinearity of Y, F, M. Apply the radical axis theorem to $\odot O$, $\odot \omega$, and $\odot T$, to derive the collinearity of X, Y, and L. In the cyclic quadrilateral $IXYF$, apply Brocard's theorem to derive the orthocentre T of $\triangle MKL$, $TK \perp ML$.

Finally, in $\triangle TFI$ and $\triangle NAD$, $TF = TI$, $NA = ND$, and

$$\angle FTI = 2\angle FXI = 2\angle AXM = \angle AND.$$

Observing $FI // BC // AD$, we conclude that $\triangle TFI$ and $\triangle NAD$ are homothetic with centre K, and moreover K, T, N are collinear.

From $TK \perp ML$ and the collinearity of K, T, N, $NK \perp ML$ follows. □

3 Let $S(k)$ denote the sum of all digits of k in base 10. Find all integers $n \geq 2$ and rational numbers $\beta \in (0,1)$, such that there exist n distinct positive integers $a_1, a_2, \ldots, a_n$ that satisfy: for any subset $l \subseteq \{1,2,\ldots,n\}$ with two or more elements,

$$S\left(\sum_{i\in I} a_i\right) = \beta \cdot \sum_{i\in I} S(a_i).$$

(Contributed by Xiao Liang)

Solution The desired are integers $n \in \{2,3,\ldots,10\}$ and all rational numbers $\beta \in (0,1)$.

First, for integer $n \in \{2,3,\ldots,10\}$ and rational $\beta \in (0,1)$, construct a_1, $a_2,\ldots,a_n$ as follows. Let c and s be positive integers that will be determined later. For $k=1,\ldots,n$, define

$$a_k = 2\cdot 10^{cn+sk} - 10^{cn+s(k-1)+1} + \sum_{\substack{i\in\{i,\ldots,n\}\\ i\neq k}} 10^{cn+s(i-1)+1} + 10^{k-1}\cdot\sum_{i=0}^{c-1} 10^{ni}.$$

Here, among the last cn places of $a_1, a_2, \ldots, a_n$, each place corresponds to a unique a_i whose digit at that place is nonzero, and that digit is 1; for the next (to the left) s places, a_1 has $199\ldots9$ ($s-1$ of 9's), while all other a_i's have $00\ldots01$; for the next s places, a_2 has $199\ldots9$ ($s-1$ of 9's), while all other a_i's have $00\ldots01$; and so on.

For every $1 \leq i \leq n$, $S(a_i) = n + 9(s-1) + c$; for every

$$I \subseteq \{1,2,\ldots,n\}(|I| \geq 2),$$

it is easy to check that

$$S\left(\sum_{i\in I} a_i\right) = n\cdot|I| + c\cdot|I|.$$

As long as s, $c \in \mathbb{N}_+$ are suitably chosen such that

$$(n+9(s-1)+c)\cdot\beta = n+c,$$

the problem requirements are all met.

Next, we prove $n \leq 10$. Assume $n = 11$. If

$$I_1, I_2 \subseteq \{1,\ldots,n\}$$

satisfy $I_1 \cap I_2 = \varnothing$ and $|I_1| \geq 2$, $|I_2| \geq 2$, then

$$S\left(\sum_{i\in I_1\cup I_2} a_i\right) = \beta \cdot \sum_{i\in I_1\cup I_2} S(a_i) = S\left(\sum_{i\in I_1} a_i\right) + S\left(\sum_{i\in I_2} a_i\right).$$

In other words, no carry occurs in the summing of $\sum_{i\in I_1} a_i$ and $\sum_{i\in I_2} a_i$. However, from $\beta < 1$ we see that a carry occurs in the summing of a_i and a_j, for every $1 \leq i < j \leq 11$.

Suppose T is the least positive integer that the summing $a_1 + a_2 + \cdots + a_{11}$ involves a carry in the Tth place (from the right). Let the Tth place digits of $a_1, a_2, \ldots, a_{11}$ be $x_1, x_2, \ldots, x_{11}$, respectively. For any subset

$$I \subseteq \{1, 2, \ldots, 11\},$$

the Tth place digit of $\sum_{i\in I} a_i$ is the ones place digit of $\sum_{i\in I} x_i$. Without loss of generality, assume

$$9 \geq x_1 \geq x_2 \geq \cdots \geq x_{11} \geq 0.$$

Then $x_1 + x_2 + \cdots + x_{11} \geq 10$. We assert that there exist I_1, $I_2 \subseteq \{1, 2, \ldots, 11\}$, $I_1 \cap I_2 = \varnothing$, $|I_1| \geq 2$, $|I_2| \geq 2$, such that the sum of the ones place digits of $\sum_{i\in I_1} x_i$ and $\sum_{i\in I_2} x_i$ is greater than or equal to 10.

If $x_{11} = 0$, starting at $x_{10} + x_9$, add one number at a time by $x_8, x_7, \ldots, x_1$. There must be a time that a carry occurs in the ones place: for some $k \in \{1, 2, \ldots, 8\}$, the summing $(x_{10} + x_9 + \cdots + x_{k+1}) + x_k$ involves a carry in the ones place, and so does $(x_{10} + x_9 + \cdots + x_{k+1}) + (x_k + x_{11})$, a contradiction.

If $x_{11} > 0$, try to make $x_1, x_2, \ldots, x_{11}$ into as many pairs as possible, such that the sum of each pair ends with a digit greater than or equal to 2. If five pairs are made, then the summing involves a carry in the ones place, a contradiction. Assume at most four pairs are made, leaving three or more unpaired numbers, which include at most one number less than 5 and one number greater than 5, as otherwise two such numbers can be paired. This indicates there is at least one 5, and furthermore, no unpaired number is less than 5, and hence there are at least two 5s, say x_a and x_b. Take the third unpaired number $x_c \geq 5$. Now, starting at any $x_d + x_e$ (not involving x_a, x_b, x_c), add one number at a time by the other numbers except x_a, x_b. Since $x_c \geq 5$, the summing must involve a carry in the ones place at some moment, say $\left(\sum_i x_i\right) + x_f$. Then we are led to contradiction by taking $\left(\sum_i x_i\right) + (x_f + x_a + x_b)$.

Based on the above argument, when $n \geq 11$, no positive integers a_1, $a_2, \ldots, a_n$ satisfy the problem condition. □

Test IV, Second Day
(8:00 am – 12:30 pm; April 12, 2021)

4 For real numbers $x_1, x_2, \ldots, x_{60} \in [-1, 1]$, find the maximum of

$$\sum_{i=1}^{60} x_i^2 (x_{i+1} - x_{i-1}),$$

where $x_0 = x_{60}$, $x_{61} = x_1$.

(Contributed by Fu Yunhao)

Solution The maximum is 40. First, notice that

$$\begin{aligned}\sum_{i=1}^{60} x_i^2(x_{i+1} - x_{i-1}) &= \sum_{i=1}^{60} x_i^2 x_{i+1} - \sum_{i=1}^{60} x_i^2 x_{i-1} \\ &= \sum_{i=1}^{60} x_i^2 x_{i+1} - \sum_{i=1}^{60} x_{i+1}^2 x_i \\ &= \sum_{i=1}^{60} x_i x_{i+1}(x_i - x_{i+1}).\end{aligned}$$

Since $3xy(x-y) = x^3 - y^3 - (x-y)^3$ for any real numbers x, y, we have

$$\begin{aligned}\sum_{i=1}^{60} x_i^2(x_{i+1} - x_{i-1}) &= \frac{1}{3}\sum_{i=1}^{60}(x_i^3 - x_{i+1}^3 - (x_i - x_{i+1})^3) \\ &= \frac{1}{3}\sum_{i=1}^{60}(x_{i+1} - x_i)^3.\end{aligned}$$

On one hand, if $x_{3k+1} = 1$, $x_{3k+2} = 0$, $x_{3k+3} = -1$ $(k = 0, 1, \ldots, 19)$,

$$\sum_{i=1}^{60}(x_{i+1} - x_i)^3 = 40 \cdot (-1)^3 + 20 \cdot 2^3 = 120;$$

on the other hand, for $a \in [-2, 2]$, $(a+1)^2(a-2) \leq 0$, or $a^3 \leq 3a + 2$, and hence

$$\sum_{i=1}^{60}(x_{i+1} - x_i)^3 \leq \sum_{i=1}^{60}(3(x_{i+1} - x_i) + 2) = 120.$$

In conclusion, the maximum of $\sum_{i=1}^{60}(x_{i+1} - x_i)^3$ is 120, and the maximum of $\sum_{i=1}^{60} x_i^2(x_{i+1} - x_{i-1})$ is 40 (when $\{x_n\} = \{1, 0, -1, 1, 0, -1, \ldots, 1, 0, -1\}$). □

5 Find the smallest real number α, such that for any convex polygon P of area 1, there exists a point M in the plane, such that the area of the convex hull of $P \cup Q$ is at most α, where Q is the central-symmetric figure of P about M.

(Contributed by Qu Zhenhua and Wu Yuchi)

Solution $\alpha = 2$.

First, we prove $\alpha \geq 2$. For a convex polygon R, let $S(R)$ be the area of R. For a plane set X, let $\bar{X}$ be the convex hull of X. Take P as $\triangle ABC$ of area 1, M as any point in the plane, and Q as the central-symmetric figure of P about M. There are two situations.

(i) If M is outside P or on the boundary. Draw a line l through M such that P, Q lie on different sides of l (they have no common interior points). Then $S(\overline{P \cup Q}) \geq S(P) + S(Q) \geq 2$.

(ii) If M is inside P. Let A', B', C' be the respective symmetric points of A, B, C about M. Then $R = \overline{P \cup Q} = \overline{\{A, B, C, A', B', C'\}}$, R has a centre of symmetry, R is a parallelogram or a hexagon. If R is a parallelogram, say $R = ABA'B'$, then C is inside or on the boundary, and

$$S(R) = S(ABA'B') \geq 2S(ABC) = 2.$$

If R is a hexagon, $R = AC'BA'CB'$, then

$$\begin{aligned} S(R) &= S(AC'BM) + S(BA'CM) + S(CB'AM) \\ &= (S(AMC) + S(BMC)) + (S(BMA) + S(CMA)) \\ &\quad + (S(CMB) + S(AMB)) = 2. \end{aligned}$$

Next, we prove $\alpha = 2$ satisfies the problem statement in two ways. □

Method 1 We begin with a lemma.

Lemma *As shown in Fig. 5.1, let $l_1 // l_2$ and P_1, P_2 be two convex polygons both contained in the shaded region, both including the points A and B. Then $S\left(\overline{P_1 \cup P_2}\right) \leq S\left(P_1\right) + S\left(P_2\right)$.*

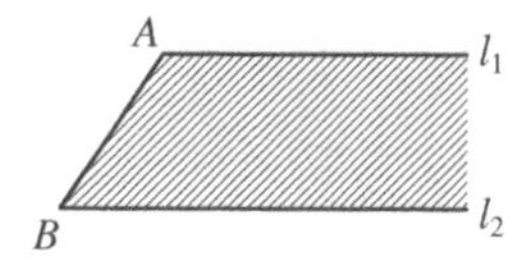

Fig. 5.1

Proof of lemma Assume $\overline{P_1 \cup P_2} = C_0C_1 \cdots C_n$, where $C_0 = A$, $C_n = B$. If there exists $0 < i < n-1$ such that C_i, $C_{i+1} \in P_1$, let

$$P_1 = \ldots UC_iC_{i+1}V \ldots.$$

Extend UC_i and VC_{i+1} to meet at point W, as shown in Fig. 5.2. Change P_1 to $P_1' = \ldots UWV \ldots$.

Obviously, the area of P_1 has increased by $S(C_iC_{i+1}W)$, and the area of $P_1 \cup P_2$ has increased by $S(C_1C_{i+1}W)$. It suffices to justify the lemma for P_1' and P_2. Notice that P_1' has one fewer vertex than P_1, and thus the above operation cannot continue infinitely. We may assume $C_1, C_3, \ldots \in P_1$, C_2, $C_4, \ldots \in P_2$.

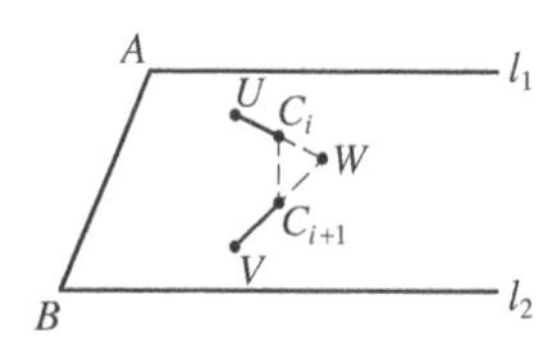

Fig. 5.2

For $1 \leq i \leq n-1$, draw a parallel line of l_1 through C_i which meets $C_{i-1}C_{i+1}$ at D_i; for $1 \leq i \leq n-2$, let $C_{i-1}C_{i+1}$ and C_iC_{i+2} intersect at E_i, as shown in Fig. 5.3. Observe that for $1 \leq i \leq n-2$,

$$\overline{P_1 \cup P_2} \setminus (P_1 \cup P_2) \subseteq \bigcup_{i=1}^{n-2} (\triangle C_iC_{i+1}E_i),$$

$$\triangle D_iD_{i+1}E_i \subseteq (P_1 \cap P_2),$$

$$S\left(C_iC_{i+1}E_i\right) = S\left(D_iD_{i+1}E_i\right).$$

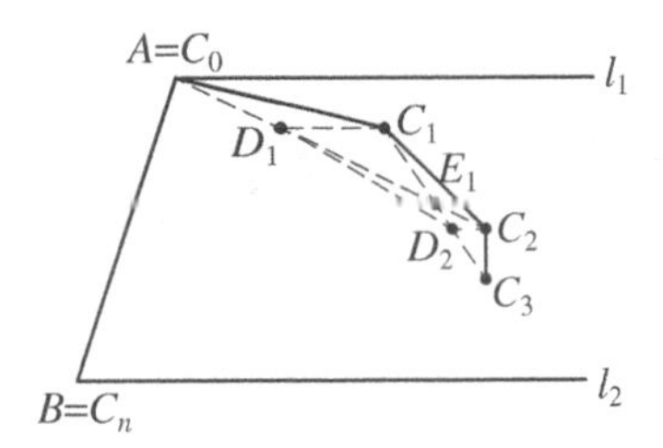

Fig. 5.3

Also, observe that for $1 \leq i < j \leq n-2$, $\triangle D_iD_{i+1}E_i \cap \triangle D_jD_{j+1}E_i = \varnothing$. Therefore,

$$S(P_1 \cap P_2) \geq \sum_{i=1}^{n-2} S(D_iD_{i+1}E_i) = \sum_{i=1}^{n-2} S(C_iC_{i+1}E_i)$$
$$\geq S\left(\overline{P_1 \cup P_2} \setminus (P_1 \cup P_2)\right),$$

and the lemma conclusion $S\left(\overline{P_1 \cup P_2}\right) \leq S(P_1) + S(P_2)$ follows immediately.

Return to the original problem. Along an arbitrary direction, draw parallel lines l_1, l_2 that bound P such that points A, B of P lie on l_1, l_2, respectively, and take the midpoint of AB as the centre of symmetry M.

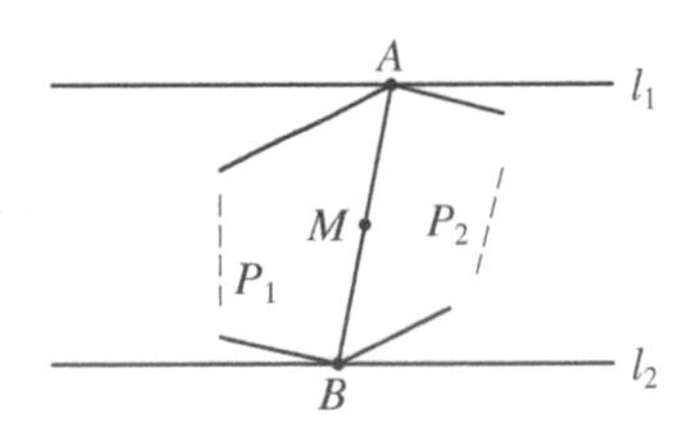

Fig. 5.4

Suppose that P is divided by AB into two convex polygons P_1, P_2, as shown in Fig. 5.4. Let P_1', P_2' be their central-symmetric figures about M. Then P_1, P_2' are on the same side of AB, and likewise for P_2, P_1'. We infer that

$$\overline{P \cup Q} = \overline{P_1 \cup P_2'} \cup \overline{P_1' \cup P_2}.$$

By the lemma,

$$S\left(\overline{P_1 \cup P_2'}\right) \leq S(P_1) + S(P_2'), \quad S\left(\overline{P_1' \cup P_2}\right) \leq S(P_1') + S(P_2).$$

Therefore,

$$\begin{aligned} S\left(\overline{P \cup Q}\right) &= S\left(\overline{P_1 \cup P_2'} \cup \overline{P_1' \cup P_2}\right) \leq S(P_1) + S(P_2') + S(P_1') + S(P_2) \\ &= 2S(P) = 2. \end{aligned}$$

Method 2 (modified from Peng Yebo's solution). Take the farthest pair of points A, B of P and draw parallel lines l_1, l_2 perpendicular to AB, as shown in Fig. 5.5. It is easy to see that A, B are vertices of P and P is entirely in the slab between l_1 and l_2.

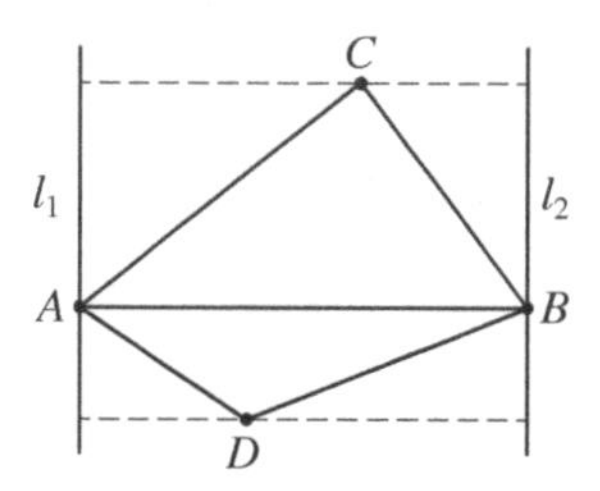

Fig. 5.5

On the two sides of AB, take points C and D of P farthest to AB (if AB is a side of P, then take C or D to be any point on the segment). Draw parallel lines of AB through C and D, respectively; the two lines, together with l_1 and l_2, enclose a rectangle R. Due to convexity, P contains $ACBD$, and hence $S(R) = 2S(ACBD) \leq 2S(P) = 2$. Choose the centre of R as the centre of symmetry M. Then Q, the central-symmetric figure of P, also lies in R. It follows that

$$S\left(\overline{P \cup Q}\right) \leq S(R) \leq 2.$$

6. There are $2n^2$ ($n \geq 2$) players in a single round-robin chess tournament. It is known that

(1) For any three players A, B, and C, if A beats B and B beats C, then A beats C.

(2) There are at most $\dfrac{n^3}{16}$ draws.

Prove: it is possible to choose n^2 players and label them P_{ij} ($1 \leq i, j \leq n$), such that for any $i, j, i', j' \in \{1, 2, \ldots, n\}$, if $i < i'$, then P_{ij} beats $P_{i'j'}$.

(Contributed by Ai Yinghua)

Solution 1 (It is reorganized from Chen Ruitao's proof; in fact it proves the conclusion when $\frac{n^3}{16}$ is replaced by $\frac{n^3}{4}$.)

Lemma *Suppose m players participate in a single round-robin tournament with possible draws, and if A beats B, B beats C, then A beats C. Then the m players can be arranged in a row such that for any two players, the one on the left either beats or draws the one on the right.*

Proof of lemma Induction on m. When $m = 1$, the lemma is trivial. Suppose it is true for $m-1$ players, and consider m players. If everyone wins a game, then we can find x_1 beats x_2, x_2 beats x_3, and so on, and someone must reappear in the sequence, say x_i beats x_j ($i \geq j$), which is contradictory. The contradiction indicates that someone has never won a game; put this player in the rightmost position. By the induction hypothesis, the other $m-1$ players can be arranged on the left such that the lemma conditions are satisfied.

For the original problem, arrange the $2n^2$ players in a row as in the lemma, and then make n groups of players as follows: set the leftmost n players as group A_1; for $i = 1, \ldots, n-2$, on the right side of A_i, set some consecutive n players as A_{i+1}; set the rightmost n players as group A_n. Moreover, between any two consecutive groups, there are $\left\lfloor \frac{n^2}{n-1} \right\rfloor$ or more players.

If players from different groups never draw, then the proof is done. Otherwise, remove two players who draw and regroup the remaining $2n^2-2$ players in the same way as before, except requiring $\left\lfloor \frac{n^2-2}{n-1} \right\rfloor$ or more players between consecutive groups. Again, if players from different groups never draw, the proof is done. Otherwise, remove two players and regroup, and so on. In general, when we regroup $2n^2-2i$ $\left(0 \leq i \leq \left\lfloor \frac{n^2}{2} \right\rfloor\right)$ players, it is required that $\left\lfloor \frac{n^2-2i}{n-1} \right\rfloor$ or more players are between consecutive groups. During the whole process, we obtain $\left\lfloor \frac{n^2}{2} \right\rfloor + 1$ groupings.

For every $0 \le i \le \left\lfloor \frac{n^2}{2} \right\rfloor$, assume in the ith grouping, x_i and y_i are from different groups and they draw. Then, there are at least $\left\lfloor \frac{n^2-2i}{n-1} \right\rfloor$ players between them, each of whom draws with either x_i or y_i. This implies that the total number of draws that x_i and y_i have is at least $\left\lfloor \frac{n^2-2i}{n-1} \right\rfloor$, and furthermore the total number of draws in the tournament is at least

$$S = \sum_{0 \le i \le \left\lfloor \frac{n^2}{2} \right\rfloor} \left\lfloor \frac{n^2-2i}{n-1} \right\rfloor.$$

When $n = 2k+1$, $S = 2k^3 + 3k^2 + 3k + 2 > \frac{n^3}{4}$; when $n = 2k$, $S > 2k^3 > \frac{n^3}{4}$. Either way, $S \le \frac{n^3}{16}$ is violated and the assumption is untrue. Hence, for $1 \le i, j \le n$, let the jth player of A_i be P_{ij}, and the conditions are satisfied.

Proof 2 Let V be the set of the $2n^2$ players. For $x, y \in V$, $x \succ y$ means x beats y. It is known from the problem that $(V, \succ)$ is a partially ordered set. Construct graph $G = (V, E)$: if x draws with y, then the edge $xy \in E$. From condition (2), $|E| \le \frac{n^3}{16}$. Let $W = \left\{ x \in V \,|\, \deg_G(x) < \frac{n}{4} \right\}$. As

$$\frac{n^3}{8} \ge 2|E| = \sum_{x \in W} \deg_G(x) + \sum_{y \in V \setminus W} \deg_G(y) \ge \frac{n}{4} \cdot |V \setminus W|,$$

we get $|V \setminus W| \le \frac{n^2}{2}$, $|W| \ge \frac{3}{2}n^2$.

Let $|W| = m \ge \frac{3}{2}n^2$. With the partial order $(W, \succ)$ induced by $(V, \succ)$, we can arrange the elements of W in a row $x_1, x_2, \ldots, x_m$ such that for $1 \le i < j \le m$, x_i either beats or draws with x_j. See proof of the lemma in the first proof.

We assert that for $1 \le i < j \le m$, $j - i \ge \left\lceil \frac{n}{2} \right\rceil + 1$, it must be $x_i \succ x_j$. Suppose to the contrary that x_i draws with x_j. Then for $i+1 \le k \le j-1$, $x_i \succ x_k$ and $x_k \succ x_j$ cannot be true simultaneously (or else $x_i \succ x_j$). Thus, x_k draws with either x_i or x_j or both, and there are $j - i - 1 \ge \left\lceil \frac{n}{2} \right\rceil \ge \frac{n}{2}$ of such k's. This implies that either x_i or x_j has $\dfrac{n}{4}$ or more draws, opposing the assumption $x_i,\ x_j \in W$.

Let $n = \left\lceil \dfrac{n}{2} \right\rceil$. Choose n groups of players

$$A_1 = \{x_1, \ldots, x_n\}, \quad A_2 = (x_{n+l+1}, \ldots, x_{2n+l}), \ldots,$$

$$A_n = \{x_{(n-1)n+(n-1)l+1}, \ldots, x_{n^2+(n-1)l}\}.$$

Notice that

$$n^2 + (n-1)\,l = n^2 + (n-1)\left\lceil \frac{n}{2} \right\rceil \le n^2 + (n-1)\,\frac{n+1}{2} < \frac{3}{2}n^2 = m.$$

Based on the previous argument, for $1 \le i < j \le n$, if $x_u \in A_i$, $x_v \in A_j$, then $v - u \ge l + 1$, $x_u \succ x_v$. Therefore, $A_1, A_2, \ldots, A_n$ satisfy the problem requirements (simply let the jth player of A_i be P_{ij}). □

China National Team Selection Test

2022

The main task of the Chinese National Training Team for the 63rd International Mathematical Olympiad (IMO) in 2022 was to select members of the Chinese national team for China to participate in the 63rd International Mathematical Olympiad held in Oslo, Norway in 2022.

Due to Covid-19, the two-round selection examinations for the training team were conducted online, and were hosted by the High School Affiliated to Renmin University of China in Beijing. The first round was held from March 22nd to March 30th, 2022. A total of 59 contestants participated in the examination. After two tests (with equal weights), 15 contestants were selected to enter the second round.

The second round was held from April 11th to April 18th, 2022. During this period, two tests were conducted, and the total scores of the four tests were compared (with equal weights). The top six scorers were identified to be the Chinese national team members for the 63rd IMO. The six members were Zhang Zhicheng (No. 1 Middle School Affiliated to Central China Normal University, 10th grade), Zhang Yiran (Shanghai High School, 12th grade), Liao Yubo (the High School Affiliated to Renmin University of China, 11th grade), Jiang Cheng (Shanghai High School, 11th grade), Qu Xiaoyu (Chongqing Bashu Secondary School, 10th grade), and Liu Jiayu (Yali High School in Changsha, 11th grade).

The coaches of the national training team are (in alphabet order): Ai Yinghua (Tsinghua University), Fu Yunhao (Guangdong Second Normal University), He Yijie (East China Normal University), Lin Tianqi (East China Normal University), Qu Zhenhua (East China Normal University), Wang Bin (Institute of Mathematics and Systems Science, Chinese Academy of Sciences), Xiao Liang (Peking University), Xiong Bin (East China Normal University), Yao Yijun (Fudan University).

Test I, First Day
(8 am – 12:30 pm; March 23, 2022)

1 As illustrated in Fig. 1.1, $ABCDEF$ is a cyclic hexagon. The extensions of AB and DC meet at G; the extensions of AF and DE meet at H. Let M and N be the circumcentres of $\triangle BCG$ and $\triangle EFH$, respectively. Prove that the lines BE, CF, and MN are concurrent.

(Contributed by Zhang Sihui)

We give four solutions as follows.

Solution 1 Let ω be the circumcircle of $\triangle BCG$. Let E' be the other intersection of the line BE and ω, F' be the other intersection of the line CF and ω, as shown in Fig. 1.2. Note that $\angle BE'F' = \angle BCF' = \angle BCF = \angle BEF$, and hence $EF // E'F'$; similarly, $\angle CF'G = \angle CBG = \angle CFA$, and hence $GF' // HF$; moreover, $GE' // HE$. It follows that $\triangle E'F'G$ and $\triangle EFH$ are homothetic triangles. Let P be the homothetic centre. Then EE' and FF' pass through P; the line connecting the circumcentres of $\triangle E'F'G$ and $\triangle EFH$, which is MN, passes through P as well. Therefore, the lines BE, CF, and MN meet at P. □

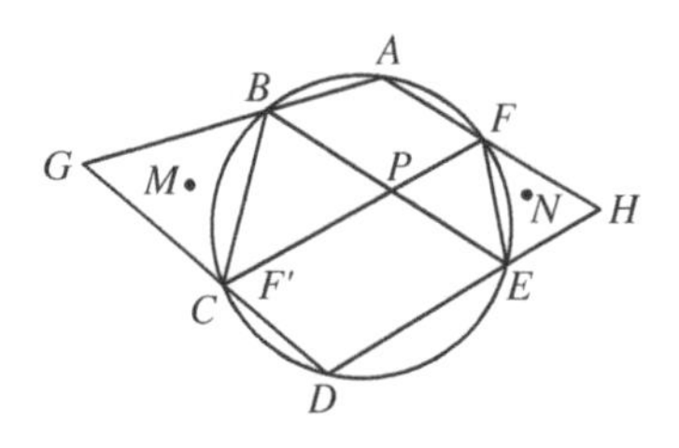

Fig. 1.1

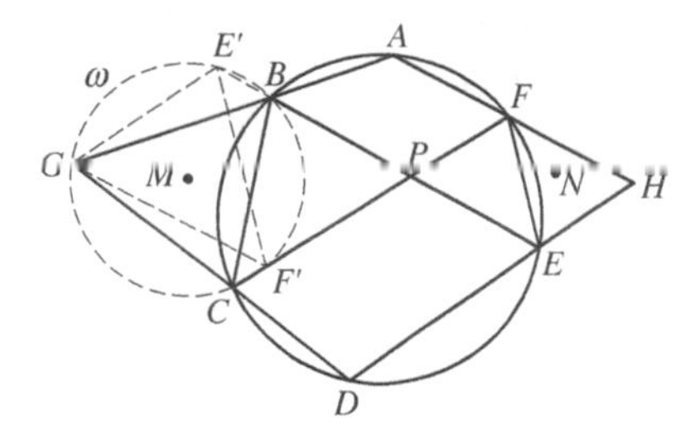

Fig. 1.2

Solution 2 (Edited from Wang Zihan's proof.) As in Fig. 1.3, we first apply Pascal's theorem to the cyclic hexagon $ABCDEF$ to deduce that G, P, H are collinear. Let A' and D' be the antipodal points of A and D, respectively. Let BA' and CD' meet at G', $D'E$ and $A'F$ meet at H'. Apply Pascal's theorem to the cyclic hexagon $BA'FCD'E$ to deduce that G', H', P are collinear.

Since A and A', D and D' are antipodal points, we have $A'B \perp AB$, $A'F \perp AF$, $CD' \perp CD$, $D'E \perp DE$. Notice that $\angle AHH' = 90^\circ - \angle FH'H = 90^\circ - \angle FAD$. Thus, the angle between AH and GG' is given by

$$\begin{aligned}180^\circ - \angle AGG' - \angle BAF &= 90^\circ - \angle BAF + \angle BG'G \\ &= 90^\circ - \angle BAF + \angle DAB \\ &= 90^\circ - \angle DAF.\end{aligned}$$

This implies $HH' \,//\, GG'$. As M is the midpoint of GG', N is the midpoint of HH', and furthermore it is shown that GH, $G'H'$ pass through P. We infer that MN passes through P, too. Therefore, BE, CF, and MN are concurrent. □

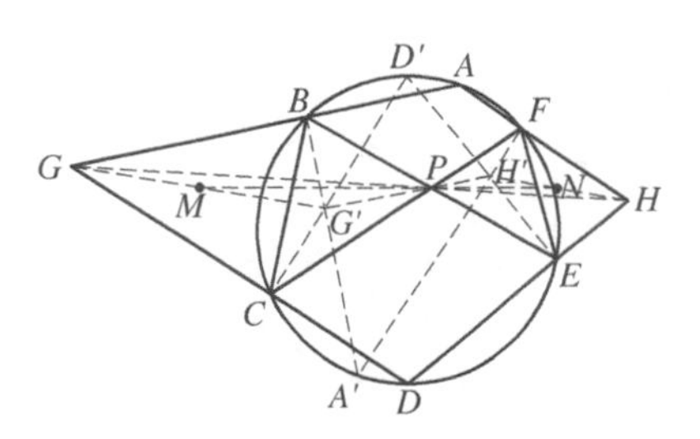

Fig. 1.3

Solution 3 As shown in Fig. 1.4, let P be the intersection of BE and CF. It suffices to show that M, P, N are collinear, or equivalently,

$$\frac{\sin\angle MPB}{\sin\angle MPC}=\frac{\sin\angle NPE}{\sin\angle NPF}.$$

Denote $\alpha=\angle MPB$, $\beta=\angle NPE$, $\gamma=\angle BPC=\angle FPE$. The above relation is equivalent to

$$\sin\alpha\sin(\gamma-\beta)=\sin\beta\sin(\gamma-\alpha),$$

or rewritten as $\sin\gamma\sin(\alpha-\beta)=0$, or just $\alpha=\beta$ (for the collinearity of M, P, N).

In $\triangle MPC$ and $\triangle MPB$, apply the law of sines to obtain

$$\frac{BM}{\sin\angle MPB}=\frac{PM}{\sin\angle PBM},\quad \frac{CM}{\sin\angle MPC}=\frac{PM}{\sin\angle PCM}.$$

As $BM=MC$, it follows that

$$\frac{\sin\angle MPB}{\sin\angle MPC}=\frac{\sin\angle PBM}{\sin\angle PCM}.$$

Similarly, in $\triangle NPF$ and $\triangle NPE$, we have

$$\frac{\sin\angle NPE}{\sin\angle NPF}=\frac{\sin\angle NEP}{\sin\angle NFP}.$$

Now it suffices to show

$$\frac{\sin\angle PBM}{\sin\angle PCM}=\frac{\sin\angle NEP}{\sin\angle NFP}. \qquad ①$$

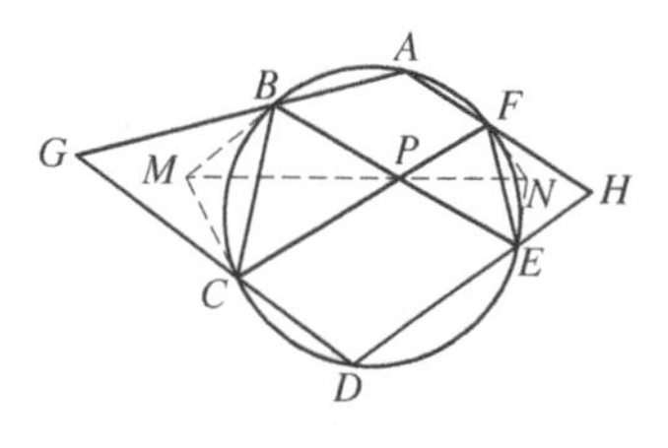

Fig. 1.4

The angles appearing in the numerators are supplementary, as

$$\begin{aligned}\angle PBM + \angle NEP &= \angle PBC + \angle MBC + \angle PEF + \angle NFE\\ &= (180^\circ - \angle CDE) + (90^\circ - \angle BGC)\\ &\quad + (180^\circ - \angle BAF) + (90^\circ - \angle AHD)\\ &= 540^\circ - (\angle CDE + \angle BGC + \angle BAF + \angle AHD)\\ &= 540^\circ - 360^\circ = 180^\circ,\end{aligned}$$

and thus $\sin\angle PBM = \sin\angle NEP$. Analogously, $\sin\angle PCM = \sin\angle NFP$. We have verified ①, and the statement. □

Solution 4 Assume that the circumcircle of $\triangle ABC$ is the unit circle in the complex plane. We use lowercase letters to represent the complex numbers corresponding to the points in the plane (for instance, a is the complex number for the point A, and so on). Let $BE \cap CF = X$. We have

$$\begin{aligned}x \in CF &\Leftrightarrow \frac{x-c}{x-f} = \overline{\left(\frac{x-c}{x-f}\right)}\\ &\Leftrightarrow x\bar{x} - \bar{f}x - c\bar{x} + c\bar{f} = x\bar{x} - \bar{c}x - f\bar{x} + \bar{c}f\\ &\Leftrightarrow (c-f)\,\bar{x} - (\bar{c} - \bar{f})\,x = c\bar{f} - \bar{c}f \Leftrightarrow \bar{x} + \bar{c}\bar{f}x = \bar{c} + \bar{f}.\end{aligned}$$

In the same manner, $\bar{x} + \bar{b}\bar{e}x = \bar{b} + \bar{e}$ Thus,

$$x = \frac{\bar{b} + \bar{e} - \bar{c} - \bar{f}}{\bar{b}\bar{e} - \bar{c}\bar{f}}. \quad ①$$

On the other hand, $\triangle BMC$ is an isosceles triangle, $\angle BMC$ is twice the angle between the lines CD and AB, and this implies that

$$\frac{c-m}{b-m} = \left(\frac{d-c}{a-b}\right)^2 \Big/ \left|\frac{d-c}{a-b}\right|^2 = \frac{d-c}{a-b} \Big/ \frac{\bar{d}-\bar{c}}{\bar{a}-\bar{b}} = \frac{cd}{ab}.$$

Consequently,

$$m = \frac{bc(a-d)}{ad-bc} = \frac{\bar{d} - \bar{a}}{\bar{c}\bar{d} - \bar{a}\bar{b}}. \quad ②$$

In the same way, we can derive

$$n = \frac{(a-d)ef}{af-bc} = \frac{\bar{d} - \bar{a}}{\bar{d}\bar{c} - \bar{a}\bar{f}}. \quad ③$$

To prove that BE, CF, and MN are concurrent, it is equivalent to prove that M, X, N are collinear, namely

$$\frac{x-n}{x-m}=\overline{\left(\frac{x-n}{x-m}\right)}\in\mathbb{R}. \qquad ④$$

From ①–③, it follows that

$$\begin{aligned} x-n &= \frac{1}{(\bar{b}\bar{e}-\bar{c}\bar{f})(\bar{d}\bar{e}-\bar{a}\bar{f})}[(\bar{b}+\bar{e}-\bar{c}-\bar{f})(\bar{d}\bar{e}-\bar{a}\bar{f})-(\bar{d}-\bar{a})(\bar{b}\bar{e}-\bar{c}\bar{f})] \\ &= \frac{1}{(\bar{b}\bar{e}-\bar{c}\bar{f})(\bar{d}\bar{e}-\bar{c}\bar{f})}[\bar{b}\bar{d}\bar{e}+\bar{d}\bar{e}^2-\bar{d}\bar{e}\bar{f}-\bar{c}\bar{d}\bar{e}-\bar{a}\bar{b}\bar{f}-\bar{a}\bar{e}\bar{f}+\bar{a}\bar{f}^2 \\ &\quad +\bar{a}\bar{c}\bar{f}-\bar{b}\bar{d}\bar{e}+\bar{a}\bar{b}\bar{e}+\bar{c}\bar{d}\bar{f}-\bar{a}\bar{c}\bar{f}] \\ &= \frac{1}{(\bar{b}\bar{e}-\bar{c}\bar{f})(\bar{d}\bar{e}-\bar{a}\bar{f})}(\bar{e}-\bar{f})(\bar{a}\bar{b}-\bar{a}\bar{f}-\bar{c}\bar{d}+\bar{d}\bar{e}), \end{aligned}$$

while

$$m-n=\frac{\bar{d}-\bar{a}}{(\bar{c}\bar{d}-\bar{a}\bar{b})(\bar{d}\bar{e}-\bar{a}\bar{f})}(\bar{a}\bar{b}-\bar{a}\bar{f}-\bar{c}\bar{d}+\bar{d}\bar{e}).$$

Hence,

$$\frac{x-n}{x-m}=\frac{(\bar{e}-\bar{f})(\bar{c}\bar{d}-\bar{a}\bar{b})}{(\bar{d}-\bar{a})(\bar{b}\bar{e}-\bar{c}\bar{f})}.$$

In the last fraction, the unit complex numbers a, b, c, d, e, and f all appear exactly once in the numerator and in the denominator; moreover, in each parenthesis, the coefficients of the homogeneous terms sum to zero (one minus one). So, the fraction is real, which justifies ④, and the argument as well.

2 Given a prime number p and an infinite set $A\subset\mathbb{Z}$. Show that one can always find a subset B of A, B contains $2p-2$ elements, and for any p distinct elements of B, their arithmetic mean does not belong to A.

(Contributed by Fu Yunhao)

Solution 1 Assume that, on the contrary, for some infinite set $A\subset\mathbb{Z}$, one cannot find a $(2p-2)$-element subset B satisfying the problem conditions. Without loss of generality, assume that A contains infinitely many positive integers; otherwise take $-A$ instead. Notice that the arithmetic mean of any p distinct elements of B must fall into the interval between the smallest and the largest elements of B. Thus, removing all elements smaller than some integer N from A, the set is still a counterexample to the problem. In particular, we take $N = 1$ and thereby assume all elements of A are positive. We may also add a constant integer to all elements of A if necessary.

Lemma *For any nonnegative integer k, there is at most one residue class modulo p^k that contains infinitely many elements of A.*

Proof of lemma Suppose the lemma is untrue and k is the smallest nonnegative integer such that two residue classes modulo p^k contain infinitely many elements of A, say with residues r_k and r'_k. Then there exists a positive integer N such that all elements of $A_{\geq N} := \{a \in A | a \geq N\}$ are congruent to r_{k-1} modulo p^{k-1}. Furthermore, in $A_{\geq N}$, those congruent to r_k and r'_k modulo p^k are both infinite. Take $p-1$ elements of $A_{\geq N}$ that are congruent to r_k, and the same number of elements congruent to $r'_k \pmod{p^k}$. Evidently, the arithmetic mean of any p elements of B is not congruent to r_{k-1} modulo p^{k-1}. This contradicts the assumption, and the lemma is verified.

Return to the original problem. For nonnegative integer k, assume the only residue class modulo p^k that contains infinitely many elements of A is $r_k \pmod{p^k}(0 \leq r_k \leq p^k - 1)$; then $r_k \equiv r_{k-1} \pmod{p^{k-1}}$. Now we use induction to select a sequence $b_1, b_2, \ldots,$ of elements of A, and verify that $2p-2$ of them constitute the desired subset B. Denote $N_0 = 0$. Let k_1 be the smallest nonnegative integer such that some element $a_1 \in A$ satisfies $a_1 \not\equiv r_{k_1} \pmod{p^{k_1}}$. Then all elements of A are congruent to r_{k_1-1} modulo p^{k_1-1}. According to the lemma, there exists a positive integer N_1 such that all elements of $A_{\geq N_1} = \{a \in A | a \geq N_1\}$ are congruent to r_{k_1} modulo p^{k_1}. Let k_2 be the smallest nonnegative integer such that some element $a_2 \in A$ satisfies $a_2 \not\equiv r_{k_2} \pmod{p^{k_2}}$. Analogously, all elements of $A_{\geq N_1}$ are congruent to r_{k_2-1} modulo p^{k_2-1}. By the lemma, we can further find N_2 such that all elements of $A_{\geq N_2} = \{a \in A | a \geq N_2\}$ are congruent to r_{k_2} modulo p^{k_2}, and so on. So, we obtain $a_1 < N_1 \leq a_2 < N_2 \leq \cdots \leq a_{2p-2}$ satisfying $p^{k_i-1} \parallel (a_i - r_{k_i})$ $(1 \leq i \leq 2p-2)$.

For convenience, subtract $r_{k_{2p-2}}$ from every element of A and from every N_i. Now $p^{k_i-1} a_i (1 \leq i \leq 2p-2)$, and every element of $A_{\geq N_i}$ is a multiple of $p^{k_{i+1}-1}$. Take the $(2p-2)$-element subset $B = \{b_1, b_2, \ldots, b_{2p-2}\}$. For a p-element subset $\{b_{i_1}, \ldots, b_{i_p}\}(i_1 < i_2 < \cdots < i_p)$ of B, the arithmetic mean is divisible by $p^{k_{i_1}-2}$ and greater than $b_{i_1} \geq N_{i_1-1}$. Hence it does not belong to A. This completes the proof. □

Solution 2 (by Feng Chenxu) We assert that for $1 \leq t \leq 2p-2$, one can always find a t-element subset B, such that for any p distinct elements of B, their arithmetic mean does not belong to A. Apply induction to t: the assertion is trivially true when $t \leq p-1$; assume the trueness when $|B| \leq t-1$ and consider t.

For $1 \leq i \leq p$, let $B_i \subseteq A$ be the residue class i modulo p, $A = \bigcup_{i=1}^{p} B_i$. If $A = B_j$ for some $1 \leq j \leq p$, then subtract j from every element of A and divide by p. This treatment will not affect the conclusion. Continue this process until at least two subsets B_j, B_k are nonempty.

If two subsets B_j and B_k are infinite, take p-1 elements from each subset to constitute B. Clearly, the arithmetic mean of any p distinct elements of B is non-integral, B satisfies the problem condition.

If only one subset, say B_j is infinite, let $C = A \backslash B_j \neq \varnothing$. Let D_+ be the set of all elements of B_j that are larger than $\max_{x \in C} x$; D_- be the set of all elements of B_j that are smaller than $\max_{x \in C} x$. Then $D_+ \cup D_-$ is infinite. Assume D_+ is infinite. By the induction hypothesis, one can find a satisfactory subset $B \subseteq D_+$, $|B| = t-1$. Since every number in D_+ is larger than every number in $A \backslash D_+$, we can infer that any p distinct numbers in B have an arithmetic mean larger than that in $A \backslash D_+$, and thus the mean is not in A. Take $x \in C$, and assume $x \in B_k$, $k \neq j$. For x and $p-1$ elements of B, their sum is $S \equiv (p-1)j + k \equiv k - j \not\equiv 0 \pmod{p}$. Thus, the mean is not an integer, nor in A. It follows that $B \cup \{x\}$ satisfies the problem condition. This completes the induction and the proof of the assertion. □

3 Given positive integers a, b, c, p, q, r, where p, q, $r \geq 2$. Let $Q = \{(x, y, z) \in \mathbb{Z}^3 | 0 \leq x \leq a,\ 0 \leq y \leq b,\ 0 \leq z \leq c\}$ be the game board.

Initially, put M game pieces on Q (no restriction on the number of pieces in each position). There are three types of legal moves:

(1) Remove p pieces from (x, y, z) and place one piece at $(x-1, y, z) \in Q$;

(2) Remove q pieces from (x, y, z) and place one piece at $(x, y-1, z) \in Q$;

(3) Remove r pieces from (x, y, z) and place one piece at $(x, y, z-1) \in Q$.

Find the minimum value of M, such that no matter how the M game pieces are placed initially, one can always take a sequence of moves to get one piece placed at $(0, 0, 0)$.

(Contributed by Ai Yinghua)

Solution 1 The minimum of M is $p^a q^b r^c$.

To prove $M = p^a q^b r^c$ is necessary, notice that if one initially places fewer than $p^a q^b r^c$ pieces at $(a,\ b,\ c)$, then the goal cannot be achieved. Indeed, let the weight of a game piece u at (x, y, z) be $w(u) = \frac{1}{p^x q^y r^z}$. According to the rule, before and after each move, the total weight of all game pieces $W = w(u)$ does not change. To have a game piece at (0,0,0), we must have $W \geq 1$; but $W < 1$ for $p^a q^b r^c$ pieces at (a, b, c), a contradiction.

It requires to prove $M = p^a q^b r^c$ is sufficient. We give a more general result on the n-dimensional game board: let $a_1, \ldots, a_n$, $p_1, \ldots, p_n$ be nonnegative integers, satisfying $2 \leq p_1 \leq \cdots \leq p_n$. Consider the game on

$$Q(a_1, \ldots, a_n) = \{(x_1, \ldots, x_n) \in \mathbb{Z}^n | 0 \leq x_i \leq a_i,\ i = 1, \ldots, n\}.$$

Initially, put M game pieces on $Q(a_1, \ldots, a_n)$. In each move, one can choose a position, remove p_i $(1 \leq i \leq n)$ pieces, and place one piece at the preceding position along the x_i direction. This is called an "X_i move". If one chooses a position and just remove q pieces, then it is called a "T_q move". For $n = 0$, $Q = \{0\}$, there is only one type of possible moves, which is the T_q move.

Divide the board into Zone-I as $Q(a_1, \ldots, a_n - 1)$ and Zone-II as $Q(a_1, \ldots, a_{n-1}) \times \{a_n\} = Q(a_1, \ldots, a_n) \backslash Q(a_1, \ldots, a_{n-1})$. Let M and K be respectively the number of pieces and the number of positions with at least one piece in $Q(a_1, \ldots, a_n)$. Let M_1, K_1 and M_2, K_2 be respectively the M, K values in Zone-I and Zone-II. In the following argument, the key idea is to take as many X_n moves as possible in Zone-II, so as to have more pieces in Zone-I. At the end of the process, there are at most $p_n - 1$ pieces at each of the K_2 positions (initially there are pieces on them) in Zone-II, and at least $\frac{M_2 - (p_n - 1)K_2}{p_n}$ pieces are added to Zone-I. We call this sequence of moves as "transmission from Zone-II to Zone-I". □

We shall justify the following three claims.

Claim $A(n)$ if $M = p_1^{a_1} \cdots p_n^{a_n}$, then a sequence of moves can make one piece at the origin.

Claim $B(n)$ let $u \geq 2$ be an integer, if $M + K > u \cdot p_1^{a_1} \ldots p_n^{a_n}$, then a sequence of $X_1, \ldots, X_n$ moves can make u pieces at the origin.

Claim $C(n)$ let $q \geq p_n$ and s be positive integers, t be a nonnegative integer, if $M + K > sq \cdot p_1^{a_1} \ldots p_n^{a_n} + t$, then one can first take $\left\lfloor \frac{t}{q} \right\rfloor$ of T_q moves, then take a sequence of $X_1, \ldots, X_n$ moves to put sq pieces at the origin.

Apply induction on n. Obviously, $A(0)$, $B(0)$, $C(0)$ hold. Suppose we have $A(n-1)$, $B(n-1)$, $C(n-1)$. Now induct on a_n: when $a_n = 0$, $Q(a_1, \ldots, a_n) = Q(a_1, \ldots, a_{n-1}) \times \{0\}$, the claims $A(n)$, $B(n)$, $C(n)$ on $Q(a_1, \ldots, a_n)$ correspond to $A(n-1)$, $B(n-1)$, $C(n-1)$ on $Q(a_1, \ldots, a_{n-1})$, respectively, and they are already justified by the induction hypothesis. In the following, consider the situations when $a_n \geq 1$.

Proof of $A(n)$

(a.1) If $M_1 \geq K_2$, one can transmit at least $\dfrac{M_2 - (p_n - 1)K_2}{p_n}$ pieces from Zone-II to Zone-I, after which the number of pieces in Zone-I is

$$\begin{aligned} M_1' &\geq M_1 + \frac{M_2 - (n-1)K_2}{p_n} \\ &= p_1^{a_1} \cdots p_{n-1}^{a_{n-1}} p_n^{a_n - 1} + \frac{(p_n - 1)(M_1 - K_2)}{p_n} \\ &\geq p_1^{a_1} \cdots p_{n-1}^{a_{n-1}} p_n^{a_n - 1}. \end{aligned}$$

In Zone-I, by the induction hypothesis, one can eventually put one piece at the origin.

(a.2) If $M_1 < K_2$, then $M_2 + K_2 > M_2 + M_1 = p_n^{a_n} \cdot p_1^{a_1} \cdots p_{n-1}^{a_{n-1}}$. Notice that $p_n^{a_n} \geq 2$. According to B$(n-1)$, in Zone-II one can transmit $p_n^{a_n}$ pieces to $(0, \ldots, 0, a_n)$, then just take X_n moves to put one piece at the origin.

Proof of $B(n)$

(b.1) If $M_1 + K_1 \geq \dfrac{p_n}{p_n - 1} K_2$, one can transmit at least $\dfrac{M_2 - (p_n - 1)K_2}{p_n}$ pieces from Zone-II to Zone-I, after which

$$\begin{aligned} M_1' + K_1' &\geq M_1 + K_1 + \frac{M_2 - (p_n - 1)K_2}{p_n} \\ &> M_1 + K_1 + \frac{u \cdot p_1^{a_1} \cdots p_n^{a_n} - M_1 - K_1 - p_n K_2}{p_n} \end{aligned}$$

$$= u \cdot p_1^{a_1} \dots p_{n-1}^{a_{n-1}} p_n^{a_n - 1} + \frac{(p_n - 1)(M_1 + K_1) - p_n K_2}{p_n}$$

$$\geq u \cdot p_1^{a_1} \dots p_{n-1}^{a_{n-1}} p_n^{a_n - 1}.$$

In Zone-I, by the induction hypothesis, one can eventually put u pieces at the origin.

(b.2) If $M_1 + K_1 < \frac{p_n}{p_n - 1} K_2$, by Bernoulli's inequality,

$$M_1 + K_1 < \frac{p_n}{p_n - 1} K_2 \leq 2(1 + a_1) \dots (1 + a_{n-1}) \leq p_n^{a_n} p_1^{a_1} \dots p_{n-1}^{a_{n-1}},$$

which implies that $p_1^{a_1} \dots p_n^{a_n} - M_1 - K_1$ is nonnegative. Since

$$M_2 + K_2 > u \cdot p_1^{a_1} \dots p_n^{a_n} - (M_1 + K_1)$$

$$= (u - 1) p_n^{a_n} \cdot p_1^{a_1} \dots p_{n-1}^{a_{n-1}} + (p_1^{a_1} \dots p_n^{a_n} - M_1 - K_1),$$

and $u - 1 \geq 1$, $p_n \geq p_{n-1}$, in Zone-II we may use the conclusion of C$(n-1)$ to take $\left\lfloor \frac{p_1^{a_1} \dots p_n^{a_n} - M_1 - K_1}{p_n} \right\rfloor$ of T_{p_n} moves, then transmit $(u - 1)p_n^{a_n}$ pieces to $(0, \dots, 0, a_n)$. The latter $(u - 1)p_n^{a_n}$ pieces at $(0, \dots, 0, a_n)$ can contribute $u - 1$ pieces at the origin by X_n moves. The former pieces removed by $\left\lfloor \frac{p_1^{a_1} \dots p_n^{a_n} - M_1 - K_1}{p_n} \right\rfloor$ of T_{p_n} moves can be transmitted by X_n moves into $\left\lfloor \frac{p_1^{a_1} \dots p_n^{a_n} - M_1 - K_1}{p_n} \right\rfloor$ pieces in Zone-I. After that, in Zone-I, there are

$$M_1' \geq M_1 + \left\lfloor \frac{p_1^{a_1} \cdots p_n^{a_n} - M_1 - K_1}{p_n} \right\rfloor$$

$$= p_1^{a_1} \dots p_{n-1}^{a_{n-1}} p_n^{a_n - 1} + \left\lfloor \frac{(p_n - 1) M_1 - K_1}{p_n} \right\rfloor$$

$$\geq p_1^{a_1} \dots p_{n-1}^{a_{n-1}} p_n^{a_n - 1}$$

pieces. Due to $A(n)$, eventually one piece can be placed at the origin. Altogether, u pieces at the origin.

Proof of $C(n)$

(c.1) If $t < q$, no T_q move is needed. Notice that $sq \geq 2$ and

$$M + K < sq \cdot p_1^{a_1} \dots p_n^{a_n} + t \geq sq \cdot p_1^{a_1} \dots p_n^{a_n}.$$

By the result of B(n), sq pieces can be transmitted to the origin.

(c.2) If $t \geq q$, then some T_q moves are needed. If there are more than q pieces at a place of Zone-II, take a T_q move, after which $M' = M - q$, $K' = K$, and the situation becomes $t' = t - q$. After a few T_q moves, it becomes either (c.1), $t < q$; or the number of pieces at every position of Zone-II is less than or equal to q. We only need to handle the latter case. Suppose there are L positions in Zone-II that contain exactly $q$pieces. By Bernoulli's inequality,

$$\begin{aligned}
M_1 + K_1 &= M + K - (M_2 + K_2) > sq \cdot p_1^{a_1} \dots p_n^{a_n} \\
&\quad + t - (q(K_2 - L) + (q+1)L) \\
&= sq \cdot p_1^{a_1} \dots p_{n-1}^{a_{n-1}} p_n^{a_n - 1} + t \\
&\quad + (sq \cdot p_1^{a_1} \dots p_{n-1}^{a_{n-1}} p_n^{a_n - 1}(p_n - 1) - qK_2 - L) \\
&\geq sq \cdot p_1^{a_1} \dots p_{n-1}^{a_{n-1}} p_n^{a_n - 1} + t \\
&\quad + (q(1 + a_1) \dots (1 + a_{n-1}) - qK_2 - L) \\
&\geq sq \cdot p_1^{a_1} \dots p_{n-1}^{a_{n-1}} p_n^{a_n - 1} + t - L.
\end{aligned}$$

As $q \geq p_n$, take X_n move once at each of these L positions with q pieces. After that,

$$M_1' + K_1' \geq M_1 + K_1 + L > sq \cdot p_1^{a_1} \dots p_{n-1}^{a_{n-1}} p_n^{a_n - 1} + t,$$

by the induction hypothesis applied to Zone-I, $\left\lfloor \frac{t}{q} \right\rfloor$ of T_q moves followed by a sequence of $X_1, \dots, X_n$ moves can send sq pieces to the origin.

This justifies $A(n)$, $B(n)$, $C(n)$, and also the problem statement.

Solution 2 (organized from Feng Chenxu's solution). The argument for $M \geq p^a q^b r^c$ is the same. We shall prove $M = p^a q^b r^c$ satisfies the problem condition.

Lemma *For a path $A_1, A_2, \dots, A_t$ in Q, A_i $(i = 2, \dots, t)$ has two coordinates equal to those of A_{i-1} and one coordinate equal to one plus that of*

A_{t-1} *Assign each A_i a number a_i such that $a_1 = 1$, $a_{i+1} = \dfrac{a_i}{x_i}$, where*

$$x_i = \begin{cases} p, & \text{if } x(A_{i+1}) = x(A_i) + 1, \\ q, & \text{if } y(A_{i+1}) = y(A_i) + 1, \\ r, & \text{if } z(A_{i+1}) = z(A_i) + 1. \end{cases}$$

If there are b_i pieces on A_i and $\sum_{i=1}^{t} b_i a_i \geq 1$, then a sequence of moves can send one piece to A_1.

Proof of lemma Every time, remove pieces on A_{i+1} and place a piece on A_i for some i, until no more move is possible. During the process, $\sum_{i=1}^{t} b_i a_i$ does not change. Suppose at the end there is no piece on A_1. There are at most $x_i - 1$ pieces on A_{i+1} or else another move can be made from A_{i+1} to A_i. It follows that $b_{i+1}a_{i+1} \leq (x_i - 1)a_{i+1} = a_i - a_{i+1}$, and

$$\sum_{i=1}^{t} b_i a_i \leq \sum_{i=1}^{t} a_{i-1} - a_i = a_1 - a_t < 1$$

is a contradiction. The lemma is verified.

For the original problem, let $f(x, y, z)$ be the number of pieces at (x, y, z) in the initial state.

First, consider $c = 0$. Apply induction on $a + b$. Assume $a \leq b$. When $a + b = 0$, the conclusion is obvious. For $a = 0$, $b > 0$, keep taking (2) moves (recall that (1)–(3) are the three types of legal moves described in the problem, in the x, y, and z directions, respectively) until there is no more move. By that time, at $(0, i, 0)$, $0 \leq i \leq b - 1$, there are at least

$$\left\lfloor \frac{f(0,b,0)}{q} \right\rfloor + M - f(0,b,0) = \frac{M}{q} + \left\lfloor \left(1 - \frac{1}{q}\right)(M - f(0,b,0)) \right\rfloor \geq \frac{M}{q}$$

pieces. The conclusion follows from the induction hypothesis. For $a = 1$, $b = 1$, assume $f(0,0,0) = 0$, and let $f(0,1,0) = a$, $f(1,0,0) = b$. Then

$$f(1,1,0) = pq - a - b \geq \min\{pq - pa, pq - qb\}.$$

Assume $pq - a - b \geq pq - pa$. Take (1) move $q - a$ times on (1, 1, 0), followed by (2) move once on (0, 1, 0). Now there is a piece at the origin.

For $a + b \geq 3$, $b \geq 2$. Assume there are t pieces in the region $0 \leq y \leq b - 1$. Take (2) moves at every position with $y = b$ until no more move is possible, at which time there are at most $q - 1$ pieces at every position. So, at least $\left\lfloor \frac{M-t-(a+1)(q-1)}{q} \right\rfloor$ moves are taken. (i) If $t \geq a+1$, then in the region $0 \leq y \leq b-1$ the number of pieces is at least $\left\lfloor \frac{M-t-(a+1)(q-1)}{q} \right\rfloor + t = \left\lfloor \frac{M+(t-a-1)(q-1)}{q} \right\rfloor \geq \frac{M}{q}$. By the induction hypothesis, another sequence of moves can send one piece to (0, 0, 0). (ii) If $t \leq a$, in the same manner, assume at most b pieces in the region $0 \leq x \leq a-1$. Then $f(a,b,0) \geq M-a-b$.

If no piece exists on the line $y=0$, then there are M pieces in the region $1 \leq y \leq b$. Using the induction hypothesis for q times, we infer that a sequence of moves can send q pieces to $(0,1,0)$. After that, one (2) move will send a piece to $(0,0,0)$.

If there are pieces on the line $y=0$, say at A. Let $B=(a,b,c)$. There is a path through the origin, A, and B; moreover, the number assigned to A is $\geq \frac{1}{p^a}$, and the number assigned to B is $\frac{1}{p^a q^b}$. We have

$$\frac{1}{p^a} + \frac{f(a,b,0)}{p^a q^b} \geq \frac{M}{p^a q^b} + \frac{1}{p^a} - \frac{a+b}{p^a q^b}.$$

Since $a \leq b$, $b \geq 2$, it follows that $a+b \leq q^b$, $\frac{M}{p^a q^b} \geq 1$, and thus $\frac{1}{p^a} + \frac{f(a,b,0)}{p^a q^b} \geq 1$. According to the lemma, a sequence of moves can send a piece to (0, 0, 0), which validates the conclusion for $c=0$. By symmetry, the conclusion holds if one of a, b, c is 0.

For a general $a \times b \times c$ board Q, let $a \leq b \leq c$. Apply induction on $a+b+c$, and assume $a \geq 1$. If $c \geq 2$, let $f(a,b,c) = M-s$, and suppose there are t pieces on the line $z=c$ but not at (a,b,c). Then there are $s-t$ pieces in the region $0 \leq z \leq c-1$. Take as many (3) moves as possible on (a,b,c), after which there are at least $\left\lfloor \frac{M-s}{r} + s - t \right\rfloor$ pieces in the region $0 \leq z \leq c-1$. If $\frac{M+(r-1)s-rt}{r} \geq \frac{M}{r}$, the conclusion is apparent from the induction hypothesis; in the following, assume it is not the case, and hence $t > \frac{r-1}{r}s$.

We can take as many (3) moves as possible on the line $z=c$, after which there at most $r-1$ pieces at each position. Then in the region $0 \leq z \leq c-1$, there are at least $\left\lceil \frac{M-t+s-(r-1)(a+1)(b+1)}{r} + s - t \right\rceil$ pieces:

if this number is $\geq \dfrac{M}{r}$, the conclusion is obvious; in the following, assume it is not the case, and thereby $(r-1)(s-t-(a+1)(b+1)) < -r$, or $s-t \leq (a+1)(b+1) - \dfrac{r}{r-1}$, which gives $s-t \leq ab+a+b-1$. This means in the region $0 \leq z \leq c-1$ there are at most $ab+a+b-1$ pieces. Similarly, we may also assume in the region $0 \leq x \leq a-1$ there are at most $bc+c+b-1$ pieces, and in the region $0 \leq y \leq a-1$ there are at most $ac+c+a-1$ pieces. The assumptions lead to $f(a,b,c) \geq M - (ab+bc+ca+2a+2b+2c-3)$.

If $f(a,b,c) \leq ap^a q^c$, then $p^a q^b(r^c - a) \leq ab+bc+ca+2a+2b+2c-3$, yet in fact $p^a q^b (r^c - a) \geq 2^{a+b}(2^c - a)$. For $c \geq 2$ and $a+b+c \geq 4$, we have

$$\begin{aligned} 2^{a+b+c} - 2^{a+b}a \geq 2^{a+b+c-1} &\geq \frac{(a+b+c)^2}{2} \\ &\geq ab+bc+ca+2a+2b+2c-3. \end{aligned}$$

In other cases, the inequalities do not hold. So, for $c \geq 2$, $f(a,b,c) \geq ap^a q^b$.

Draw $(a+1)$ paths from $(0,0,c)$ to (a,b,c), the ith of which traverses all positions on the line $x = i-1$. Send pieces from (i,y,c) to the $(i+1)$th path, $((x,y) \neq (a,b))$. As the assigned number at (a,b,c) along every path is $\dfrac{1}{p^a q^b}$, and $f(a,b,c) > ap^a q^b$, we may take moves to distribute the pieces on (a,b,c) such that along each of the a paths (excluding the path along $x = a$), the total weight is an integer. By the lemma, a sequence of moves can result in $\left\lfloor \sum_{0\leq x\leq a, 0\leq y\leq b} \dfrac{f(x,y,c)}{p^x q^y} \right\rfloor$ pieces on $(0,0,c)$. Since

$$\sum_{0\leq x\leq a,\ 0\leq y\leq b} \frac{f(x,y,c)}{p^x q^y} \geq \frac{M-s+2t}{p^a q^b}$$

and $t \geq \dfrac{r-1}{r} s \geq \dfrac{s}{2}$, the above quantity is $\geq \dfrac{M}{p^a q^b} = r^c$. Take as many (3) moves as possible towards the pieces on the plane $x = y = 0$. By induction, eventually one piece is placed at $(0,0,0)$.

Finally, consider $a = b = c = 1$, and assume $f(0,0,1) \geq 1$. If $f(0,1,1) + f(1,0,1) + f(1,1,1) \geq pq(r - f(0,0,1))$, focusing on the plane $z = 1$, we may send $r - f(0,0,1)$ pieces to $(0,0,1)$, then take a (3) move to put one piece at (0, 0, 0). If the inequality does not hold, then there are at least $(pq-1)f(0,0,1) + 1 \geq pq$ pieces on the plane $z = 0$. Again, just like $c = 0$

we see that one piece can be sent to (0, 0, 0). For the same reason, the conclusion is true when $f(0,1,0) \geq 1$, or $f(1,0,0) \geq 1$.

In the following, assume $f(1,0,0) = f(0,1,0) = f(0,0,1) = 0$. Let

$$f(0,1,1) = a', \quad f(1,0,1) = b', \quad f(1,1,0) = c'.$$

Repeatedly take (3) moves on the plane $z = 1$ until no more move is possible, by then there are at least $\left\lceil \dfrac{M - c' - 3(r-1)}{r} \right\rceil + c'$ pieces on the plane $z = 0$. If this number is $\geq \dfrac{M}{r}$, the conclusion follows from the induction hypothesis for $c = 0$; otherwise, $\left\lceil \dfrac{M - c' - 3(r-1)}{r} \right\rceil + c' \leq \dfrac{M}{r} - 1$, yielding

$$(r-1)(c'-3) + r \leq 0,$$

and thus $c' \leq 1$. Similarly, assume a', $b' \leq 1$. If $c' = 0$, the plane $z = 1$ includes all the pieces. Like $c = 0$, one can make r pieces at $(0,0,1)$ and take a (3) move to get one piece at $(0,0,0)$. The same conclusion if $a' = 0$ or $b' = 0$. For $a' = b' = c' = 1$, on $(1,1,1)$, take $q-1$ of (1) moves and $(r-1)p - 1$ of (2) moves, reaching

$$f(0,1,1) = q, \quad f(1,0,1) = (r-1)p, \quad f(1,1,1) = p + q - 3 \geq 0.$$

Take a (2) move on $(0,1,1)$, followed by $(r-1)$ of (1) moves on $(1,0,1)$, then a (3) move on $(0,0,1)$. Now, a piece is placed at $(0,0,0)$. This justifies that $M = p^a q^b r^c$ is sufficient. □

Test I, Second Day
(8 am – 12:30 pm; March 24, 2022)

4 As shown in Fig. 4.1, in the acute $\triangle ABC$, $\angle ACB > 2\angle ABC$. Let I be the incentre of $\triangle ABC$, K be the symmetric point of I about BC. The extensions of BA and KC meet at D. The line through B and parallel to CI crosses the minor arc $\overset{\frown}{BC}$ of the circumcircle of $\triangle ABC$ at $E(E \neq B)$. The line through A and parallel to BC meets the line BE at F. Prove: if $BF = CE$, then $AD = FK$.

(Contributed by He Yijie)

Solution 1 Extend CI to meet the circumcircle of $\triangle ABC$ say at T. Clearly, T is the midpoint of $\overset{\frown}{AB}$, $TI = TA = TB$. As $BE // CT$, we have

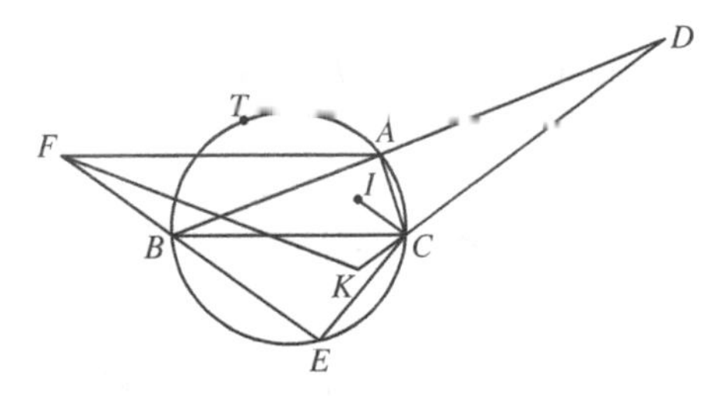

Fig. 4.1

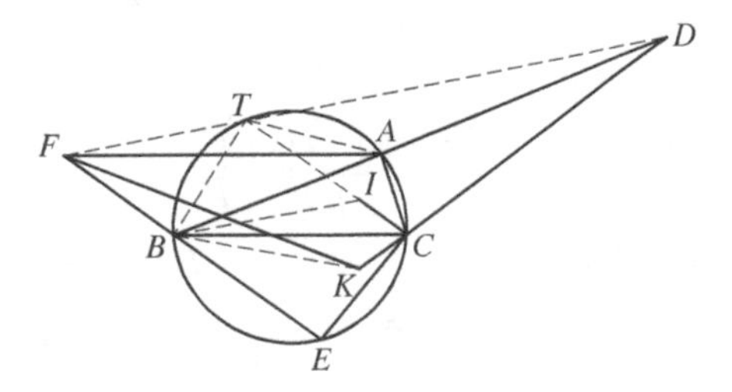

Fig. 4.2

$BT = CE$, and $BF = CE = BT = AT = TI$. Hence $BFTI$ is a parallelogram, $BI // FT$, $BI = FT$. As $BI = BK$, thus $FT = BK$. Moreover,

$$\angle FBK = \angle FBI + \angle IBK = \angle FTI + \angle ATI = \angle ATF,$$

hence $\triangle FBK \cong \triangle ATF$, $AF = FK$.

Denote the interior angles of $\triangle ABC$ as $A = 2\alpha$, $B = 2\beta$, $C = 2\gamma$. Since $BI // FT$, $BC // FA$, thus $\beta = \angle IBC = \angle TFA = \angle BKF$. Also, $\angle EBK = \angle EBC - \angle CBK = \angle BCI - \beta = \gamma - \beta$, yielding $\angle BFK = \angle KBE - \angle BKF = \gamma - 2\beta$.

Notice that $\angle BDK = \angle BCK - \angle CBA = \gamma - 2\beta = \angle BFK$. Therefore, B, F, D, K are concyclic, implying that $\angle ADF = \angle BDF = \angle BKF = \beta = \frac{1}{2}\angle ABC = \frac{1}{2}\angle BAF$, and thus $AF = AD$.

Altogether, we obtain $FK = AF = AD$. Done. □

Solution 2 Let the circumcircle of $\triangle ABC$ be the unit circle in the complex plane, $C = 1$, $A = x^2$, and $B = x^2y^2$. Then the complex number for I is

$$x + x^2y + x^2y^2z = x + (xyz)xy + x^2y = x - xy + x^2y = x(1 - y + xy).$$

As K and I are symmetric about BC, we have $\dfrac{\overline{I-C}}{B-C} = \dfrac{K-C'}{B-C}$, or equivalently

$$\begin{aligned} K &= C + \frac{B-C}{\bar{B}-\bar{C}}(\bar{I}-\bar{C}) = 1 - BC(\bar{I}-\bar{C}) \\ &= 1 - x^2y^2\left[\frac{1}{x}\left(1-\frac{1}{y}+\frac{1}{xy}\right)-1\right], \end{aligned}$$

that is, $K = x^2y^2 - xy^2 + xy - y + 1$. Furthermore, $BE \parallel CI$ implies $E = \dfrac{1}{y}(B \neq E \Rightarrow x^2y^3 = 1)$, $CE = \left|1 - \dfrac{1}{y}\right| = |y-1|$. As F lies on the line BE, we find

$$\begin{aligned} &\frac{B-F}{E-F} = \frac{\overline{B}-\overline{F}}{\overline{E}-\overline{F}} \\ \Leftrightarrow\ & B\bar{E} - \bar{E}F - B\bar{F} + F\bar{F} = \bar{B}E - \bar{B}F - E\bar{F} + F\bar{F} \\ \Leftrightarrow\ & (B-E)\bar{F} = (\bar{B}-\bar{E})F + B\bar{E} - \bar{B}E \\ \Leftrightarrow\ & \bar{F} + \frac{1}{BEF} = \frac{B+E}{BE} \\ \Leftrightarrow\ & F + x^2y\bar{F} = x^2y^2 + \frac{1}{y}. \end{aligned}$$

Hence, $yF + x^2y^2\bar{F} = x^2y^3 + 1$. Meanwhile,

$$\begin{aligned} AF // BC \Leftrightarrow R \ni \frac{A-F}{B-C} = \frac{\bar{A}-\bar{F}}{\bar{B}-\bar{C}} \\ \Leftrightarrow A - F = -BC(\overline{A}-\overline{F}) \Leftrightarrow F + BC\overline{F} = A + \frac{BC}{A}. \end{aligned}$$

Hence, $F + x^2y^2\overline{F} = x^2 + y^2$. Combine the two equations, obtaining

$$F = \frac{1}{y-1}[x^2(y^3-1) - (y^3-1)] = x^2(y^2+y+1) - (y+1).$$

This leads to $BF = |x^2(y+1) - (y+1)| = |(x^2-1)(y+1)|$, and

$$FK = |x^2y^2 + x^2y + x^2 - y - 1 - x^2y^2 - xy + y - 1| = |x^2(y+1) + xy^2 - xy - 2|.$$

Let $D = BA \cap KC$. Then D is on the line $BA \Leftrightarrow D + AB\bar{D} = A + B$, namely

$$D + x^4y^2\bar{D} = x^2y^2 + x^2.$$

Also, D is on the line KC, the extension of CK meets the unit circle at $x^2y^3 \Leftrightarrow D$ is on the line through C and x^2y^3, namely

$$D + 1 \cdot x^2y^3\bar{D} = x^2y^3 + 1.$$

Combine the above two equations, obtaining $x^2y^2(x^2 - y)\bar{D} = x^2y^2 + x^2 - x^2y^3 - 1$, or

$$\bar{D} = \frac{x^2y^2 + x^2 - x^2y^3 - 1}{x^2y^2(x^2 - y)}, \quad D = \frac{x^2y + x^2y^3 - x^2 - x^4y^3}{y - x^2},$$

$$DA = |\bar{D} - \bar{A}| = \left|\frac{x^2y^2 + x^2 - x^2y^3 - 1 - x^2y^2 + y^3}{x^2y^2(x^2 - y)}\right| = \left|\frac{(x^2 - 1)(1 - y^3)}{x^2 - y}\right|.$$

Essentially, it is the following question.

Question. Two unit complex numbers x and y with $\arg x$, $\arg y \in \left(0, \frac{\pi}{2}\right)$, $\arg y > 2\arg x$ ($\triangle ABC$ is acute, $\angle ACB > 2\angle ABC$), satisfy

$$|y - 1| = |(x^2 - 1)(y + 1)|. \quad ①$$

Prove

$$|x^2(y + 1) + (y^2 - y)x - 2| = \left|\frac{(x^2 - 1)(1 - y^3)}{x^2 - y}\right|. \quad ②$$

□

Proof. Notice that

$$① \Leftrightarrow \left|\frac{(x^2 - 1)(y + 1)}{y - 1}\right| = 1 \Leftrightarrow \frac{(\bar{x}^2 - 1)(\bar{y} + 1)}{\bar{y} - 1} = \frac{y - 1}{(x^2 - 1)(y + 1)}$$

$$\Leftrightarrow \frac{\frac{1 - x^2}{x^2}\frac{1 + y}{y}}{\frac{1 - y}{y}} = \frac{y - 1}{(x^2 - 1)(y + 1)}$$

$$\Leftrightarrow \left(\frac{x^2 - 1}{x}\right)^2 = \left(\frac{y - 1}{y + 1}\right)^2,$$

that is, $\frac{x^2 - 1}{x} = \pm\frac{y - 1}{y + 1}$. The left-hand side equals pi for some $p > 0$; the right-hand side satisfies $\operatorname{Im}\frac{y - 1}{y + 1} > 0$ as $\arg y \in \left(0, \frac{\pi}{2}\right)$. Hence,

$\dfrac{x^2-1}{x}=\dfrac{y-1}{y+1}$, which is $(y+1)x^2=(y-1)x+(y+1)$, or simply

$$x^2-1=\left(\frac{y-1}{y+1}\right)x.$$

On the other hand, in ②,

$$\text{left}=|(y^2-1)x+(y-1)|=|y-1|\cdot|(y+1)x+1|;$$

$$\text{right}=\left|\frac{\left(\frac{y-1}{y+1}\right)x(1-y^3)}{\left(\frac{y-1}{y+1}x+1-y\right)}\right|=\left|\frac{(y-1)x(1-y^3)}{(y-1)(x-y-1)}\right|$$

$$=\left|\frac{(1-y)(1+y+y^2)}{x-(y+1)}\right|.$$

Now, ② is equivalent to $|[(y+1)x+1][x-(y+1)]|=|1+y+y^2|$. Expand all terms and use $(y+1)x^2=(y-1)x+(y+1)$, obtaining $|-(y^2+y+1)x|=|1+y+y^2|$, which is true as $|x|=1$. The proof is complete. □

5 Let $C=\{z\in\mathbb{C}||z|=1\}$ be the unit circle in the complex plane. 240 complex numbers $z_1, z_2, \ldots, z_{240}\in C$ (can be repeated) satisfy the following conditions:

(1) for any open arc Γ of length π on C, there are at most 200 j's $(1\le j\le 240)$ such that $z_j\in\Gamma$;

(2) for any open arc γ of length $\dfrac{\pi}{3}$ on C, there are at most 120 j's $(1\le j\le 240)$ such that $z_j\in\gamma$.

Find the maximum of $|z_1+z_2+\cdots+z_{240}|$.

(Contributed by Qu Zhenhua)

Solution The maximum is $80+40\sqrt{3}$. Take 80 of 1's, 40 of each of $\exp\left(\dfrac{\pi}{6}\mathrm{i}\right)$, $\exp\left(-\dfrac{\pi}{6}\mathrm{i}\right)$, i, −i. It is straightforward to check that they satisfy (1), (2), and their sum equals $80+40\sqrt{3}$. We need to show this is the maximum value.

Let $z_1, z_2, \ldots, z_{240}$ satisfy (1), (2). By rotation if necessary, we may assume that $S=z_1+z_2+\cdots+z_{240}$ is a nonnegative real number; starting from -1 (included) along C, in the counterclockwise order, let the numbers be $z_1, z_2, \ldots, z_{240}$.

Condition (1) indicates that for $1 \le j \le 40$, to go from z_j to z_{j+200} along C, at least an arc of length π is needed, that is to say, there exists $2\pi \ge \alpha \ge \pi$ such that $z_{j+200} = z_j \cdot \exp(\alpha i)$. Let $z_j = (-1) \cdot \exp(\beta i)$, $\beta \in [0, 2\pi)$. From $\arg z_{j+200}$ we see $\beta + \alpha < 2\pi$, and thus

$$\operatorname{Re}(z_j + z_{j+200}) = -\cos\beta - \cos(\beta + \alpha) = -2\cos\frac{\alpha}{2}\cos\left(\beta + \frac{\alpha}{2}\right) \le 0$$

$\left(\text{because } \frac{\alpha}{2} \in \left[\frac{\pi}{2}, \pi\right], \beta + \frac{\alpha}{2} \in \left[\frac{\pi}{2}, \frac{3\pi}{2}\right]\right)$. Summing up $1 \le j \le 40$, we obtain

$$\operatorname{Re}(z_1 + z_2 + \cdots + z_{40} + z_{201} + z_{202} + \cdots + z_{240}) \le 0. \quad ①$$

Similarly, condition (2) indicates that for $41 \le j \le 80$, to go from z_j to z_{j+120} along C, at least an arc of length $\frac{\pi}{3}$ is needed. So there exists $2\pi \ge \alpha \ge \frac{\pi}{3}$, such that $z_{j+120} = z_j \cdot \exp(\alpha i)$. Let $z_j = (-1) \cdot \exp(\beta i)$, $\beta \in [0, 2\pi)$. From $\arg z_{j+120}$ we see $\beta + \alpha < 2\pi$, and thus

$$\operatorname{Re}(z_j + z_{j+120}) = -\cos\beta - cos(\beta + \alpha) = -2\cos\frac{\alpha}{2}\cos\left(\beta + \frac{\alpha}{2}\right).$$

If $\alpha \ge \pi$, the above quantity is non-positive; if $\alpha \in \left[\frac{\pi}{3}, \pi\right)$, then

$$\left|2\cos\frac{\alpha}{2}\cos\left(\beta + \frac{\alpha}{2}\right)\right| \le 2 \times \cos\frac{\pi}{6} \times 1 = \sqrt{3},$$

so the real part is $\le \sqrt{3}$. Summing up for $41 \le j \le 80$, we obtain

$$\operatorname{Re}(z_{41} + z_{42} + \cdots + z_{80} + z_{161} + z_{162} + \cdots + z_{200}) \le 40\sqrt{3}. \quad ②$$

The inequalities ① and ② lead to an estimate of the sum

$$\begin{aligned}
&|z_1 + z_2 + \cdots + z_{240}| \\
&= \operatorname{Re}\left(\sum_{j=1}^{40}(z_j + z_{200+j})\right) + \operatorname{Re}\left(\sum_{j=41}^{80}(z_j + z_{120+j})\right) + \operatorname{Re}\left(\sum_{j=81}^{120} z_j\right) \\
&\le 0 + 40\sqrt{3} + 80 = 80 + 40\sqrt{3}.
\end{aligned}$$

Here, the maximum is attained only when the numbers are: 80 of 1's, 40 of each of $\exp\left(\frac{\pi}{6}\mathrm{i}\right)$, $\exp\left(-\frac{\pi}{6}\mathrm{i}\right)$, i, $-i$ (or by any rotation). □

6 Let m be a positive integer and A be a finite set. Let $A_1, A_2, \ldots, A_m$ be subsets of A (not necessarily distinct). It is known that for any nonempty set $I \subseteq \{1, 2, \ldots, m\}$,

$$\left|\bigcup_{i \in I} A_i\right| \geq |I| + 1.$$

Prove: the elements of A can be coloured black or white, such that every of $A_1, A_2, \ldots, A_m$ contains both black and white elements.

(Contributed by Fu Yunhao)

We give three solutions as follows.

Solution 1 Construct a bipartite graph G whose two parts are $X = \{A_1, A_2, \ldots, A_m\}$ and $Y = A$: for $1 \leq i \leq m$ and $a \in A$, $A_i \in X$ and $a \in Y$ are adjacent if and only if $a \in A_i$. Since for each nonempty $I \subseteq \{1, 2, \ldots, m\}$, $\left|\bigcup_{i \in I} A_i\right| \geq |I| + 1$, it follows that for any k vertices of X, the number of vertices of Y that are neighbouring to one or more of them, is at least $k + 1$. According to Hall's theorem, there exists a transversal f from X to Y. For $1 \leq i \leq m$, let $a_i = f(A_i)$. Clearly, $a_i \in A_i$, and $a_1, a_2, \ldots, a_m$ are distinct elements of A. Colour all the other elements $A \backslash \{a_1, \ldots, a_m\}$ white, and determine the colours of $a_1, a_2, \ldots, a_m$ as follows.

Every time, choose an i such that a_i is uncoloured and A_i has a coloured neighbour, say b. Colour a_i the opposite colour to b. Suppose, during the process, some elements say $a_1, a_2, \ldots, a_k$ $(1 \leq k \leq m)$ are uncoloured, but we cannot find another i and colour a_i. Based on the algorithm, this means that all the neighbours of $A_1, A_2, \ldots, A_k$ are in $\{a_1, \ldots, a_k\}$, yet it contradicts $|A_1 \cup A_2 \cup \ldots \cup A_k| \geq k + 1$. Hence, this process can continue until all of $a_1, a_2, \ldots, a_m$ are coloured. Moreover, for $1 \leq i \leq m$, when a_i is coloured, A_i is guaranteed to have black and white neighbours, that is, A_i contains both black and white elements. This colouring satisfies the problem requirements. □

Solution 2 We begin with a lemma.

Lemma *The finite sets $A_1, A_2, \ldots, A_m$ are called "nice", if the union of any $k \leq m$ (k is arbitrary) of them contains at least $k + 1$ elements. If $A_1, A_2, \ldots, A_m$ are nice, then there exist 2-element sets $B_1, B_2, \ldots, B_m$, such that $B_i \subseteq A_i$, $i = 1, 2, \ldots, m$, and $B_1, B_2, \ldots, B_m$ are nice.*

Proof of lemma Assume the lemma is untrue. Let $A_1, A_2, \ldots, A_m$ be a counterexample with the smallest m value and the smallest $|A_1| + |A_2| + \cdots + |A_m|$ for such m. There are three situations.

Case 1, if the union of any $k \le m$ (k is arbitrary) sets does not contain exactly $k+1$ elements. Take $k = 1$, and apparently every set contains 3 or more elements. Choose an arbitrary set and remove any element from it. The sets are still nice (as a counterexample), yet $|A_1| + |A_2| + \cdots + |A_m|$ is smaller. A contradiction. Case 2, if the union of some $k \le m-1$ sets contains exactly $k+1$ elements. Assume $|A_1 \cup A_2 \cup \cdots \cup A_k| = k+1$. Since $k < m$, there existc 2-element sets $B_1, B_2, \ldots, B_k$ such that $B_i \subseteq A_i$, $i = 1, 2, \ldots, k$, and $B_1, B_2, \ldots, B_k$ are nice.

Let $C_i := A_i \backslash (A_1 \cup A_2 \cup \cdots \cup A_k)$, $i = k+1, \ldots, m$. It is easy to see that, among $C_{k+1}, C_{k+2}, \ldots, C_m$, the union of any d ($1 \le d \le m-k$, d is arbitrary) sets contains d or more elements. By Hall's theorem, there exist $x_i \in C_i$ ($i = k+1, \ldots, m$), such that $x_{k+1}, x_{k+2}, \ldots, x_m$ are all distinct. We construct $B_{k+1}, B_{k+2}, \ldots, B_m$ in the following way: if for some $i \in \{k+1, k+2, \ldots, m\}$, B_i has not been made yet, but A_i contains an element y_i that belongs to some B set already constructed, then let $B_i = \{x_i, y_i\}$. Clearly, x_i is a new element in the constructed B sets, hence these B sets are still nice. Suppose, after the construction of several B_i's, no more set can be constructed in the above way. Then the remaining A sets do not contain any element in the constructed B_i's. Since m is minimal, we may construct B_j's from those A sets such that they are nice. Furthermore, B_j's and B_i's are disjoint, when combined, $B_1, B_2, \ldots, B_m$ are nice. So, case 2 is not possible.

Case 3, if the union of any $k (1 \le k \le m-1$, k is arbitrary) sets contains $k+2$ or more elements, while the union of all m sets contains exactly $m+1$ elements. Observe that any m-1 sets have the same union as that of all m sets. Hence, every element must belong to at least 2 sets. Meanwhile, each set contains at least 3 elements. Choose any set and remove any element from it. The sets are still nice, but they have a smaller $|A_1| + |A_2| + \cdots + |A_m|$, contradiction.

This verifies the lemma.

Return to the original problem. According to the lemma, we may find 2-element subsets $B_1, B_2, \ldots, B_m$ of $A_1, A_2, \ldots, A_m$, respectively, such that the union of any k of them has $k+1$ or more elements. Treat every element of A as a vertex, and then for every $1 \le i \le m$, connect the two elements in B_i by an edge. Note that this graph has no cycle (otherwise, the edges in a cycle correspond to B sets the number of which is the same as the number of elements in their union), and thus it is a forest, which is a bipartite graph. Colour all elements in one part black, and all elements in the other part white. This colouring satisfies the problem condition. □

Solution 3 Use induction on the number of sets m. When $m = 1$, $|A_1| \geq 2$, choose two elements of A_1 and colour one black and one white. Suppose the conclusion holds for all $m \leq n-1$. Consider $m = n$.

Case 1. If for any $I \subseteq \{1, \ldots, n\}$, $|\cup_{i \in I} A_i| \geq |I| + 2$. Take $a \in A_n$. By the induction hypothesis, there is a colouring of $A_1 \backslash \{a\}, A_2 \backslash \{a\}, \ldots, A_{n-1} \backslash \{a\}$ for which each $A_i \backslash \{a\}$ contains black and white elements. As $A_n \backslash \{a\}$ has been coloured, we can colour a such that A_n has elements of both colours.

Case 2. If for some $I \neq \{1, 2, \ldots, n\}$, $|\cup_{i \in I} A_i| = |I| + 1$. Let I be the largest, $J = \{1, 2, \ldots, n\} \backslash I$, and $B = \cup_{i \in I} A_i$, $C = \cup_{j \in J} A_j$.

Case 2a. If $B \cap C = \varnothing$, then colour B and C separately. By induction, there is a colouring that guarantees each A_i contains black and white elements.

Case 2b. If $B \cap C \neq \varnothing$, take $b \in B \cap C$, and define $A'_j = A_j \backslash (B \backslash \{b\})$ $(j \in J)$. For any $T \subseteq J$, since I is the largest, it follows that

$$\left|\bigcup_{t \in T} A'_t\right| \geq \left|\bigcup_{t \in I \cup T} A_t\right| - \left|\bigcup_{i \in I} A_i\right| \geq (|I \cup T| + 2) - (|I| + 1) = |T| + 1.$$

Similarly,

$$\left|\bigcup_{j \in J} A'_j\right| = \left|\bigcup_{i=1} A_i\right| - |B \setminus \{b\}| \geq n + 1 - |I| = |J| + 1.$$

By induction, we may colour $C \backslash (B \backslash \{b\}) = \cup_{j \in J} A'_j$ and $B = \cup_{i \in I} A_i$ such that $A'_j (j \in J)$ and $A_i (i \in I)$ contain elements of both colours. If b has different colours in the two colourings, then reverse the colouring of B. Now they are compatible and give a colouring of A which meets the problem requirements. □

Test II, First Day
(8 am – 2:30 pm; March 28, 2022)

1 In an $m \times n$ grid (with $m + 1$ horizontal lines and $n + 1$ vertical lines), it is possible to add diagonals to some unit squares, that is, each unit square □ becomes one of □, ⧄, or ⧅, such that the grid becomes an Eulerian cycle (there exists a path that visits every edge exactly once, and that starts and ends on the same vertex). Find all such $m \times n$ grids.

(Contributed by Xiao Liang)

Solution An $m \times n$ grid can be made into an Eulerian cycle if and only if $m = n$. We give two solutions as follows. □

Solution 1 If $m = n$, add a diagonal (top left to bottom right) to every unit square not in the main diagonal of the grid. It is easy to verify the new graph is an Eulerian cycle.

It requires to prove when $m \neq n$, the grid cannot be made into an Eulerian cycle. Let $n > m$ and suppose it is possible. Notice in the original grid, every lattice point has an even degree of edges except for those on the four sides but not at the corners. Let us call them even and odd points. There are $m - 1$ or $n - 1$ odd points on each side of the grid. To make the grid into an Eulerian cycle, the diagonals must turn each odd point into an even point while maintain all the even points. We focus on all the diagonals. If three or four diagonals meet at a lattice point, separate them like in the figure.

This divides the diagonals into several cycles and non-intersecting paths. Remove all the cycles. The remaining paths of diagonals turn all the odd points into even points, and hence a path must start at an odd point and end at another odd point. Colour all the lattice points alternately black and white. Evidently, a path can only pass through lattice points of the same colour. If a path starts and ends on the same side of the grid, then there are an odd number of odd points on this side between the two ends, and they cannot be connected by paths. If a path starts and ends on opposite sides of the grid, say from row a and column 1, to row b and column $n+1$. Then $(a, 1)$ and $(b, n+1)$ are of the same colour, $a+1$ and $b+n+1$ are of the same parity, which indicates that there are $(a-2)+(n-1)+(b-2) = a+n+b-5$, an odd number of odd points between the two ends, which cannot be connected by paths. The above argument shows that a path must start and end at neighbouring sides, which requires the $2(m-1)$ odd points on two opposite sides match the $2(n-1)$ odd points on the other two sides. Hence, $2(m-1) = 2(n-1)$, or $m = n$. This completes the proof. □

Solution 2 (organized from Liu Yexu's proof). The solution for $m = n$ is the same as in the first proof. Now assume that the grid can be made into an Eulerian cycle. It requires to show $m = n$.

We claim that the Eulerian cycle can be decomposed into figures of five types ◺, ◺, ◿, ◿ and □, any two figures sharing no common sides. Once proved, the numbers of unit length horizontal and vertical segments must agree, which gives $m(n+1) = n(m+1)$ and $m = n$.

To show such decomposition is possible, we find and remove one figure at a time from the graph in the following way. Before and after each removal, every lattice point has an even degree of edges. Every time, take the leftmost column and pick the top lattice point A. Then, no upward, upper left, leftward, lower left edges can emanate from A. Since A has degree 2 or 4, these are the possible configurations at A:

For the first one, remove the two triangles including the dash lines; for the second, third and fourth, there is only one choice to remove a triangle including the dash line as follows.

For the fifth configuration, if there is a diagonal from top right to bottom left in the unit square, remove the triangle shown in the following; otherwise, remove the square.

Based on the above algorithm, the last two configurations never occur. Eventually, the graph can be thoroughly decomposed, and we deduce that $m = n$. □

2 In the oblique $\triangle ABC$, $BC > AC > AB$. Let P_1, P_2 be two points in the plane satisfying: for $i = 1, 2$, the lines AP_i, BP_i, and CP_i intersect with the circumcircle of $\triangle ABC$ at D_i, E_i, and F_i, respectively (other than A, B, C), with $D_iE_i \perp D_iF_i$ and $D_iE_i = D_iF_i \neq 0$. It is known that the line P_1P_2 crosses the circumcircle of $\triangle ABC$ at two points Q_1 and Q_2. For $i = 1, 2$, let the projections of Q_i on the lines AB, AC be X_i and Y_i, respectively. The lines X_1Y_1 and X_2Y_2 meet at W. Prove: W lies on the nine-point circle of $\triangle ABC$.

(Contributed by Yao Yijun)

Solution There are two major steps in the proof. First, we need to identify the positional relationship between P_1 and P_2. This can be done by manipulation of angles; however, it relies heavily on the positions and shapes, which require a lot of case-by-case discussions. To prevent this, we have to use directed angles or avoid any addition or subtraction of angles. In fact, it is always true that $\triangle P_iAB \sim \triangle P_iE_iD_i$, and thus

$$\frac{P_iA}{P_iE_i} = \frac{P_iB}{P_iD_i} = \frac{AB}{D_iE_i}.$$

Similarly,

$$\frac{P_iA}{P_iF_i} = \frac{P_iC}{P_iD_i} = \frac{AC}{D_iF_i}, \quad \frac{P_iC}{P_iE_i} = \frac{P_iB}{P_iF_i} = \frac{BC}{E_iF_i}.$$

It follows that

$$\frac{P_iA}{P_iB} = \left(\frac{P_iA}{P_iF_i}\right) \Big/ \left(\frac{P_iB}{P_iF_i}\right) = \left(\frac{AC}{D_iF_i}\right) \Big/ \left(\frac{BC}{E_iF_i}\right) = \frac{AC}{BC}\frac{E_iF_i}{D_iF_i} = \sqrt{2}\frac{AC}{BC},$$

$$\frac{P_iA}{P_iC} = \left(\frac{P_iA}{P_iE_i}\right) \Big/ \left(\frac{P_iC}{P_iE_i}\right) = \left(\frac{AB}{D_iE_i}\right) \Big/ \left(\frac{BC}{E_iF_i}\right) \frac{AB}{BC} \cdot \frac{E_iF_i}{D_iE_i} = \sqrt{2}\frac{AB}{BC},$$

and hence $P_1A : P_1B : P_1C = P_2A : P_2B : P_2C$, indicating

$$\frac{P_1A}{P_2A} = \frac{P_1B}{P_2B} = \frac{P_1C}{P_2C}.$$

Therefore, A, B, C lie on an Apollonius circle of foci P_1 and P_2, which implies that P_1, P_2 and the circumcentre O of $\triangle ABC$ are collinear. (Note: here we use the given condition $P_1 \neq P_2$. It cannot be derived from the above argument.)

We can also use complex numbers. Let the circumcircle of $\triangle ABC$ be the unit circle in the complex plane. The lowercase letter of a point represents the corresponding complex number of that point (for instance, point A corresponds to $a \in \mathbb{C}$, and so on). For a point P on the plane, denote D, E, and F as the second intersections of AP, BP, CP and $\triangle ABC$, respectively. Then,

$$\frac{p-a}{p-d} = \overline{\left(\frac{p-a}{p-d}\right)} \Leftrightarrow \frac{p-a}{p-d} = \frac{\bar{p}-\bar{a}}{\bar{p}-\bar{d}},$$

or equivalently,

$$\bar{p} + \bar{a}\bar{d}p = \bar{a} + \bar{d},$$

that is,

$$\bar{d} = \frac{\bar{a} - \bar{p}}{\bar{a}p - 1}, \quad d = \frac{a - p}{a\bar{p} - 1}.$$

Likewise,

$$e = \frac{b - p}{b\bar{p} - 1}, \quad f = \frac{c - p}{c\bar{p} - 1}.$$

On the other hand, $DE \perp DF$ and $ED = DF \Leftrightarrow \frac{e-d}{f-d} = \pm \mathrm{i} \Leftrightarrow \left(\frac{e-d}{f-d}\right)^2 = -1$. Together, we obtain

$$\left[\frac{(p\bar{p} - 1)(a - b)(c\bar{p} - 1)}{(p\bar{p} - 1)(c - b)(a\bar{p} - 1)}\right]^2 = -1.$$

Simplify it to an equation about $\bar{p}$:

$$\begin{aligned}&[c^2(a-b)^2 + b^2(c-a)^2]\bar{p}^2 - 2[c(a-b)^2 + b(c-a)^2]\bar{p} \\ &\quad + [(a-b)^2 + (c-a)^2] = 0. \qquad ①\end{aligned}$$

The leading coefficient is 0 if and only if $|a - b| = |c - a|$ ($\triangle ABC$ is isosceles with vertex angle A) and $\frac{b}{c} = \pm \mathrm{i}\frac{a-b}{c-a}$ (the central angle subtended by BC is $\frac{\pi}{2}$ larger than the inscribed angle), for that $\triangle ABC$ is an isosceles right triangle, which is known to be untrue. Therefore, ① is indeed a quadratic equation. The discriminant is

$$\begin{aligned}\triangle &= 4\{[c(a-b)^2 + b(c-a)^2]^2 \\ &\quad - 4[c^2(a-b)^2 + b^2(c-a)^2][(a-b)^2 + (c-a)^2]\} \\ &= 4\{2bc(a-b)^2(c-a)^2 - (b^2 + c^2)(a-b)^2(c-a)^2\} \\ &= -4(a-b)^2(b-c)^2(c-a)^2 \neq 0,\end{aligned}$$

implying that ① has distinct roots. Furthermore, P_1P_2 passes through the circumcentre O of $\triangle ABC$. It suffices to prove

$$\begin{aligned}\frac{c(a-b)^2 + b(c-a)^2}{(a-b)(b-c)(c-a)} \in \mathrm{i}\mathbb{R} \Leftrightarrow &\frac{c(a-b)^2 + b(c-a)^2}{(a-b)(b-c)(c-a)} \\ &+ \overline{\frac{c(a-b)^2 + b(c-a)^2}{(a-b)(b-c)(c-a)}} = 0.\end{aligned}$$

Since

$$\frac{c(a-b)^2+b(c-a)^2}{(a-b)(b-c)(c-a)}=\frac{c(a-b)}{(b-c)(c-a)}+\frac{b(c-a)}{(a-b)(b-c)}$$
$$=-\frac{c}{b-c}-\frac{c}{c-a}-\frac{b}{a-b}-\frac{b}{b-c},$$

and

$$\overline{\frac{b+c}{b-c}}=\frac{b+c}{c-b},\quad \frac{\bar{c}}{\bar{c}-\bar{a}}+\frac{\bar{b}}{\bar{a}-\bar{b}}=\frac{a}{a-c}+\frac{a}{b-a}$$
$$=\left(1-\frac{c}{c-a}\right)+\left(-1-\frac{b}{a-b}\right),$$

the conclusion follows.

The second part is to use properties of Simson line to verify the problem statement. First, by Steiner's theorem, for any point on the circumcentre, its corresponding lines X_1Y_1 and X_2Y_2 pass through a pair of antipodal points of the nine-point circle of $\triangle ABC$, respectively.

Let Q be a point on the circumcircle of $\triangle ABC$. It is well known that the closest points to P on lines AB, AC, BC are collinear (the Simson line of P), and this line bisects the segment connecting Q and the orthocentre H of $\triangle ABC$. Here we give a simple proof by complex numbers. Let the projections of Q (on the unit circle of the complex plane) on AB, AC be X and Y, respectively. Then X is the midpoint of QQ' where Q and Q' are symmetric about AB. Notice $\angle QBA=\angle Q'BA$ and they are on the opposite sides of AB, and we have

$$\frac{(q'-b)(q-b)}{|q-b|^2}=\frac{(a-b)^2}{|a-b|^2},$$

that is,

$$q'=b+\frac{(a-b)(\bar{q}-\bar{b})}{\bar{a}-\bar{b}}$$

For unit complex numbers a and b, $\bar{a}=\frac{1}{a}$, $\bar{b}=\frac{1}{b}$. Therefore,

$$q'=b-ab\left(\bar{q}-\frac{1}{b}\right)=a+b-ab\bar{q},$$

yielding

$$x=\frac{1}{2}\left(q-ab\bar{q}+a+b\right),\quad y=\frac{1}{2}\left(q-ca\bar{q}+c+a\right).$$

To show they are collinear with $\frac{1}{2}(q+a+b+c)$, it suffices

$$\frac{c+\dfrac{ab}{q}}{b+\dfrac{ca}{q}}=\frac{cq+ab}{bq+ac}\in\mathbb{R},$$

which is obvious (here, the Simson line equation is not necessary). Now, as the orthocentre H is the homothetic centre of the circumcircle and the nine-point circle, we infer that X_1Y_1 and X_2Y_2 pass through a pair of antipodal points of the nine-point circle of $\triangle ABC$, respectively.

Finally, we show $X_1Y_1 \perp X_2Y_2$. Clearly, this is due to

$$\angle CY_1X_1 = \angle CQ_1X_1 = 90^\circ - \angle X_1CQ_1,$$
$$\angle BAQ_2 = \angle Q_2Y_2X_2 = 90^\circ - \angle CY_2X_2.$$

Based on the above argument, in conclusion, the intersection of X_1Y_1 and X_2Y_2 lies on the nine-point circle of $\triangle ABC$. □

3 Let $a_1, a_2, \ldots, a_n$ be n positive integers that are not divisible by each other, namely $a_i \nmid a_j$ for $i \neq j$. Prove $a_1 + a_2 + \cdots + a_n \geq 1.1n^2 - 2n$. Note: some credit will be given to a proof of the inequality for sufficiently large n.

(Contributed by Wang Bin)

Solution Consider the set of all positive integers coprime with 6: $B = \{1, 5, 7, 11, 13, \ldots\}$. The sum of the smallest k elements satisfies

$$f(k)=\begin{cases}\dfrac{3k^2}{2}, & \text{if } k \text{ is even},\\[2mm] \dfrac{3k^2-1}{2}, & \text{if } k \text{ is odd}.\end{cases}$$

$$f(k+1)-f(k)=3k+1 \text{ or } 3k+2;\quad f(0)=0,\quad f(1)=1,$$
$$f(2)=6,\quad f(3)=13,\quad f(4)=24,\quad f(5)=37,\ldots.$$

Every positive integer a has a uniquely representation $a = 2^\alpha 3^\beta b$ where $b \in B$, α, β are nonnegative integers. Define $h(a) = b$ or we say b is the kernel of a. If m positive integers $a_1, \ldots, a_m$ are not divisible by each other and are of the same kernel b, write $a_k = 2^{\alpha_k}3^{\beta_k}b$, $k = 1, \ldots, m$. It is easy to see that $\alpha_1, \ldots, \alpha_m$ are distinct (or else some $2^\alpha 3^\beta b$ is divisible by $2^\alpha 3^{\beta'} b$). Assume $\alpha_1 < \alpha_2 < \cdots < \alpha_m$; accordingly, $\beta_1 > \beta_2 > \cdots > \beta_m \geq 0$, and we have $a_k \geq 2^{k-1}3^{m-k}b$ and their sum

$$a_1 + a_2 + \cdots + a_m \geq 2^0 3^{m-1} b + 2^1 3^{m-2} b + \cdots + 2^{m-1} 3^0 b = (3^m - 2^m)b.$$

Consider the mapping h from $A = \{a_1, a_2, \ldots, a_n\}$ to B, $h(a) = b$ where $b \in h(A)$ is the kernel of one or more elements of A. Define

$$B_k = \{b \in B : \#\{a \in A : h(a) = b\} \geq k\},$$

that is to say, every number in B_k is the kernel of k or more elements of $\{a_1, \ldots, a_n\}$. Then $B \supseteq B_1 \supseteq B_2 \supseteq \ldots$, and the numbers in $B_k \backslash B_{k+1}$ are kernels of exactly k elements of $\{a_1, \ldots, a_n\}$, giving

$$n = \sum_{k=1}^{\infty} k \times |B_k \backslash B_{k+1}| = |B_1| + |B_2| + \cdots .$$

Let $S(\cdot)$ be the number of elements in a set. We have

$$S(A) = a_1 + \cdots + a_n \geq \sum_{k=1}^{\infty} (3^k - 2^k) \times S(B_k \backslash B_{k+1})$$

$$= \sum_{k=1}^{\infty} [(3^k - 2^k) - (3^{k-1} - 2^{k-1})] \times S(B_k),$$

$$S(A) \geq (2 \times 3^{k-1} - 2^{k-1}) \times f(|B_k|).$$

Define sequence $c_k = 2 \times 3^{k-1} - 2^{k-1}$, $c_1 = 1$, $c_2 = 4$, $c_3 = 14$, $c_4 = 46$, $c_5 = 146, \ldots$.

We are led to consider the optimization problem ♠: under the constraint $x_1 + x_2 + \cdots = n$ (x_i nonnegative), minimize $T = c_1 f(x_1) + c_2 f(x_2) + \cdots$.

Suppose $X = (x_1, x_2, \ldots, x_K, 0, 0, \ldots)$ minimizes T, where $x_1 \geq x_2 \geq \cdots \geq x_K \geq 1$ and after that $x_{K+1} = x_{K+2} = \cdots = 0$. If $K \leq 2$, then

$$T = c_1 f(x_1) + c_2 f(x_2) \geq \frac{3x_1^2 - 1}{2} + 4 \times \frac{3x_2^2 - 1}{2} \geq \frac{6}{5}(x_1 + x_2)^2$$

$$- \frac{5}{2} \geq 1.1n^2 - 2n.$$

Assume $K \geq 3$. As X is a minimizer, if X is changed to $X' = (x_1 + 1, x_2, \ldots, x_{K-1}, x_K - 1, 0, \ldots)$, then T does not decrease. Thus

$$\begin{aligned} 0 \leq \triangle T &= c_1(f(x_1+1) - f(x_1)) - c_K(f(x_K) - f(x_K - 1)) \\ &\leq 3x_1 + 2 - c_K \Rightarrow c_K \leq 3x_1 + 2. \end{aligned}$$

Also, if X is changed to $X'' = (x_1, x_2+1, \ldots, x_{K-1}, x_K - 1, 0, \ldots)$, then T does not decrease, either:

$$\begin{aligned} 0 \leq \triangle T &= c_2(f(x_2+1) - f(x_2)) - c_K(f(x_K) - f(x_K - 1)) \\ &\leq 4(3x_2 + 2) - c_K \Rightarrow c_K \leq 12x_2 + 8. \end{aligned}$$

Therefore, $n = x_1 + x_2 + \cdots + x_K \geq x_1 + x_2 + 1 \geq \dfrac{c_K - 2}{3} + \dfrac{c_K - 8}{12} + 1 = \dfrac{5c_K - 4}{12}$, which implies

$$c_1 + c_2 + \cdots + c_K \leq c_K\left(1 + \frac{1}{3} + \cdots + \frac{1}{3^{K-1}}\right) \leq \frac{12n+4}{5} \times \frac{3}{2} \leq 4n.$$

Now we have,

$$\begin{aligned} T(X) &= c_1 f(x_1) + \cdots + c_K f(x_K) \\ &\geq c_1 \frac{3x_1^2 - 1}{2} + \cdots + c_K \frac{3x_K^2 - 1}{2}, \\ T(X) &\geq \frac{3}{2}[c_1 x_1^2 + c_2 x_2^2 + \cdots + c_K x_K^2] - \frac{1}{2}[c_1 + c_2 + \cdots + c_K] \\ &\geq \frac{3}{2} \frac{(x_1 + x_2 + \cdots + x_K)^2}{\dfrac{1}{c_1} + \dfrac{1}{c_2} + \cdots + \dfrac{1}{c_K}} - 2n. \end{aligned}$$

As $c_{k+1} \geq 3c_k$ always holds, it follows that

$$\begin{aligned} \frac{1}{c_1} + \frac{1}{c_2} + \cdots + \frac{1}{c_K} &\leq 1 + \frac{1}{4} + \frac{1}{14} + \frac{1}{14 \times 3} + \frac{1}{14 \times 3^2} + \cdots \\ &= 1 + \frac{1}{4} + \frac{1.5}{14} \leq 1.36, \end{aligned}$$

and thus

$$T \geq \frac{3}{2} \times \frac{n^2}{1.36} - 2n > 1.1n^2 - 2n.$$

We have shown that the minimum value of problem ♠ must satisfy $T_{\min} \geq 1.1n^2 - 2n$. For the original problem,

$$S(A) \geq \sum_{k=1}^{\infty} c_k \times f(|B_k|) \geq 1.1n^2 - 2n.$$

This completes the proof. □

Remark 1 For problem ♠, let $X = (x_1, x_2, \ldots, x_K, 0, \ldots)$ be the optimal solution. X corresponds to the n-element set

$$A = \{a_1, \ldots, a_n\} = \bigcup_{k=1}^{K} \bigcup_{b \in [3x_{k+1}, 3x_k] \bigcap B} \{2^{k-1}3^0 b,\ 2^{k-2}3^1 b, \ldots, 2^0 3^{k-1} b\}.$$

This set $A = \{a_1, \ldots, a_n\}$ satisfies $a_1 + \cdots + a_n = T = \sum_{k=1}^{K} c_k f(x_k)$, and all the elements are not divisible by each other. Indeed, it suffices to prove two numbers in $[3x_{k+1}, 3x_k] \cap B$ are not multiple of each other. As X minimizes $T(X)$, replacing X by $X' = (x_1, \ldots, x_k - 1,\ x_{k+1} + 1, \ldots, x_K, 0, \ldots)$, T does not decrease. Thus

$$0 \leq \triangle T = c_k(f(x_k) - f(x_k - 1)) - c_{k+1}(f(x_{k+1} + 1) - f(x_{k+1})).$$

Observe that $f(x_k) - f(x_k - 1)$ is the largest number in $[3x_{k+1}, 3x_k] \cap B$, while $f(x_{k+1} + 1) - f(x_{k+1})$ is the smallest number in $[3x_{k+1}, 3x_k] \cap B$. Since $\frac{c_{k+1}}{c_k} \leq 4$, the largest cannot exceed 4 times of the smallest; hence, the numbers in $[3x_{k+1}, 3x_k] \cap B$ are not divisible by each other (as they are all coprime with 6, the multiple between two numbers must be at least 5). It further implies that all elements of $A = \{a_1, \ldots, a_n\}$ are not divisible by each other.

In fact, the solution of the minimization problem ♠ (the minimal sum of n non-divisible numbers) is the sum of the smallest n numbers in the following table:

1	5	7	11	13	17	19	⋯
4	20	28	44	52	68	76	⋯
14	70	98	154	182	238	266	⋯
46	230	322	506	598	782	874	⋯
146	730	1022	1606	1898	2482	2774	⋯
⋯	⋯	⋯	⋯	⋯	⋯	⋯	⋯

Remark 2 An estimate of the series gives

$$\lambda = \frac{1}{c_1} + \frac{1}{c_2} + \frac{1}{c_3} + \cdots = 1 + \frac{1}{4} + \frac{1}{14} + \frac{1}{46} + \cdots \approx 1.3533, \quad \frac{3}{2\lambda} \approx 1.1084.$$

We can show that the solution of the minimization problem ♠ satisfies $T_{min} \leq \frac{3}{2\lambda} n^2 + O(n)$: consider

$$\min_{x_1+x_2+\cdots=n} H = c_1 x_1^2 + c_2 x_2^2 + = \cdots,$$

in which every x_k is an integer. First, relax the constraint to real numbers and take $y_k = \frac{n}{\lambda} \frac{1}{c_k}$, satisfying

$$\sum_{k=1}^{\infty} c_k y_k^2 = \frac{1}{\lambda} n^2.$$

Then use integer $x_k = y_k + \delta_k$ to approximate the optimal solution and estimate the error

$$c_k (y_k + \delta_k)^2 - c_k y_k^2 = c_k \delta_k^2 + \frac{2n}{\lambda} \delta_k.$$

As consecutive terms in $\{c_k\}$ have ratio $3 \leq \frac{c_{k+1}}{c_k} \leq 4$, we may choose c_K such that

$$\frac{n}{4} < c_K \leq n,$$

and set $y_{K+1}, y_{K+2}, \ldots$ all to 0. The decrement of this part is

$$y_{K+1} + y_{K+2} + \cdots = \frac{n}{\lambda} \left(\frac{1}{c_{K+1}} + \frac{1}{c_{K+2}} + \cdots \right) \leq \frac{n}{\lambda} \times \frac{1}{2c_K} \leq \frac{4}{2\lambda} < 2.$$

Finally, set $y_1, y_2, \ldots, y_K$ to their nearest integers while maintaining their sum as n (if necessary, change y_1 to $y_1] + 1$). We have

$$\begin{aligned} H - \frac{1}{\lambda} n^2 &= \sum_k c_k (y_k + \delta_k)^2 - c_k y_k^2 \\ &= \sum_k c_k \delta_k^2 + \frac{2n}{\lambda} \delta_k = c_k \delta_k^2 \\ &\leq 4c_1 + \sum_{k=2}^{K} c_k + \sum_{k=K+1}^{\infty} c_k y_k^2 \end{aligned}$$

$$\leq 3 + \frac{3}{2}c_K + \frac{n^2}{\lambda^2} \times \frac{1}{2c_K}$$
$$\leq 3 + 1.5n + \frac{2}{\lambda^2}n \leq 3 + 2.6n,$$

and thus $H_{\min} \leq \frac{n^2}{\lambda} + 2.6n + 3$. Furthermore, $T_{\min} \leq \frac{3}{2}H_{\min} \leq \frac{3}{2\lambda}n^2 + 4n + 5$. A sharper estimate of the linear coefficient in $T_{\min}$ is left to the interested readers.

Test II, Second Day
(8 am – 12:30 pm; March 29, 2022)

4 Given n as a positive integer, find all points $(x_1, x_2, \ldots x_n) \in \mathbb{R}^n$ that minimize $f : \mathbb{R}^n \to \mathbb{R}$,

$$f(x_1, x_2, \ldots, x_n) = \sum_{k_1=0}^{2}\sum_{k_2=0}^{2}\cdots\sum_{k_n=0}^{2} |k_1x_1 + k_2x_2 + \cdots + k_nx_n - 1|.$$

(Contributed by Wang Bin)

Solution f is minimized when $x_1 = \cdots = x_n = \frac{1}{n+1}$. Partition $A = \{0,1,2\}^n$ into $2n+1$ subsets, $A = A_0 \cup A_1 \cup \cdots \cup A_{2n}$ where

$$A_k = \{\beta = (i_1, i_2, \ldots, i_n) \in A : i_1 + i_2 + \cdots + i_n = k\}, \quad k = 0, 1, \ldots, 2n.$$

Let A_k have $|A_k| = a_k$ elements. Clearly,

$$(1 + t + t^2)^n = a_0 + a_1t + \cdots + a_nt^n + \cdots + a_{2n}t^{2n}, a_{2n-k} = a_k.$$

Denote $X = (x_1, \ldots, x_n)$, $y = (x_1 + \cdots + x_n)/n$. Then

$$f(X) = \sum_{k=0}^{2n}\sum_{\beta\in A} |\beta \cdot X - 1|.$$

First, remove the absolute value signs and sum by partition, where

$$B_k = \sum_{\beta\in A_k} (\beta \cdot X - 1) = \frac{k|A_k|}{n}(x_1 + \cdots + x_n) - |A_k| = ka_ky - a_k.$$

We estimate $\sum_{k=0}^{2n} |B_k|$ by canceling positive and negative terms. To identify large and small ka_k's, observe that

$$\sum_{k=0}^{2n} ka_kt^{k-1} = \frac{d}{dt}[(1 + t + t^2)^n] = n(1 + 2t)(1 + t + t^2)^{n-1}$$

and in

$$(1+t+t^2)^{n-1} = \sum_{k=0}^{2n-2} c_k t^k$$

the coefficients $(c_0, c_1, \ldots, c_{n-1}, \ldots, c_{2n-2})$ are unimodal and symmetric. From

$$ka_k = \frac{c_{k-1} + 2c_{k-2}}{n}$$

we have

$$\begin{aligned}(n+1)a_{n+1} &\geq na_n \geq (n+2)a_{n+2} \geq (n-1)a_{n-1} \\ &\geq (n+3)a_{n+3} \geq \cdots \geq 2a_2 \geq 2na_{2n} \geq 1a_1,\end{aligned}$$

which means $U_1 := \sum_{k=0}^{n} ka_k < \sum_{k=n+1}^{2n} ka_k$, $U_2 := \sum_{k=n+2}^{2n} ka_k < \sum_{k=0}^{n+1} ka_k$.

Define $\lambda = \dfrac{U_1 - U_2}{(n+1)a_{n+1}} \in (-1, 1)$. Then

$$f(X) \geq \sum_{k=0}^{2n} |B_k| \geq \left[\sum_{k=0}^{n}(-B_k) + \lambda B_{n+1} + \sum_{k=n+2}^{2n} B_k\right] + (1-|\lambda|)|B_{n+1}|.$$

In the square brackets, the coefficient of y equals $-U_1 + \lambda(n+1)a_{n+1} + U_2 = 0$; hence it is a constant

$$C = \sum_{k=0}^{n} a_k - \lambda a_{n+1} + \sum_{k=n+2}^{2n} a_k.$$

We infer that $f(X) \geq C$ always holds, with equality if and only if $x_1 = \cdots = x_n = \dfrac{1}{n+1}$. In the meantime, the a_{n+1} terms in the sum of B_{n+1} must all vanish, that is, the inner product of $X = (x_1, \ldots, x_n)$ and any vector in A_{n+1} equals 1, so $x_1 = \cdots = x_n = \dfrac{1}{n+1}$. □

5 For a fixed positive integer n, let D be the set of all positive factors of n. Prove that for a mapping $f : D \to \mathbb{Z}$, the following two assertions are equivalent:

(A) For any positive factor m of n,

$$n \Big| \sum_{d|m} f(d)\mathrm{C}_{n/d}^{m/d};$$

(B) For any positive factor k of n,

$$k \Big| \sum_{d|k} f(d).$$

(Contributed by Ai Yinghua)

Solution For the given mapping $f : D \to \mathbb{Z}$, define mapping $g : D \to \mathbb{Z}$ as

$$g(k) = \sum_{d|k} f(d), \quad \forall k \in D.$$

According to the Mobius transformation, f is uniquely determined by g,

$$f(k) = \mu\left(\frac{k}{d}\right) g(d), \quad \forall k \in D,$$

which further gives

$$\begin{aligned}
\sum_{d|m} f(d)\mathrm{C}_{n/d}^{m/d} &= \sum_{d|m}\sum_{x|d} \mu\left(\frac{d}{x}\right) g(x)\mathrm{C}_{n/d}^{m/d} \\
&= \sum_{x|m} g(x) \left(\sum_{x|d,d|m} \mu\left(\frac{d}{x}\right) \mathrm{C}_{n/d}^{m/d} \right) \\
&= \sum_{x|m} g(x) \left(\sum_{S|\frac{m}{x}} \mu(s)\mathrm{C}_{n/(xs)}^{m/(xs)} \right). \qquad ①
\end{aligned}$$

We claim

Lemma *If $b|a$, then $\sum\limits_{k|b} \mu(k)\mathrm{C}_{a/k}^{b/k} \equiv 0 \pmod a$.*

Assuming the lemma is true, we prove that assertions (A) and (B) are equivalent. The proof of the lemma is at the end.

"$(B) \Rightarrow (A)$": suppose $x|g(x)$ holds for every $x \in D$. By the lemma, we have

$$\frac{n}{x} \Big| \sum_{S|\frac{m}{x}} \mu(s)\mathrm{C}_{n/(xs)}^{m/(xs)},$$

plugged into ① to derive (A).

"$(A) \Rightarrow (B)$": assume (A) is true. We use induction to verify $k|g(k)$ for every $k \in D$. Suppose $k|g(k)$ holds for all $k < m$, $k \in D$ and consider $k = m$. From the lemma and the induction hypothesis, it follows that

$$\begin{aligned} 0 &\equiv \sum_{d|m} f(d)\mathrm{C}_{n/d}^{m/d} \equiv \sum_{x|m} g(x)\left(\sum_{S|\frac{m}{x}} \mu(s)\mathrm{C}_{n/(xs)}^{m/(xs)}\right) \\ &\equiv g(m)\cdot\frac{n}{m} + \sum_{x|m,x<m} g(x)\left(\sum_{S|\frac{m}{x}} \mu(s)\mathrm{C}_{n/(xs)}^{m/(xs)}\right) \\ &\equiv g(m)\cdot\frac{n}{m} \pmod n, \end{aligned}$$

and hence $m|g(m)$. This completes the induction.

Finally, we need to prove the lemma.

Proof of the lemma Note that

$$\sum_{k|b} \mu(k)\mathrm{C}_{a/k}^{b/k} = \frac{a}{b}\sum_{k|b} \mu(k)\mathrm{C}_{\frac{a}{k}-1}^{\frac{b}{k}-1}.$$

Suppose b has the prime factorization $b = \prod_{i=1}^{t} p_i^{\beta_i}$. It suffices to prove for every $1 \le i \le t$,

$$\sum_{k|b} \mu(k)\mathrm{C}_{\frac{a}{k}-1}^{\frac{b}{k}-1} \equiv 0 \pmod{p_i^{\beta_i}}.$$

Let $i = 1$, $p_1^{\beta_1} = p^{\beta}$, and rewrite C_u^v as $\mathrm{C}(u, v)$ (for other i's, it can be proved in the same manner). Notice that

$$\begin{aligned} &\sum_{k|b} \mu(k)\mathrm{C}_{\frac{a}{k}-1}^{\frac{b}{k}-1} \\ &= \sum_{I\subset\{1,\ldots,t\}} (-1)^{|I|}\mathrm{C}\left(\frac{a}{\prod_{i\in I} p_i} - 1, \frac{b}{\prod_{i\in I} p_i} - 1\right) \\ &= \sum_{J\subset\{2,\ldots,t\}} (-1)^{|J|}\left(\mathrm{C}\left(\frac{a}{\prod_{j\in J} p_j} - 1, \frac{b}{\prod_{j\in J} p_j} - 1\right)\right. \\ &\quad \left. -\mathrm{C}\left(\frac{a}{p\prod_{j\in J} p_j} - 1, \frac{b}{p\prod_{j\in J} p_j} - 1\right)\right). \end{aligned}$$

Now it only requires to prove: if u and v are multiples of p^β, then

$$C(u-1, v-1) - C\left(\frac{u}{p}-1, \frac{v}{p}-1\right) \equiv 0 \pmod{p^\beta}. \qquad ②$$

Let $w = u - v$. We have

$$C(u-1, v-1) = \frac{\prod\limits_{0 \le x \le \frac{v}{p}-1, 1 \le r \le p-1} (xp + r + w) \cdot \prod\limits_{1 \le y \le \frac{v}{p}-1} (yp + w)}{\prod\limits_{0 \le x \le \frac{v}{p}-1, 1 \le r \le p-1} (xp + r) \cdot \prod\limits_{1 \le y \le \frac{v}{p}-1} (yp)}$$

$$= \frac{\prod\limits_{0 \le x \le \frac{v}{p}-1, 1 \le r \le p-1} (xp + r + w)}{\prod\limits_{0 \le x \le \frac{v}{p}-1, 1 \le r \le p-1} (xp + r)} \cdot C\left(\frac{u}{p}-1, \frac{v}{p}-1\right).$$

As w is a multiple of p^β,

$$\begin{aligned} &C(u-1, v-1) \prod_{0 \le x < \frac{v}{p}-1, 1 \le r < p-1} (xp + r) \\ &= C\left(\frac{u}{p}-1, \frac{v}{p}-1\right) \prod_{0 < x < \frac{v}{p}1, 1 \le r \le p-1} (xp + r + w) \\ &\equiv C\left(\frac{u}{p}-1, \frac{v}{p}-1\right) \prod_{0 < x < \frac{v}{p}1, 1 \le r \le p-1} (xp + r) \pmod{p^\beta}, \end{aligned}$$

and ② follows. The lemma is now verified.

Another proof of the lemma We use the following proposition.

Let X be a finite set, $h : X \to X$ be a bijection satisfying $h^{(a)} = \mathrm{Id}_X$. Under the action of h, the set X can be decomposed into several orbits, each orbit as $\{x, h(x), h^{(2)}(x), \ldots, h^{(l)}(x) = x\}$, where l, the length of the orbit, is the smallest positive integer satisfying $h^{(l)}(x) = x$. As $h^{(a)} = \mathrm{Id}_X$, every l must be a factor of a. Suppose for each factor l of n, there are $N(l)$ orbits of length l. For each positive integer $k|a$, let

$$m(k) = \#\{x \in X : h^{(k)}(x) = x\}.$$

Clearly, $m(k) = \sum\limits_{l|k} N(l)l$, and

$$\begin{aligned}\sum_{k|a}\mu(k)m\left(\frac{a}{k}\right) &= \sum_{k|a}\sum_{l|\frac{a}{l}}\mu(k)N(l)l \\ &= \sum_{l|a}N(l)l\sum_{k|\frac{a}{l}}\mu(k) \\ &= N(a)a,\end{aligned}$$

in particular,

$$\sum_{k|a}\mu(k)m\left(\frac{a}{k}\right) \equiv 0 \pmod a. \qquad ③$$

We apply the above result to the following model: let X consist of all b-element subsets of $\mathbb{Z}/a$, define $h : X \to X$ as

$$h(\{y_1, \ldots, y_b\}) = \{y_1 + 1, \ldots, y_b + 1\}, \quad \forall \{y_1, \ldots, y_b\} \in X.$$

If k is a factor of a, it is easy to see that

$$m(k) = \mathrm{C}_k^{b/(a/k)}.$$

From ③, we arrive at

$$\sum_{k|a}\mu(k)\mathrm{C}_{a/k}^{b/k} \equiv 0 \pmod a.$$

□

6 Let m and n be positive integers, $m \geq n \geq 2022$. Prove: for arbitrary real numbers $a_1, a_2, \ldots, a_n$, $b_1, b_2, \ldots, b_n$, the number of ordered pairs $(i, j)(1 \leq i, j \leq n)$ satisfying $|a_i + b_j - ij| \leq m$ does not exceed $3n\sqrt{m}\ln n$.

(Contributed by Zhang Ruixiang)

Solution Colour all (i, j) satisfying $|a_i + b_j - ij| \leq m$ red.

Lemma *If (i_1, j_1), (i_1, j_2), (i_2, j_1), (i_2, j_2) are all red, then*

$$|(i_2 - i_1)(j_2 - j_1)| \leq 4m.$$

Proof of lemma By definition of red points,

$$|a_{i_1} + b_{j_1} - i_1 j_1| \leq m, \quad |a_{i_1} + b_{j_2} - i_1 j_2| \leq m,$$

$$|a_{i_2} + b_{j_1} - i_2 j_1| \leq m, \quad |a_{i_2} + b_{j_2} - i_2 j_2| \leq m.$$

Taking two differences to eliminate a_{i_1}, a_{i_2}, b_{j_1}, and b_{j_2}, we obtain $|i_1j_1 - i_1j_2 - i_2j_1 + i_2j_2| \leq 4m$ and the lemma conclusion. $\square$

Return to the original problem. First, for given $1 \leq i_1 < i_2 \leq n$, we need to find the number of j's such that both (i_1, j) and (i_2, j) are red points. Let $d = i_2 - i_1$. Apparently, any two such j's differ by $\frac{4m}{d}$ at most; hence, there are at most $\frac{4m}{d}+1$ of such j's.

In the meantime, if d is fixed, there are $n - d$ ways to choose i_1 and i_2. As a result, the number of red pairs (i_1, j) and (i_2, j) with $1 \leq i_1 < i_2 \leq n$, $1 \leq j \leq n$ cannot exceed

$$\sum_{d=1}^{n-1}(n-d)\left(\frac{4m}{d}+1\right) = \sum_{d=1}^{n-1}(n-d) - 4mn + 4m\sum_{d=1}^{n-1}\frac{n}{d}$$
$$< \frac{n(n-1)}{2} + 4mn\ln n.$$

On the other hand, for $1 \leq j \leq n$, let the number of red points in $(1, j)$, $(2, j), \ldots, (n, j)$ be x_j. Then

$$\sum_{j=1}^{n} \mathrm{C}_{x_j}^2 < \frac{n(n-1)}{2} + 4mn\ln n,$$

and thus

$$\sum_{j=1}^{n}\left(x_j - \frac{1}{2}\right)^2 < n^2 - \frac{1}{4}n + 8mn\ln n.$$

By Cauchy's inequality, it follows that

$$\sum_{j=1}^{n}\left(x_j - \frac{1}{2}\right) \leq \sqrt{n\left(n^2 - \frac{1}{4}n + 8mn\ln n\right)},$$

or

$$\sum_{j=1}^{n} x_j \leq \frac{n}{2} + n\sqrt{n - \frac{1}{4} + 8m\ln n}\,.$$

When $m \geq n \geq 2022$, a simple estimate reveals that the right hand side is $\leq 3n\sqrt{m\ln n}$. $\square$

Remark This result originates from a new class of methods in harmonic analysis, which has the following variation: there exists a constant $C >$

0, such that for any positive integers $m \geq n^2$ and any functions f, g, $h: \{1, 2, \ldots, n\} \to \mathbb{R}$, the number of ordered pairs $(x, y)(x, y \in \{1, 2, \ldots, n\})$ satisfying

$$|f(x) + yg(x) + h(y) - xy^2| \leq m,$$

cannot exceed

$$Cm^{\frac{1}{3}}n(\ln n + 1)^{\frac{2}{3}}.$$

In the above statement, the leading order term $m^{1/3}n$ is the optimal bound (consider $f(x)+yg(x)+h(y)-xy^2 = \frac{1}{3}(y-x)^3$). If it is approaching $f(x) + g(y) + yh(x) + xl(y) - x^2y^2$ even with sufficiently large m, whether a bound of leading order term $m^{1/4}n^{1+\delta}$ exists or not is still an open problem.

Test III, First Day
(8 am – 12:30 pm; April 12, 2022)

1. Given that circle Γ_2 is inside circle Γ_1 on the plane, prove that there exists a point P on the plane with the following property: if l is a line not through P, l intersects Γ_1 and Γ_2 at distinct points A and B, C and D, respectively (A, C, D, and B are consecutive on l), then $\angle APC = \angle DPB$.

(Contributed by Fu Yunhao)

Solution As shown in Fig. 1.1, let O_1, O_2, r_1, r_2 be the centres and the radii of circles Γ_1 and Γ_2, respectively, $r_1 > r_2$. We find two points P and Q

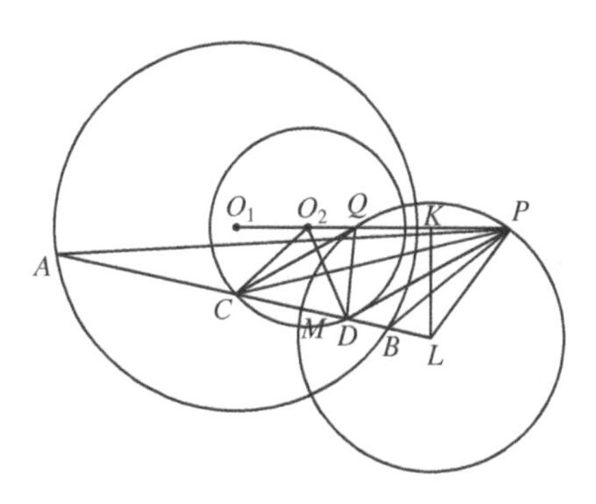

Fig. 1.1

on the ray O_1O_2 such that

$$O_1P \cdot O_1Q = r_1^2, \quad O_2P \cdot O_2Q = r_2^2.$$

First, take Kon the ray O_1O_2 with

$$O_1K = \frac{O_1O_2^2 + r_1^2 - r_2^2}{2O_1O_2}.$$

As $O_1O_2 \leq r_1 - r_2$, $O_1K \geq r_1$. Then take P and Q on the ray O_1O_2 with

$$KP = KQ = \sqrt{O_1K^2 - r_1^2},$$

which leads to $O_1P \cdot O_1Q = r_1^2$; meanwhile,

$$\begin{aligned} O_2P \cdot O_2Q - O_1P \cdot O_1Q &= O_2K^2 - O_1K^2 = (O_1K - O_1O_2)^2 - O_1K^2 \\ &= O_1O_2^2 - 2O_1O_2 \cdot O_1K = r_2^2 - r_1^2, \end{aligned}$$

which gives $O_2P \cdot O_2Q = r_2^2$.

For an arbitrary line l not through P, if l is perpendicular to O_1O_2, due to symmetry the conclusion is obvious. Otherwise, from $O_2C^2 = O_2Q \cdot O_2P$ we have $\triangle O_2CQ \backsim \triangle O_2PC$, and $\frac{CQ}{CP} = \frac{r_2}{O_2P}$. Similarly, $\frac{DQ}{DP} = \frac{r_2}{O_2P}$, and hence $\frac{CQ}{CP} = \frac{DQ}{DP}$, indicating that the bisectors of $\angle CPD$ and $\angle CQD$ meet l at the same point, say M. In the same way, we can infer that the bisectors of $\angle APB$ and $\angle AQB$ meet l at the same point as well, say M'.

Note that P, Q, and M all lie on the Apollonian circle which is the locus of points X such that $\frac{CX}{CD} = \frac{CQ}{DQ}$, hence the centre L must be on l, more precisely, at the intersection of the perpendicular bisector of PQ and l. In the same way, P, Q, and M' all lie on the Apollonian circle, $\frac{AP}{AQ} = \frac{AQ}{BQ} = \frac{AM'}{BM'}$, with the same centre L. As K lies on the boundary or outside of Γ_1, we infer that L is outside Γ_1, $M = M'$, and

$$\angle APC = \angle APM - \angle CPM = \angle BPM - \angle DPM = \angle BPD.$$

This completes the proof. □

2 Let α and β be positive real numbers satisfying: for any positive integers k_1 and k_2, it is always true that $\lfloor k_1\alpha \rfloor \neq \lfloor k_2\beta \rfloor$, in which

$\lfloor x \rfloor$ represents the largest integer not exceeding x. Prove: there exist positive integers m_1, m_2 such that $\dfrac{m_1}{\alpha} + \dfrac{m_2}{\beta} = 1$.

(Contributed by Wang Bin)

Solution 1 First, note that $\dfrac{\beta}{\alpha}$ is irrational, or else there exist positive integers k_1 and k_2 such that $k_1\alpha = k_2\beta$, contradicting the problem condition.

If $\alpha = \dfrac{q}{p}$ is rational, then there exists a positive integer k_2 such that the decimal part of $\dfrac{k_2\beta}{q}$ is less than $\dfrac{1}{q}$. Take $k_1 = p\left\lfloor \dfrac{k_2\beta}{q} \right\rfloor$ to obtain

$$k_1\alpha = q \cdot \left\lfloor \frac{k_2\beta}{q} \right\rfloor < k_2\beta < q\left\lfloor \frac{k_2\beta}{q} \right\rfloor + 1,$$

contradicting the problem condition. Hence, α is irrational; for the same reason, β is also irrational.

We call a pair of positive integers (a, b) "great" if $0 < b\beta - a\alpha < 1$. For a great pair (a, b), there exists a unique positive integer t such that $a\alpha < t < b\beta$. Let $u = t - a\alpha$, $v = b\beta - t$. We say that the great pair (a, b) corresponds to the intermediate number t and the differences (u, v).

Lemma *If the great pairs (a_1, b_1), (a_2, b_2) correspond to the differences (u_1, v_1), (u_2, v_2), respectively, then $\dfrac{u_1}{v_1} = \dfrac{u_2}{v_2}$.*

Proof of Lemma Let the great pairs correspond to intermediate numbers t_1, t_2, respectively. Assume the conclusion is untrue, $\dfrac{u_1}{v_1} > \dfrac{u_2}{v_2}$; let $\varepsilon = \dfrac{u_1v_2 - u_2v_1}{2} > 0$.

As $\dfrac{\beta}{\alpha}$ is irrational, there exist positive integers a_0, b_0 such that $0 < a_0\alpha - b_0\beta < \varepsilon$. Moreover, it is known that some positive integer t_0 satisfies $b_0\beta < t_0 < a_0\alpha$. Let $u_0 = a_0\alpha - t_0$, $v_0 = t_0 - b_0\beta$. In case $\dfrac{u_1}{u_0} - \dfrac{v_1}{v_0} > 1$, we take $L = \left\lfloor \dfrac{v_1}{v_0} \right\rfloor + 1$ such that $\dfrac{u_1}{u_0} > L > \dfrac{v_1}{v_0}$, and thus

$$u_1 - Lu_0 = (t_1 + Lt_0) - (a_1 + La_0)\alpha > 0,$$

$$v_1 - Lv_0 = (t_1 + Lt_0) - (b_1 + Lb_0)\beta < 0.$$

Now choose $k_1 = a_1 + La_0$, $k_2 = b_1 + Lb_0$, and $\lfloor k_1\alpha \rfloor = \lfloor k_2\beta \rfloor = t_1 + Lt_0 - 1$ leads to a contradiction. So, it must be $\dfrac{u_1}{u_0} - \dfrac{v_1}{v_0} \leq 1$; similarly,

$\frac{u_2}{u_0} - \frac{v_2}{v_0} \geq -1$. Then,

$$u_1 - \frac{u_0}{v_0}v_1 \leq u_0, \quad u_2 - \frac{u_0}{v_0}v_2 \geq -u_0,$$

yielding $u_1v_2 - u_2v_1 \leq u_0(v_1 + v_2) < 2\varepsilon$, which is inconsistent with the definition of ε. The lemma is proved.

According to the lemma, every great pair (a, b) corresponds to the same ratio of the differences $\frac{u}{v} = \frac{t - a\alpha}{b\beta - t}$. Denote the common ratio $\frac{v}{u+v}$ by λ, $\lambda \in (0, 1)$ and $\lambda \cdot a\alpha + (1 - \lambda) \cdot b\beta = t \in \mathbb{Z}$. For two great pairs, a linear combination of them with integer coefficients is called a "nice" pair. It is easy to see that each nice pair (c, d) satisfies

$$\lambda \cdot c\alpha + (1 - \lambda) \cdot d\beta = c \cdot \lambda\alpha + d \cdot (1 - \lambda)\beta \in \mathbb{Z}.$$

Take a great pair (a, b) and let $\delta = b\beta - a\alpha \in (0, 1)$. For any $M > 0$, every integer pair (c, d) satisfying $0 < d\beta - c\alpha < M$ and $c > \frac{a}{\delta}M$ is nice. (Indeed, choose $L = \left\lfloor \frac{d\beta - c\alpha}{\delta} \right\rfloor < \frac{M}{\delta} < \frac{c}{a}$ such that $(d\text{-}Lb)\beta\text{-}(c\text{-}La)\alpha = (d\beta - c\alpha) - L\delta \in (0, \delta) \subset (0, 1)$. Then, $(c - La, d - Lb)$ is great, while $(c, d) = (c - La,\, d - Lb) + L \cdot (a, b)$ is nice.)

Take $M = \alpha + 2\beta$, integers $c_0 > \frac{a}{\delta}M + 1$ and $d_0 = \left\lceil \frac{c_0\alpha}{\beta} \right\rceil$ that satisfy $0 < d_0\beta - c_0\alpha < \beta$. According to the above argument, (c_0, d_0), $(c_0, d_0 + 1)$, $(c_0 - 1, d_0)$ are all nice; so are their linear combinations $(0, 1)$ and $(1, 0)$. This indicates that both $\lambda\alpha$ and $(1 - \lambda)\beta$ are (positive) integers. Then we just choose

$$m_2 = \lambda\alpha, \quad m_1 = (1 - \lambda)\beta,$$

so that $\frac{m_2}{\alpha} + \frac{m_1}{\beta} = \lambda + (1 - \lambda) = 1$, or $m_1\alpha + m_2\beta = \alpha\beta$. Done. □

Solution 2 Let the decimal part of x be $\{x\} = x - \lfloor x \rfloor$.

Evidently, a positive integer n-1 appears in the sequence $\lfloor k\alpha \rfloor$ if and only if $n - 1 \leq k\alpha < n$ for some integer k, or equivalently $k < \frac{n}{\alpha} \leq k + \frac{1}{\alpha}$, that is, $\left\{\frac{n}{\alpha}\right\} \leq \frac{1}{\alpha}$. From the problem condition, we infer that the ordered decimal pairs

$$\left(\left\{\frac{n}{\alpha}\right\}, \left\{\frac{n}{\beta}\right\}\right) \notin \left[0, \frac{1}{\alpha}\right] \times \left[0, \frac{1}{\beta}\right], \quad \forall n = 2, 3, \ldots. \qquad ①$$

Consider the linear dependence relation of $\frac{1}{\alpha}$, $\frac{1}{\beta}$, and 1 in $\mathbb{Q}$. If $\left\{\frac{1}{\alpha}, \frac{1}{\beta}, 1\right\}$ is a linearly independent set in $\mathbb{Q}$, then by the 2-dimensional Kronecker's theorem, the sequence $\left(\left\{\frac{n}{\alpha}\right\}, \left\{\frac{n}{\beta}\right\}\right)_{n=1,2,\dots}$ is dense in the unit square $[0,1] \times [0,1]$, and there exists $n \geq 2$ such that ① does not hold. Hence, $\left\{\frac{1}{\alpha}, \frac{1}{\beta}, 1\right\}$ is linearly dependent. Suppose $\frac{a}{\alpha} + \frac{b}{\beta} = c$, in which a, b, $c \in \mathbb{Z}$ and $(a, b, c) = 1$. Then α, $\beta \notin \mathbb{Q}$, $a, b = 0$.

If a, b have different signs, assume $a > 0$, $b < 0$. Then for an arbitrary positive integer k,

$$a\frac{k}{\alpha} + b\frac{k}{\beta} = kc \Rightarrow a \times \left\{\frac{k}{a}\right\} - |b| \times \left\{\frac{k}{\beta}\right\} \in \mathbb{Z}.$$

Choose $k \geq 2$ that satisfies $0 < \left\{\frac{k}{\beta}\right\} < \min\left\{\frac{1}{|b|\alpha}, \frac{1}{\alpha\beta}\right\}$, and we find

$$\left\{\frac{ak}{\alpha}\right\} = \left\{a \times \left\{\frac{k}{\alpha}\right\}\right\} = \left\{|b| \times \left\{\frac{k}{\beta}\right\}\right\} < \frac{1}{\alpha},$$

$$\left\{\frac{ak}{\beta}\right\} = \left\{a \times \left\{\frac{k}{\beta}\right\}\right\} < \frac{1}{\beta}.$$

So, $n = ak$ makes ① untrue.

In this case, a, b must have the same sign, say $a > 0$, $b > 0$. Since $(a, b, c) = 1$, there exist positive integers u, v such that $uc - vb \equiv 1 \pmod{a}$. From $a\frac{n}{\alpha} + b\frac{n}{\beta} = nc$, it follows that

$$a \times \left\{\frac{n}{\alpha}\right\} + b \times \left\{\frac{n}{\beta}\right\} = nc - b\left\lfloor\frac{n}{\beta}\right\rfloor - a\left\lfloor\frac{n}{\alpha}\right\rfloor.$$

We hope to find n such that $n \equiv v$, $\left\lfloor\frac{n}{\beta}\right\rfloor \equiv v \pmod{a}$ and $\left\{\frac{n}{\beta}\right\}$ is very small, in other words,

$$a \times \left\{\frac{n}{\alpha}\right\} + b \times \left\{\frac{n}{\beta}\right\} \equiv 1 \pmod{a},$$

and thus $\left\{\frac{n}{\alpha}\right\} < \frac{1}{a}$ (which further implies $\frac{a}{\alpha} < 1$).

To this end, let $n = ak + u$, and take positive integer k such that

$$t = \frac{\frac{n}{\beta} - v}{a} = \frac{ak + u - v\beta}{a\beta} = \frac{k}{\beta} - \frac{v\beta - u}{a\beta}$$

has decimal part $\{t\} < \min\left\{\frac{1}{ab}, \frac{1}{a\beta}\right\}$ (this only requires $\frac{v\beta - u}{a\beta} < \left\{\frac{k}{\beta}\right\} < \frac{v\beta - u}{a\beta} + \min\left\{\frac{1}{ab}, \frac{1}{a\beta}\right\}$; as $\left\{\frac{k}{\beta}\right\}$ is dense in $[0, 1]$, such k always exists). It follows that

$$\left\{\frac{n}{\alpha}\right\} = -\left\lfloor\frac{n}{\alpha}\right\rfloor + \frac{c(ak + u)}{a} - \frac{b}{a} \times \frac{ak + u}{\beta}$$

$$= \left\{\frac{uc - bv}{a} - b \times \frac{ak + u - v\beta}{a\beta}\right\} < \frac{1}{a},$$

$$\left\{\frac{n}{\beta}\right\} = \left\{a \times \frac{ak + u - v\beta}{a\beta} + v\right\}$$

$$= \left\{a \times \left\{\frac{ak + u - v\beta}{a\beta}\right\}\right\} < \frac{1}{\beta},$$

and hence $\frac{1}{a} > \frac{1}{\alpha}$ (otherwise $n = ak + u$ makes ① untrue), or $\frac{a}{\alpha} < 1$; likewise, $\frac{b}{\beta} < 1$. Thus, the positive integer $c = \frac{a}{\alpha} + \frac{b}{\beta} < 2$, $c = 1$. We arrive at $\frac{a}{\alpha} + \frac{b}{\beta} = 1$ which is the problem conclusion. □

3 Given positive integer $n \geq 2$, find all $(a_1, a_2, \ldots, a_n) \in \mathbb{Z}^n$ satisfying the following conditions:

(1) a_1 is odd, $1 < a_1 \leq a_2 \leq \cdots \leq a_n$, and $M = \frac{1}{2^n}(a_1 - 1)a_2 \ldots a_n$ is an integer;

(2) One can find M of $(c_{i,1}, c_{i,2}, \ldots, c_{i,n}) \in \mathbb{Z}^n$ $(i = 1, 2, \ldots, M)$, such that for any $1 \leq i < j \leq M$, there exists $k \in \{1, 2, \ldots, n\}$,

$$c_{i,k} - c_{j,k} \not\equiv -1, 0, 1 \pmod{a_k}.$$

(Contributed by Ding Jian and Xiao Liang)

Solution The desired n-tuples satisfy:

If there are exactly r odd numbers among $a_2, \ldots, a_n$, then $2^r | (a_1 - 1)$.

First, we show that ① is necessary. Temporarily drop the constraint "$a_n \geq a_{n-1} \geq \cdots \geq a_1$" and assume $a_1, \ldots, a_r$ to be odd; $a_{r+1}, \ldots, a_n$ to be even. Suppose we have found M of $(c_{i,1}, c_{i,2}, \ldots, c_{i,n}) \in \mathbb{Z}^n$. For each $s \in \mathbb{Z}$, define $B_s = \{i | 1 \leq i \leq M,\ c_{i,n} \equiv s \pmod{a_n}\}$. We have

$$|B_1| + |B_2| + \cdots + |B_{a_n}| = M,$$

and for some s, $|B_s| + |B_{s+1}| \geq \dfrac{M}{a_n/2}$, indicating that we can select at least $\dfrac{M}{a_n/2}$ of n-tuples whose nth coordinates $c_{i,n}$ have pairwise differences $\equiv 0, \pm 1 \pmod{a_n}$.

Continue the above operation on these n-tuples for their $(n-1)$th coordinates modulo a_{n-1}, their $(n-2)$th coordinates modulo a_{n-2}, and so on. At the end, we obtain at least $\dfrac{M}{\frac{a_n}{2} \cdots \frac{a_2}{2}} = \dfrac{a_1 - 1}{2}$ of n-tuples whose kth coordinates $(2 \leq k \leq n) c_{i,k}$ have pairwise differences $\equiv 0, \pm 1 \pmod{a_k}$. From the problem condition, their first coordinates have pairwise differences not congruent to $0, \pm 1 \pmod{a_1}$, which requires that at most $\dfrac{a_1 - 1}{2}$ of n-tuples are left. Consequently, in each of the preceding operations, the equality holds, namely for each t, we can select exactly $\dfrac{M}{\frac{a_n}{2} \cdots \frac{a_t}{2}}$ of n-tuples; in particular, for $t = r + 1$,

$$\frac{M}{\frac{a_n}{2} \cdots \frac{a_{r+1}}{2}} = \frac{1}{2^r}(a_1 - 1)a_2 \cdots a_r \in \mathbb{Z},$$

implying $2^r | (a_1 - 1)$.

Next, we try to find $(a_1, a_2, \ldots, a_n)$ for $2^r | (a_1 - 1)$. Temporarily, weaken the condition "$a_n \geq a_{n-1} \geq \cdots \geq a_1$" to be $a_1 = \min\{a_1, \ldots, a_n\}$. First, if there are even number(s) in $a_2, \ldots, a_n$, assume it is a_n. Suppose for $a_1, \ldots, a_{n-1}$ we have found $(c_{i,1}, c_{i,2}, \ldots, c_{i,n-1}) \left(1 \leq i \leq M' = \dfrac{1}{2^{n-1}}(a_1 - 1)a_2 \cdots a_n = \dfrac{2}{a_n} M\right)$. Then just take $(c_{i,1}, c_{i,2}, \ldots, c_{i,n-1}, 2c) \left(1 \leq i \leq M', 1 \leq c \leq \dfrac{a_n}{2}\right)$.

Now consider $a_2, a_3, \ldots, a_n$ are odd numbers. Let $a_1 = 2^n t + 1$. Suppose $a_1 = a_2 = \cdots = a_n$, $M = t a_1^{n-1}$. Define the function

$$f(x_1, x_2, \ldots, x_{n-1}) = \sum_{i=1}^{n-1} 2^i x_i,$$

and select $M = ta_1^{n-1}$ of n-tuples

$$(x_1, x_2, \ldots, x_{n-1}, f(x_1, \ldots, x_{n-1}) + 2^n s),$$

where $x_1, \ldots, x_{n-1} \in \{1, 2, \ldots, a_1\}$, $s \in \{1, \ldots, t\}$. We are to justify that they satisfy the problem condition. For the above n-tuples and another n-tuple

$$(x_1', x_2', \ldots, x_{n-1}', f(x_1', \ldots, x_{n-1}') + 2^n s'),$$

in case that the differences of their kth coordinates $\equiv 0, \pm 1 \pmod{a_1}$ for any k, then

$$x_i - x_i' \equiv 0, \pm 1 \pmod{a_1}, \quad i = 1, \ldots, n-1,$$

and

$$\sum_{i=1}^{n-1} 2^{i-1}(x_i - x_i') + 2^n(s - s') \equiv 0, \quad \pm 1 \pmod{a_1}. \qquad ②$$

The former congruence relations indicate that

$$\sum_{i=1}^{i-1} 2^i(x_i - x_i') \text{ can only } \equiv 2^{n-1} - 1,\ 2^{n-2} - 1, \ldots, 1 - 2^{n-1} \pmod{a_1},$$

while $s - s' \in \{1 - t, 2 - t, \ldots, t - 1\}$. If ② holds, then it must be $s = s'$ and

$$\sum_{i=1}^{n-1} 2^{i-1}(x_i - x_i') = 0 \pmod{a_1}.$$

By the uniqueness of the binary representation, we obtain $x_i = x_i'$ (mod a_1), that is, they are the same n-tuple, and the problem conditions are all met.

Finally, consider $a_1, \ldots, a_n$ are all odd numbers. We demonstrate that if $a_2 > a_1$, it can be made into a_1, $a_2 - 2$, $a_3, \ldots, a_n$ and further into the situation that all a_i's are identical simply by induction, which has been covered in the previous discussion. Suppose for $a_1' = a_1$, $a_2' = a_2 - 2$, $a_3' = a_3, \ldots, a_n' = a_n$ we have $M' = \dfrac{1}{2^n}(a_1 - 1)(a_2 - 2)a_3 \ldots a_n$ of n-tuples satisfying the problem conditions and their kth coordinates always take values in $\{0, 1, \ldots, a_k' - 1\}$. Now define the $M = \dfrac{1}{2^n}(a_1 - 1)a_2 a_3 \ldots a_n$ of n-tuples as follows: first, choose the M' of them that we already have, (i) for $x_2 = a_2 - 2$, choose $(x_1, a_2 - 2, x_3, \ldots, x_n)$ if and only if we have chosen

$(x_1, a_2-4, x_3, \ldots, x_n)$; (ii) for $x_2 = a_2-1$, choose $(x_1, a_2-1, x_3, \ldots, x_n)$ if and only if we have chosen $(x_1, a_2-3, x_3, \ldots, x_n)$. Clearly, these n-tuples satisfy the conditions. Furthermore, from the previous inductive proof of $2^r|(a_1-1)$ and equality holding, it follows that among the M' of n-tuples we already have, there are $M' \cdot \dfrac{2}{a_2-2}$ of them whose second coordinates are a_2-1 or a_2-2. Therefore, the total number of n-tuples is $M'+M'\cdot\dfrac{2}{a_2-2} = M$, and the proof is complete. □

Test III, Second Day
(8 am – 12:30 pm; April 13, 2022)

4 In the rectangular coordinate system, there exist finitely many triangles such that: their centroids are all integer points; the intersection of any two triangles is either empty, or a common vertex, or a common side (connecting the two common vertices); the union of all the triangles is a square with integer side length k (the vertices of the square are not necessarily integer points and the sides are not necessarily parallel to the axes). Find all integers k such that this could happen.

(Contributed by Qu Zhenhua)

Solution The answer is all multiples of 3.

On one hand, if $3|k$, let $k = 3t$. Consider the square whose vertices are at $(0,0)$, $(3t, 3t)$, $(3t, 0)$, and $(0, 3t)$. Divide the square by parallel lines

$$x = 3i \ (i = 1, \ldots, t) \quad \text{and} \quad y = 3j \ (j = 1, \ldots, t)$$

into t^2 of 3×3 small squares, and use diagonals to cut each small square into two right isosceles triangles. Evidently, the centroids of the right isosceles triangles are integer points, and the conditions are all met.

On the other hand, assume that a square of side length k has the desired triangulation. Let V be the set of all vertices of the triangles. Define the binary relation $A \sim_0 A \sim B$ if there are two triangles $\triangle ACD$, $\triangle BCD$ in the triangulation, and say $A \sim B$ (they are in the same equivalent class) if and only if there exist vertices $A_1, \ldots, A_r$ such that $A \sim_0 A_1 \sim_0 \cdots \sim_0 A_r \sim_0 B$. For an arbitrary point P with coordinates x_P, y_P, we have:

(i) If $A \sim B$, then $3|(x_A - x_B)$, $3|(y_A - y_B)$. Indeed, by transitivity, we may assume $A \sim_0 B$. Then there exist $\triangle ACD$ and $\triangle BCD$ whose centroids are integer points, which implies $3|(x_A + x_C + x_D)$, $3|(x_B + x_C + x_D)$, and hence $3|(x_A - x_B)$; similarly, $3|(y_A - y_B)$.

(ii) There are at most three equivalent classes. Fix a triangle T_0 and let A be an arbitrary vertex in V. There exist a sequence of triangles $T_0, \ldots, T_r$ such that T_{i-1} and T_i share a common side ($i = 1, \ldots, r$) and A is a vertex of T_r. By definition of the equivalent classes, each vertex of T_{i-1} is equivalent to a vertex of T_i. Then by transitivity, A is equivalent to some vertex of T_0.

According to (ii) and the pigeonhole principle, two vertices of the large square must be in the same equivalent class. By (i), it follows that $3|k^2$ or $3|2k^2$, and $3|k$ is verified. □

5 Prove: there exist $C > 0$ and $\alpha > \frac{1}{2}$, such that for any positive integer n, there exists a subset $A \subseteq \{1, 2, \ldots, n\}$, $|A| \geq Cn^{\alpha}$, and the difference of any two elements of A is not a perfect square.

(Contributed by the contest committee)

Solution For $n \geq 25$, let $5^{2t} \leq n < 5^{2t+2}$ ($t \in \mathbb{N}_+$). Take

$$A = \{(\alpha_{2t}, \ldots, \alpha_1)_5 | \alpha_{2i} \in \{0, 1, 2, 3, 4\}, \quad i = 1, \ldots, t;$$
$$\alpha_{2i-1} \in \{1, 3\}, \quad i = 1, \ldots, t\}$$

where $(\alpha_{2t}, \ldots, \alpha_1)_5$ is the base-5 representation of $m = 5^{2t-1}\alpha_{2t} + \cdots + 5\alpha_2 + \alpha_1$. Obviously, $A \subset \{1, 2, \ldots, n\}$. For $u_1, u_2 \in A$, $u_1 = (a_{2t}, \ldots, a_1)$, $u_2 = (b_{2t}, \ldots, b_1)$, $u_1 > u_2$, we claim that $u_1 - u_2$ is not a perfect square. Suppose $a_s \neq b_s$ is the first pair of distinct digits of u_1 and u_2 under base 5, namely, $a_1 = b_1, \ldots, a_{s-1} = b_{s-1}$, $a_s \neq b_s$. Then

$$u_1 - u_2 = (a_{2t} - b_{2t})5^{2t-1} + \cdots + (a_s - b_s)5^{s-1}.$$

If $2|s$, then $5^{s-1} \| (u_1 - u_2)$ ($a_s - b_s \neq 0$ is between -4 and 4), $u_1 - u_2$ cannot be a perfect square.

If $2+$, then $\frac{u_1 - u_2}{5^{s-1}} = (a_{2t} - b_{2t})5^{2t-1} + \cdots + (a_s - b_s) \in \mathbb{Z}$. Suppose $u_1 - u_2$ is a perfect square. From $2|(s-1)$, we know that $\frac{u_1 - u_2}{5^{s-1}}$ is also a perfect square; however, from $2|(s-1)$ and $a_s \neq b_s$, it must be true that $\{a_s, b_s\} = \{1, 3\}$, and $\frac{u_1 - u_2}{5^{s-1}} \equiv 2$ or $3 \pmod 5$, which indicate that

$\dfrac{u_1 - u_2}{5^{s-1}}$ cannot be a perfect square. This proves the claim and A satisfies the problem requirement.

Since $|A| = 10^t$, we take $\alpha = \log_{25} 10 > \dfrac{1}{2}$ and arrive at

$$n^\alpha < 5^{(2t+2)} \log_{25} 10 = 10^{t+1} = 10|A|.$$

Now $C = \dfrac{1}{24}$, $\alpha = \log_{25} 10$ ($\alpha \in (0,1)$) will suffice.

For $n \leq 24$, take $A = \{1\} : |A| \geq \dfrac{1}{24}n \geq \dfrac{1}{24}n^\alpha$ holds as well.

In conclusion, for any $n \in \mathbb{N}_+$, we can find suitable A, such that $|A| \geq Cn^\alpha$. □

Remark One can also consider the residues modulo 16 and take

$$A = \{(\alpha_t, \ldots, \alpha_1)_{16} | \alpha_i \in \{2, 5, 7, 13, 15\},\ i = 1, \ldots, t\}. \qquad \square$$

6 (1) Prove: on the complex plane, the convex hull of the set of zeros of $z^{20} + 63z + 22 = 0$ has area greater than π.

(2) Let n be a positive integer, $1 \leq k_1 < k_2 < \cdots < k_n$ be n odd integers. Prove: for any n complex numbers $a_1, a_2, \ldots, a_n$, $\sum\limits_{i=1}^{n} a_i = 1$, and for any complex number w, $|w| \geq 1$, the equation

$$a_1 z^{k_1} + a_2 z^{k_2} + \cdots + a_n z^{k_n} = w$$

has at least one root whose magnitude does not exceed $3n|w|$.

(Contributed by Yao Yijun)

Solution (1) To begin, we prove

Lemma (Gauss-Lucas Theorem)[1] *If $f(z)$ is a polynomial with complex coefficients, then all zeros of $f'(z)$ belong to the convex hull of the set of zeros of $f(z)$.*

[1] This result was first used by Gauss implicitly in 1836, and was first proved by French mathematician Édouard Lucas in 1874. Here, we adopt the main idea of Lucas; the detailed explanation can be found in Hungarian mathematician Lipót Fejér's paper, *Sur la racine de moindre module d'une équation algébrique* (*C.R.A.S.* *145*(*1907*), *459–461*).

Proof of lemma By the fundamental theorem of algebra, $f(z)$ can be written as

$$f(z) = A(z - z_1)^{\alpha_1}(z - z_2)^{\alpha_2} \dots (z - z_n)^{\alpha_n}.$$

We infer that $f'(z)$ has a zero z_1 of multiplicity $(\alpha_1 - 1)$, a zero z_2 of multiplicity $(\alpha_2 - 1), \dots$, and a zero z_n of multiplicity $(\alpha_n - 1)$. For any other zero Z of $f'(z)$,

$$\frac{f'(Z)}{f(Z)} = \frac{\alpha_1}{Z - z_1} + \frac{\alpha_2}{Z - z_2} + \dots + \frac{\alpha_n}{Z - z_n} = 0,$$

that is,

$$\frac{\alpha_1}{|Z - z_1|^2}(\bar{Z} - \bar{z}_1) + \frac{\alpha_2}{|Z - z_2|^2}(\bar{Z} - \bar{z}_2) + \dots + \frac{\alpha_n}{|Z - z_n|^2}(\bar{Z} - \bar{z}_n) = 0.$$

Take the complex conjugate and shift the terms, obtaining

$$Z = \frac{\dfrac{\alpha_1}{|Z - z_1|^2} z_1 + \dfrac{\alpha_2}{|Z - z_2|^2} z_2 + \dots + \dfrac{\alpha_n}{|Z - z_n|^2} z_n}{\dfrac{\alpha_1}{|Z - z_1|^2} + \dfrac{\alpha_2}{|Z - z_2|^2} + \dots + \dfrac{\alpha_n}{|Z - z_n|^2}}.$$

Hence, Z is a convex combination of $z_1, z_2, \dots, z_n$, and the lemma is verified.

Return to the original problem. According to the lemma, the desired convex hull contains the convex hull of the zeros of

$$20z^{19} - 63 = 0,$$

which is a regular 19-gon of radius $\left(\frac{63}{20}\right)^{1/19}$, whose incircle has radius

$$\left(\frac{63}{20}\right)^{1/19} \cos\frac{\pi}{19} > \left(\frac{63}{20}\right)^{1/19} \left(1 - \frac{1}{2}\left(\frac{\pi}{19}\right)^2\right) > \left(\frac{63}{20}\right)^{1/19} \left(1 - \frac{1}{2}\left(\frac{1}{6}\right)^2\right)$$
$$= \frac{71}{72}\left(\frac{63}{20}\right)^{1/19}.$$

Since

$$\left(\frac{72}{71}\right)^{19} = \left(1 + \frac{1}{71}\right)^{19} < 1 + \frac{19}{71} + 18 \cdot \mathrm{C}_{19}^2 \frac{1}{72^2} < 3 < \frac{63}{20},$$

the incircle is larger than the unit circle, and thc conclusion follows.

(2) **Assertion** For $a_1, a_2, \ldots, a_n$ and w that satisfy the problem conditions, the minimum magnitude of the zeros of

$$a_1 z^{k_1} + a_2 z^{k_2} + \cdots + a_n z^{k_n} = w \qquad ①$$

does not exceed $3mn$.

Proof of the assertion Taking the reciprocal of z, it is equivalent by showing that the maximum magnitude M_0 of the zeros of

$$wz^{k_n} - a_1 z^{k_n - k_1} - a_2 z^{k_n - k_2} - \cdots - a_{n-1} z^{k_n - k_{n-1}} - a_n = 0 \qquad ②$$

satisfies $M_0 \geq \dfrac{1}{3mn}$.

By the lemma in (1), the maximum magnitude M_0 of ② is no less than that of the derivative equation

$$wk_n z^{k_n - 1} - a_1(k_n - k_1) z^{k_n - k_1 - 1} - \cdots - a_{n-1}(k_n - k_{n-1}) z^{k_n - k_{n-1}} - 1 = 0, \qquad ③$$

say M_1 of

$$wk_n z^{k_{n-1}} - a_1(k_n - k_1) z^{k_{n-1} - k_1} - a_2(k_{n-1} \\ -k_2) z^{k_{n-1}} - k_2 - \cdots - a_{n-1}(k_n - k_{n-1}) = 0. \qquad ④$$

Starting from (4), repeat the above operation n-s times. Every time, the maximum magnitude of the resulting equation is decreasing. Hence,

$$M_0 \geq M_1 \geq \cdots \geq M_{n-s},$$

where M_{n-s} is the maximum magnitude of

$$wk_n k_{n-1} \ldots k_{s+1} z^{k_s} - a_1(k_n - k_1)(k_{n-1} - k_1) \cdots (k_{s+1} - k_1) z^{k_s - k_1} \\ - \cdots - a_s(k_n - k_s) \cdots (k_{s+1} - k_s) = 0.$$

By Vieta's theorem,

$$M_0 \geq M_1 \geq \cdots \geq M_{n-s} \geq \left[\frac{(k_n - k_s) \cdots (k_{s+1} - k_s)}{k_n k_{n-1} \cdots k_{s+1}} \cdot \left| \frac{a_s}{w} \right| \right]^{1/k_s}.$$

Now it suffices to prove, there exists $s \in \{1, 2, \ldots, n\}$ such that

$$\left[\frac{k_n k_{n-1} \cdots k_{s+1}}{(k_n - k_s)(k_{n-1} - k_s) \cdots (k_{s+1} - k_s)} \right]^{1/k_s} \cdot \left[\left| \frac{w}{a_s} \right| \right]^{1/k_s} \leq 3mn.$$

Notice that

$$\frac{k_p}{k_p - k_s} = 1 + \frac{k_s}{k_p - k_s} \leq \left(1 + \frac{1}{k_p - k_s}\right)^{k_s}$$

and it suffices to have

$$\left[\prod_{p=s+1}^{n}\left(1 + \frac{1}{k_p - k_s}\right)\right] \cdot \left(\frac{m}{|a_s|}\right)^{1/k_s} \leq 3mm$$

for some s.[2] From the given conditions

- that all k_j's are odd, it follows that $k_p - k_s \geq 2(p-s)$ and
$$1 + \frac{1}{k_p - k_s} \leq 1 + \frac{1}{2(p-s)} < \frac{p-s+1}{p-s};$$
- that $\sum\limits_{j=1}^{n} a_j = 1$, it follows that $\sum\limits_{j=1}^{n} |a_j| \geq 1$, and there exists s, such that $|a_s| \geq \frac{1}{2^s}$, or $\frac{1}{|a_s|} \leq 2^s$.

There are two cases requiring final clarifications.

Case 1, there exists s, such that $k_s \geq 3$ (s could be 1) and $|a_s| \geq \frac{1}{2^s}$. Then (due to $s \leq 2s - 1 \leq k_s$)

$$\left[\prod_{p=s+1}^{n}\left(1 + \frac{1}{k_p - k_s}\right)\right] \cdot \left(\frac{m}{|a_s|}\right)^{1/k_s} < \left[\prod_{p=s+1}^{n} \frac{p-s+1}{p-s}\right] \cdot m^{1/k_s} \cdot 2^{s/k_s}$$
$$< n \cdot \sqrt[3]{m} \cdot 2 < 3mn.$$

Case 2, if case 1 does not occur, then $k_1 = 1$, $|a_1| \geq \frac{1}{2}$, and

$$\left[\prod_{p=2}^{n}\left(1 + \frac{1}{k_p - k_1}\right)\right] \times \frac{m}{|a_s|} \leq \left[\prod_{p=2}^{n} \frac{2(p-1)+1}{2(p-1)}\right] \times m \times 2$$
$$< \sqrt{\prod_{k=1}^{2n-2} \frac{k+1}{k}} \times 63 \times 2$$
$$= \sqrt{2n-1} \times m \times 2 < 3mn.$$

This completes the proof of the assertion. □

---

[2]The above operations come from Hungarian mathematician Michael Fekete's paper, *Analoga zu den Sätze von Rolle und Bolzano fÿr komplexe Polynome und Potenzreihen mit Lüchen* (*Jahr. der deutschen Math. Verieiningung, 32(1924), 299—306*).

Test IV, First Day
(8 am – 12:30 pm; April 16, 2022)

1 Initially, each unit square of an $n \times n(n \geq 2)$ grid is coloured red, yellow, or blue. In each second, the colours of the unit squares simultaneously change in the following way:

(1) if A is red and A shares a common side with a yellow square, then A turns yellow;
(2) if B is yellow and B shares a common side with a blue square, then B turns blue;
(3) if C is blue and C shares a common side with a red square, then C turns red;
(4) in all other cases, the colour does not change.

Prove: if the grid does not become monochromatic after $2n$-2 seconds, then it will never be so in finite time.

(Contributed by Leng Fusheng)

Solution 1 We use 0, 1, 2 for red, yellow, and blue colours, respectively. For two squares u, v (or two colours), define

$$w(u,v)=\begin{cases}-1, & \text{if } (u,v)=(0,1),(1,2) \text{ or } (2,0);\\ 1, & \text{if } (u,v)=(1,0),(2,1) \text{ or } (0,2);\\ 0, & \text{if } u=v.\end{cases}$$

In other words, $w(u,v) \equiv u-v \pmod 3$. Consider the simple graph G whose vertices are the n^2 unit squares, and two vertices are adjacent if and only if the two squares share a common side. For each unit square v, let $v^{(t)}$ represent the colour of v at t-th seconds. For every directed cycle $\alpha = v_1 \dots v_k v_1$ of G, define

$$w_t(\alpha)=\sum_{k=1}^{k} w(v_j^{(t)}, v_{j+1}^{(t)})$$

where the subscripts are taken modulo k. We claim that $w_t(\alpha)$ does not depend on t during the process. It suffices to prove

$$\sum_{j=1}^{k}\left(w\left(v_j^{(t+1)},\ v_{j+1}^{(t+1)}\right)-w\left(v_j^{(t)},v_{j+1}^{(t)}\right)\right)=0$$

or

$$w\left(v_j^{(t+1)}, v_{j+1}^{(t+1)}\right) - w\left(v_j^{(t)}, v_{j+1}^{(t)}\right) = w\left(v_j^{(t+1)}, v_j^{(t)}\right) - w\left(v_{j+1}^{(t+1)}, v_{j+1}^{(t)}\right) \quad ①$$

for each j.

First, notice that the two sides of ① are congruent modulo 3. Second, according to the rule, each term on the right-hand side must belong to $\{0,1\}$, and hence their difference belongs to $\{-1,0,1\}$. We show that the left-hand side cannot be ± 2: if it equals 2, then $w(v_j^{(t+1)}, v_{j+1}^{(t+1)} = 1)$ and $w(v_j^{(t)}, v_{j+1}^{(t)}) = -1$. Suppose $v_j^{(t)} = 0$, $v_{j+1}^{(t)} = 1$. By the rule, it must be $v_j^{(t+1)} = 1$, $v_{j+1}^{(t+1)} = 0$. However, a yellow square cannot turn red in the next second, a contradiction. In the same manner, it cannot be equal to -2, either. So, (*) is verified and $w_t(\alpha)$ is independent of t.

Evidently, if $w_0(\alpha) \neq 0$ for some directed cycle α, then the squares in α cannot ever become monochromatic. Assume that the grid becomes monochromatic in finite time; necessarily, assume $w_0(\alpha) = 0$ for every α. Then at $t = 0$, for each unit square v, we may assign a value $h_0(v) \in \mathbb{Z}$ to it such that for any vertices u, v of G and any directed path $\rho = uv_1 \ldots v_k v$, the following equation holds

$$w_0(u, v_1) + \left(\sum_{j=1}^{k-1} w_0(v_j, v_{j+1})\right) + w_0(v_k, v) = h_0(u) - h_0(v).$$

Suppose among all squares, u has the largest h_0 value. Due to the maximum value at u, u does not change colour in the next second. Since $w_t(\alpha)$ is independent of t, we have $w_1(\alpha) = 0$ for every directed cycle α. At $t = 1$, we may assign a value $h_1(v)$ to each square v in a similar way, satisfying $h_1(u) = h_0(u)$. We assert that $h_1(u)$ has the largest h_1 value as well. Suppose otherwise, then there exists v, $h_1(v) > h_1(u)$. Choose a path $uv_1 \ldots v_k v$, apply (*) to every edge, and take the summation of the equations to obtain

$$\begin{aligned}(h_1(u) - h_1(v)) - (h_0(u) - h_0(v)) = w(u^{(1)},\\ u^{(0)}) - w(v^{(1)}, v^{(0)}) = -w(v^{(1)}, v^{(0)})\end{aligned}$$

which implies $h_0(u) = h_0(v)$ and $v^{(1)} \neq v^{(0)}$, namely v also has the largest h_0 value. Yet by definition of h_0, the colour of v does not change, which is a contradiction.

Moreover, observe that for any v adjacent to u, $h_0(v) = h_0(u)$ or $h_0(u) - 1$. Either way, v must turn to u's colour in the next second.

Based on the above argument, we can use induction to show that u (whose h_0 value is maximal) never changes colour, and for each time t, h_t can be defined such that $h_t(u) = h_0(u)$. Now for any square v, if there is a path $\rho = uv_1 \dots v_k v$, then v turns to u's colour in at most $k+1$ seconds. As the distance between u and any other square is at most $2n-1$, if the grid indeed becomes monochromatic in finite time, then it will be so in at most $2n-2$ seconds. □

Solution 2 First, we provide and justify the following claim.

Claim If the grid becomes monochromatic in finite time, say blue, then there exists a unit square whose colour is always blue during the process.

Proof of claim Suppose the claim is untrue: the grid eventually becomes all blue, but every square has changed the colour. Construct a directed graph G as follows: let every unit square be a vertex, and draw $A \to B$ if (i) A and B are adjacent; (ii) there exists time t_0 such that A is not blue at $t = t_0 - 1$ but is blue at all $t \geq t_0$, and B is blue at $t = t_0 - 1$ (in other words, the last colour change of A is due to B). Since every square has changed the colour (by our assumption), the outdegree is ≥ 1 for each vertex, and G contains a cycle, say $A_1 \to A_2 \to \cdots \to A_k \to A_1$. Suppose A_i finally becomes blue at $t = t_i$, $1 \leq i \leq k$, and $t_1 = \max\{t_1, \dots, t_k\}$. As $A_1 \to A_2$, we infer that A_1 is yellow at $t = t_1 - 1$; as $A_k \to A_1$, we see that A_1 is blue at $t = t_k - 1$. Hence, $t_k < t_1$. Note that a blue square must turn red first, then turn yellow. So, there exists $t_k < t' < t_1$ such that A_1 is red at $t = t' - 1$. However, as $t' - 1 \geq t_k$, the colour of A_k at $t = t' - 1$ is blue, and A_1 will force A_k to turn red at $t = t'$. This is contradictory as we assumed A_k to be blue from the time $t = t_k$ on. The claim is now verified.

For the original problem, suppose the grid eventually turns all blue. We will prove it becomes so in $2n - 2$ seconds. First, define the distance of two squares as the sum of their vertical and horizontal distances. The farthest distance is $2n-2$. By the claim, there exists a unit square A which remains blue during the whole process. Consider any square B at distance 1 from A: B cannot be red; either B is originally blue, or B (yellow) turns blue at $t = 1$. Now we use induction to show any square B at distance k $(1 \leq k \leq 2n-2)$ from A will be blue at $t \geq k$. Suppose the conclusion holds for $k - 1$ and take B' adjacent to B and at distance $k - 1$ from A. By the hypothesis, B' is blue at $t \geq k - 1$. Then B turns blue no later than the instant $t = k$. This completes the induction and the proof of the statement. □

 As Fig. 2.1 shows, in the convex quadrilateral $ABCD$, I and J are the incentres of $\triangle ABC$ and $\triangle ADC$, respectively. It is known that IJ, AC, and BD meet at P; the line through P and perpendicular to BD meets the exterior angle bisectors of $\angle BAD$ and $\angle BCD$ at E and F, respectively. Prove: $PE = PF$.

(Contributed by Lin Tianqi)

Solution 1 If $AB // CD$ and $AD // BC$, $ABCD$ is a parallelogram, P is the midpoint of AC and $AE // CF$, hence $PE = PF$. In the following, assume AB is not parallel to CD.

First, we prove $AB + AD = CB + CD$. As illustrated in Fig. 2.2, let the extensions of BA and CD meet at T; the B-escribed circle $\odot K$ of $\triangle TBC$ touches the lines AB, BC, and CD at X, Y, and Z, respectively. Since the internal homothetic center of $\odot I$ and $\odot J$ lies on the line IJ and AC is a common internal tangent of the two circles, it follows that the internal homothetic centre of $\odot I$ and $\odot J$ is P.

It is well known that the internal homothetic centre P of $\odot I$ and $\odot J$, the external homothetic centre B of $\odot I$ and $\odot K$, and the internal homothetic centre say G of $\odot J$ and $\odot K$, the above three points are collinear. Therefore, G lies on the line BP. Furthermore, CD is a common internal tangent of $\odot J$ and $\odot K$. We conclude that the internal homothetic centre G of $\odot J$ and $\odot K$ is D, AD is tangent to $\odot K$, say the point of tangency is at W.

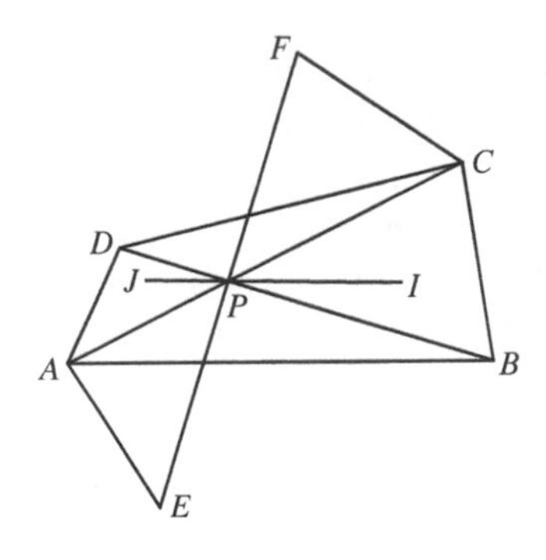

Fig. 2.1

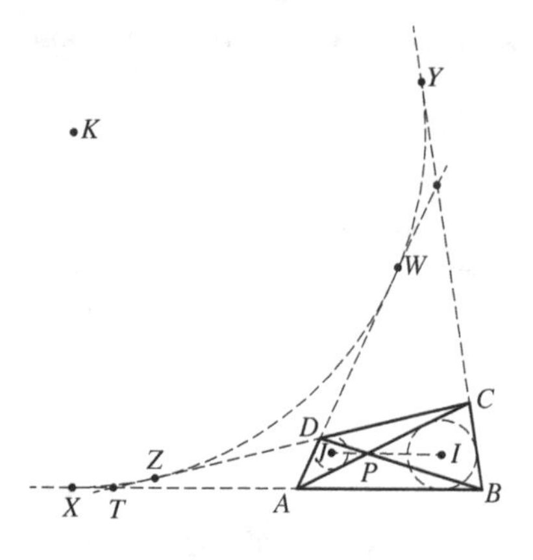

Fig. 2.2

By theorem of length of tangent,

$$\begin{aligned} AB + AD &= (BX - AX) + (AW - DW) \\ &= BX - DW = BY - DZ = (CB + CY) - (CZ - CD) \\ &= CB + CD, \end{aligned}$$

that is,

$$AB + AD = CB + CD. \qquad ①$$

Let $\odot U$ and $\odot V$ be the B-escribed and D-escribed circles of $\triangle ABD$, respectively. Let $\odot Q$ and $\odot R$ be the B-escribed and D-escribed circles of $\triangle BCD$, respectively. We prove UR and VQ meet at P.

Consider $\odot K$, $\odot U$, and $\odot R$. It is well known that the external homothetic centre A of $\odot K$ and $\odot U$, the internal homothetic centre C of $\odot K$ and $\odot R$, and the internal homothetic centre of $\odot U$ and $\odot R$, the above three points are collinear. As BD is a common internal tangent of $\odot U$ and $\odot R$, we conclude that the internal homothetic centre of $\odot U$ and $\odot R$ is P, UR passes through P; similarly, VQ passes through P, too.

Now, we show $UQ \perp BD$ and $VR \perp BD$. As Fig. 2.3 shows, $\odot U$ and $\odot Q$ touch the extension of BD at L and L', respectively. Clearly,

$$BL = \frac{1}{2}(AB + AD + BD),$$

$$BL' = \frac{1}{2}(CB + CD + BD).$$

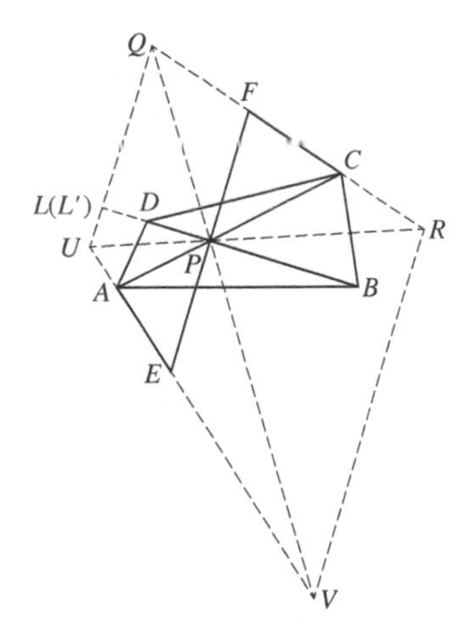

Fig. 2.3

From ①, it follows that $BL = BL'$, $L = L'$, and hence $UQ \perp BD$. Likewise, $VR \perp BD$. As $EF \perp BD$, we arrive at $UQ // EF // VR$ and

$$\frac{PE}{UQ} = \frac{VP}{VQ} = \frac{RF}{RQ} = \frac{PF}{UQ},$$

and thus $PE = PF$. □

Solution 2 (organized from Chen Ziqing's proof). As shown in Fig. 2.4, if $AB // CD$ and $AD // BC$, $ABCD$ is a parallelogram, P is the midpoint of AC and $AE // CF$, hence $PE = PF$. In the following, assume AB is not parallel to CD.

Similar to the first proof, we obtain $\odot O$, the excircle of the ex-tangential quadrilateral $ABCD$; suppose $\odot O$ touches the lines AB, BC, CD, and DA at X, Q, Y, and Z, respectively. Let $AD \cap BC = S$, $AB \cap CD = R$.

According to Newton's theorem, in every tangential quadrilateral other than a rhombus, the centre of the incircle lies on the Newton line. It follows that the four lines AC, BD, XY, and ZQ meet at P; the four lines AC, RS, YZ, and XQ meet at N; the four lines BD, RS, XZ, QY meet at M.

Now consider the cyclic quadrilateral $XYZQ$: clearly, both M and P lie on the polar line l_N of N with respect to $\odot O$, namely $MP = l_N$, which leads to $ON \perp PM$, or $ON \perp BD$. In the complete quadrangle $BARDSC$, A, C; P, N form a harmonic range of points, implying that ON, OP; OA, OC form a harmonic pencil of straight lines. Since $EF \perp BD$, it follows that $EF // NO$ and P is the midpoint of EF. □

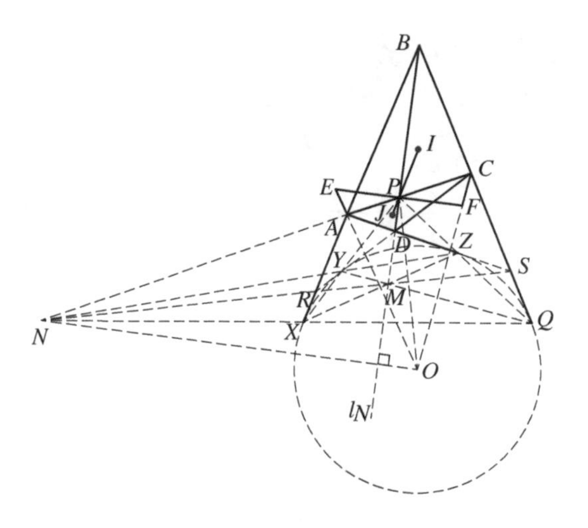

Fig. 2.4

3 Find all functions $f:\mathbb{R}\to\mathbb{R}$, such that for any real numbers x and y, the following multisets are identical:

$$\{f(xf(y)+1), f(yf(x)-1)\}=\{xf(f(y))-1, yf(f(x))+1\}. \quad (*)$$

Note: two multisets $\{a,b\}$ and $\{c,d\}$ are identical if and only if $a=c$, $b=d$, or $a=d$, $b=c$.

(Contributed by Xiao Liang)

Solution The desired functions are $f(x)=x$ and $f(x)=-x$. It is straightforward to check that both satisfy (*).

Throughout the argument, a set refers to a multiset. Taking $x=y=0$ in (*), we obtain $\{f(1), f(-1)\}=\{1,-1\}$. First, we assume $f(1)=1$ and prove $f(x)=x (x\in\mathbb{R})$.

Step 1. $f(n)=n (n\in\mathbb{Z})$.

We show $f(0)=0$ first. Taking $x=0$ in (*) to obtain

$$\{f(1), f(yf(0)-1)\}=\{-1, yf(f(0))+1\}.$$

If $f(0)\neq 0$, take $y=\dfrac{2}{f(0)}$. Then $\{1,1\}$ on the left-hand side cannot include -1, a contradiction.

Now use induction to prove $f(n) = n (n \in \mathbb{N})$. For $n = 1$, it is true. Suppose $f(m) = m$ for all $m \leq n$. Taking $x = 1$, $y = n$ in (*) to get

$$\{f(n+1), f(n-1)\} = \{n-1, n+1\},$$

and hence $f(n+1) = n+1$, completing the induction. In a similar way, taking $x = 1$, $y = -n$, we can use induction to verify $f(-n) = -n$ for $n \in \mathbb{N}$.

Step 2. f is a bijection.

First, take $y = 1$ in (*): on the right-hand side, $xf(f(1)) - 1 = x - 1$ can be any real number and it is in the image set of f. Hence, f is onto.

Next, assume $f(y_0) = 0$. If $y_0 \neq 0$, take $y = y_0$ in (*) and obtain

$$\{f(1), f(y_0 f(x) - 1)\} = \{-1, y_0 f(f(x)) + 1\}.$$

We must have $1 = f(1) = y_0 f(f(x)) + 1$, or $f(f(x)) = 0$ for all x, which is absurd as $f(1) = 1$. Therefore, $y_0 = 0$.

Now we prove f is one-to-one. If $y_1 \neq y_2$ satisfy $f(y_1) = f(y_2) \neq 0$, take $y = y_1$ and $y = y_2$ respectively in (*) to get

$$\{f(xf(y_1) + 1), f(y_1 f(x) - 1)\} = \{xf(f(y_1)) - 1, y_1 f(f(x)) + 1\},$$

$$\{f(xf(y_2) + 1), f(y_2 f(x) - 1)\} = \{xf(f(y_2)) - 1, y_2 f(f(x)) + 1\}.$$

Observe in the above two relations that $f(xf(y_1) + 1) = f(xf(y_2) + 1)$, and $xf(f(y_1)) - 1 = xf(f(y_2)) - 1$. If, for some $x_0 \in \mathbb{R}$, $f(x_0 f(y_1) + 1) \neq x_0 f(f(y_1)) - 1$, then

$$y_1 f(f(x_0)) + 1 = f(xf(y_1) + 1) = f(xf(y_2) + 1) = y_2 f(f(x_0)) + 1.$$

It follows that $y_1 f(f(x_0)) = y_2 f(f(x_0))$, further $f(f(x_0)) = 0$, which gives $x_0 = 0$. Consequently, for $x \neq 0$, $f(xf(y_1) + 1) = xf(f(y_1)) - 1$. Now

$$f(x) = ax + b \ (a, b \in \mathbb{R},\ x \neq 1)$$

is linear. From step 1, $f(n) = n (n \in \mathbb{Z})$, we get $a = 1$, $b = 0$. Evidently, $f(x) = x$ is one-to-one. Hence, f is a bijection.

Step 3. For any $n \in \mathbb{Z}$, $y \in \mathbb{R}$,

$$f\left(f\left(\frac{n}{f(y)}\right)\right) = \frac{n}{y}. \qquad ①$$

It suffices to consider $n \neq 0$ and $y \neq 0$. Take $x = \dfrac{n}{f(y)}$ in (*) to obtain

$$\left\{n+1, f\left(yf\left(\frac{n}{f(y)}\right)-1\right)\right\} = \left\{n\frac{f(f(y))}{f(y)}-1, yf\left(f\left(\frac{n}{f(y)}\right)\right)+1\right\}. \quad ②$$

Replace x by y and replace y by $\dfrac{n}{f(y)}$ in (*) to obtain

$$\left\{n-1, f\left(yf\left(\frac{n}{f(y)}\right)+1\right)\right\} = \left\{n\frac{f(f(y))}{f(y)}+1, yf\left(f\left(\frac{n}{f(y)}\right)\right)-1\right\}. \quad ③$$

If, for some $y_0 \in \mathbb{R}\backslash\{0\}$, $n \neq y_0 f\left(f\left(\dfrac{n}{f(y_0)}\right)\right)$, then ② and ③ give

$$n+1 = n\frac{f(f(y_0))}{f(y_0)}-1, \quad n-1 = n\frac{f(f(y_0))}{f(y_0)}+1$$

respectively. Taking the difference to yield $2 = -2$, a contradiction. This verifies ①.

Step 4. For $\alpha \in \mathbb{Q}$ and $y \in \mathbb{R}$, $f(\alpha y) = \alpha f(y)$.

It suffices to consider $\alpha \neq 0$ and $y \neq 0$. In ①, replace y by $f\left(\dfrac{m}{f(y)}\right)$ $(m \in \mathbb{Z}\backslash\{0\})$ to find

$$f\left(f\left(\frac{n}{f\left(f\left(\frac{m}{f(y)}\right)\right)}\right)\right) = \frac{n}{f\left(\frac{m}{f(y)}\right)}.$$

Plug it into ①, the denominator of the left-hand side:

$$f\left(f\left(\frac{n}{m}y\right)\right) = \frac{n}{f\left(\frac{m}{f(y)}\right)}.$$

In the above, replace y by $\dfrac{t}{f(y)}$ $(t \in \mathbb{Z}\backslash\{0\})$ to find

$$\frac{nt/m}{y} = f\left(f\left(\frac{n}{m}\frac{t}{f(y)}\right)\right) = \frac{n}{f\left(\frac{m}{f\left(\frac{t}{f(y)}\right)}\right)}.$$

After simplification, $f\left(\dfrac{m}{f\left(\dfrac{t}{f(y)}\right)}\right)=\dfrac{m}{t}y$. Now apply f on both sides to derive

$$\frac{m}{t}f(y)=f\left(f\left(\frac{m}{f\left(\frac{t}{f(y)}\right)}\right)\right)=f\left(\frac{m}{t}y\right).$$

This justifies $f(\alpha y)=\alpha f(y)$.

Step 5. For $y\in\mathbb{R}$ and $a\in\mathbb{Q}$, $f(y+a)=f(y)+a$.

Take $x=\dfrac{1}{r}\in\mathbb{Q}\backslash\{0\}$ in (*) to derive

$$\left\{f\left(\frac{1}{r}f(y)+1\right),\quad f\left(\frac{1}{r}y-1\right)\right\}=\left\{\frac{1}{r}f(f(y))-1,\frac{1}{r}y+1\right\}.$$

By the result of step 4, multiply on both sides by r to get

$$\{f(f(y)+r),f(y-r)\}=\{f(f(y))-r,y+r\}. \qquad ④$$

We shall prove $f(y-r)=f(f(y))-r$. If, for some $y_0\in\mathbb{R}$, $r_0\in\mathbb{R}\backslash\{0\}$, the following equation holds,

$$f(y_0-r_0)=y_0+r_0,$$

then take $y=y_0-r_0$, $r=-2r_0$ in ④ to obtain

$$\{f(f(y_0-r_0)-2r_0),f(y_0-r_0+2r_0)\}=\{f(f(y_0-r_0))+2r_0,y_0-3r_0\}.$$

After simplification, it becomes

$$\{y_0+r_0,f(y_0+r_0)\}=\{f(y_0+r_0)+2r_0,y_0-3r_0\}.$$

Since $y_0+r_0\neq y_0-3r_0$, it must be $y_0+r_0=f(y_0+r_0)+2r_0$, namely

$$f(y_0+r_0)=y_0-r_0.$$

However, we arrive at $\{y_0+r_0,y_0-r_0\}=\{y_0+r_0,y_0-3r_0\}$, which cannot hold, a contradiction. Therefore, $f(y-r)=f(f(y))-r$ holds for all $y\in\mathbb{R}$, $r\in\mathbb{Q}\backslash\{0\}$. Let $r=-a$ and $r=0$ respectively, then take subtraction to deduce $f(y+a)=f(y)+a$ for all $y\in\mathbb{R}$, $a\in\mathbb{Q}$.

Step 6. $f(y) = y(y \in \mathbb{R})$.

Combine step 5 and ④ to derive

$$\{f(f(y)) + r, f(y) - r\} = \{f(f(y)) - r, y + r\}.$$

Taking the sum of the two elements of each multiset, we are led to $f(f(y)) + f(y) = f(f(y)) + y$, or $f(y) = y$. So, $f(1) = 1$ indeed implies that $f(x) = x(x \in \mathbb{R})$.

Now, we assume $f(1) = -1$ and prove $f(x) = -x$ $(x \in \mathbb{R})$.

Step 1. $f(n) = -n(n \in \mathbb{Z})$.

In (*), take $y = 1$ and $x = 1, -1$, respectively, to get

$$\{f(0), f(-2)\} = \{0, 2\}, \quad \{f(2), f(0)\} = \{-2, 0\}.$$

It follows that $f(0) = 0$, $f(2) = -2$, $f(-2) = 2$.

Now take induction on $|n|$ to show $f(n) = -n$. For $|n| = 1, 2$, the conclusion is already verified; suppose $f(n) = -n$ for $|n| \leq n_0$ $(n_0 \geq 2)$. In (*), take $y = 1$ and $x = n_0, -n_0$, respectively to obtain

$$\{f(-n_0 + 1), f(f(n_0) - 1)\} = \{n_0 - 1, f(f(n_0)) + 1\},$$
$$\{f(n_0 + 1), f(f(-n_0) - 1)\} = \{-n_0 - 1, f(f(-n_0)) + 1\}.$$

Together, they give $f(n_0+1) = -n_0-1$, $f(-n_0-1) = n_0+1$, completing the induction for step 1.

Step 2. For all $x \in \mathbb{R}$, $f(x) = -x$.

First, similar to $f(1) = 1$, we can show that f is onto.

For nonzero integer n and $z \in \mathbb{R}\backslash\{0\}$, take

$$x = \frac{n}{f(z)}, \quad y = z,$$

and $x = z$, $y = \dfrac{n}{f(z)}$, respectively, to obtain

$$\left\{-n-1, f\left(zf\left(\frac{n}{f(z)}\right) - 1\right)\right\} = \left\{n\frac{f(f(z))}{f(z)} - 1, zf\left(f\left(\frac{n}{f(z)}\right)\right) + 1\right\},$$
$$\left\{-n+1, f\left(zf\left(\frac{n}{f(z)}\right) + 1\right)\right\} = \left\{n\frac{f(f(z))}{f(z)} + 1, zf\left(f\left(\frac{n}{f(z)}\right)\right) - 1\right\}.$$

⑤

If, for some $z = z_0 \in \mathbb{R}\backslash\{0\}$, $f(f(z_0)) \neq -f(z_0)$, then the two relations in ⑤ give

$$-n-1 = zf\left(f\left(\frac{n}{f(z)}\right)\right)+1,$$

$$-n+1 = zf\left(f\left(\frac{n}{f(z)}\right)\right)-1.$$

Take the subtraction to arrive at $2 = -2$, a contradiction.

Therefore, $f(f(z_0)) = -f(z_0)$. Since f is onto, $f(x) = -x$ holds for all $x \in \mathbb{R}\backslash\{0\}$. As we have $f(0) = 0$, it must be $f(x) = -x$ $(x \in \mathbb{R})$.

In conclusion, $f(x) = x(x \in \mathbb{R})$ and $f(x) = -x$ $(x \in \mathbb{R})$ are the only two functions that satisfy the problem conditions. □

Test IV, Second Day
(8 am – 12:30 pm; April 17, 2022)

4 Find all positive integers a, b, c and prime p satisfying that

$$2^a p^b = (p+2)^c + 1.$$

(Contributed by Qu Zhenhua)

Solution Clearly, p is odd, $p \geq 3$. If $c = 1$, then $p + 3 = 2^a p^b \geq 2p \geq p + 3$, equality holds only when $p = 3$, $a = b = 1$. We obtain a solution $(p, a, b, c) = (3, 1, 1, 1)$. In the following, assume $c \geq 2$.

Case 1: c is odd. Let q be a prime factor of c. Since

$$(p+2)^q + 1 | (p+2)^c + 1,$$

$$\text{hence } (p+2)^q + 1 = 2^\alpha p^\beta. \qquad ①$$

Obviously, $\alpha > 0$. Observe that $(p+2)^q + 1 = (p+3)A$, in which

$$\begin{aligned} A &= (p+2)^{q-1} - (p+2)^{q-2} + \cdots + 1 \\ &> (p+2)^{q-1} - (p+2)^{q-2} \\ &= (p+2)^{q-2}(p+1) > p^{q-1}, \end{aligned}$$

and A is odd. Hence, A is a power of p, $A \geq p^q$, $\beta \geq q$. Taking ① modulo p, we have

$$2^q \equiv -1 \pmod{p},$$

indicating that the order of 2 modulo p is 2 or $2q$.

If the order of 2 modulo p is 2, then $p = 3$. Now ① becomes $5^q + 1 = 2^\alpha 3^\beta$. As $5^q + 1 \equiv 2 \pmod 4$, $\alpha = 1$. By the lifting-the-exponent lemma, $v_3(5^q + 1) = v_3(5 + 1) + v_3(q) \leq 2$, and hence $\beta \leq 2$. Checking $\beta = 1, 2$, neither satisfies the desired equation.

If the order of 2 modulo p is $2q$, then $2q|(p-1)$, implying that $q \leq \dfrac{p-1}{2} < \dfrac{p}{2}$. In ①, divide p^q on both sides and use the inequality $\left(1 + \dfrac{1}{x}\right)^x < \mathrm{e}$ for $x \geq 1$, to derive

$$2^\alpha p^{\beta - q} = \left(1 + \frac{2}{p}\right)^q + p^{-q} < \left(1 + \frac{2}{p}\right)^{\frac{p}{2}} + p^{-q} < \mathrm{e} + 3^{-3} < 3.$$

Hence, $\beta = q$, $\alpha = 1$. Then, from $2 \cdots p^q = (p+2)^q + 1 = (p+3)A$, we get

$$A = p^q, \quad p + 3 = 2,$$

a contradiction.

Case 2: c is even, $2^d \| c$, $d \geq 1$. By the assumption, $(p+2)^{2^d} + 1 | (p+2)^c + 1$, and thus

$$(p+2)^{2d} + 1 = 2^\alpha p^\beta.$$

Since $(p+2)^{2^d} + 1 \equiv 2 \pmod 4$, it must be $\alpha = 1$,

$$(p+2)^{2^d} + 1 = 2 \cdot p^\beta. \tag{2}$$

Taking ② modulo p, $2^{2^d} \equiv -1 \pmod p$, we infer that the order of 2 modulo p is 2^{d+1}, and hence $2^d < \dfrac{p}{2}$. As

$$p^{\beta+1} > 2 \cdot p^\beta = (p+2)^{2^d} + 1 > p^{2^d},$$

it follows that $\beta \geq 2^d$. In ②, divide both sides by p^{2^d} to find

$$\begin{aligned} 2 \cdot p^{\beta \cdot 2^d} &= \left(1 + \frac{2}{p}\right)^{2^d} + p^{-2^d} \\ &< \left(1 + \frac{2}{p}\right)^{\frac{p}{2}} + p^{-2^d} \\ &< \mathrm{e} + 3^{-2} < 3, \end{aligned}$$

and so $\beta = 2^d$.

If $d \geq 2$, then from $2^{d+1}|(p-1)$ it follows that $p \equiv 1 \pmod 8$. By ②, we have

$$p^{2^d} - 1 = (p+2)^{2^d} - p^{2^d}.$$

Analyze the 2-adic orders on both sides: $v_2(p^{2^d} - 1) = v_2(p^2 - 1) + d - 1 \geq d + 3$, while

$$\begin{aligned} v_2((p+2)^{2^d} - p^{2^d}) &= v_2((p+2)^2 - p^2) + d - 1 \\ &= v_2(2) + v_2(2p+2) + d - 1 = d + 2, \end{aligned}$$

a contradiction. Thus, $d = 1$, $2p^2 = (p+2)^2 + 1$, yielding $p = 5$. Now return to the original equation, where $c = 2k$, k being odd, and $a = 1$. We have

$$2 \cdot 5^b = 7^{2k} + 1.$$

By the lifting-the-exponent lemma, $b = v_5(7^{2k} + 1) = v_5(7^2 + 1) + v_5(k) \leq 2 + \frac{k}{5}$. If $k \geq 3$, then

$$2 \cdot 5^b \leq 2 \cdot 5^{2+\frac{k}{5}} = (7^2 + 1) \cdot 5^{\frac{k}{5}} < 7^{2k} + 1,$$

which is impossible. Consequently, $k = 1$, $c = 2$, $b = 2$, and we derive another solution $(p, a, b, c) = (5, 1, 2, 2)$.

In all, there are two solutions (p, a, b, c): (3, 1, 1, 1) and (5, 1, 2, 2).

□

5 Let n be a positive integer, $x_1, x_2, \ldots, x_{2n}$ be nonnegative real numbers satisfying that $x_1 + x_2 + \cdots + x_{2n} = 4$. Prove: there exist nonnegative integers p and q such that $q \leq n - 1$ and

$$\sum_{i=1}^{q} x_{p+2i-1} \leq 1, \quad \sum_{i=q+1}^{n-1} x_{p+2i} \leq 1.$$

Note 1: the subscripts are modulo $2n$, that is, $k \equiv l \pmod{2n}$ implies $x_k = x_l$.

Note 2: if $q = 0$, the first sum is 0; if $q = n-1$, the second sum is 0.

(Contributed by the contest committee)

Solution 1 Divide $x_1, x_2, \ldots, x_{2n}$ into two groups by the parity of the subscripts: $A = x_1 + x_3 + \cdots + x_{2n-1}$, $B = x_2 + x_4 + \cdots + x_{2n}$, and define the partial sums $A(0) = B(0) = 0$,

$$\begin{cases} A(2k+1) = A(2k) + x_{2k+1}, \\ B(2k+1) = B(2k), \end{cases} \quad \begin{cases} A(2k+2) = A(2k+1), \\ B(2k+2) = B(2k+1) + x_{2k+2}, \end{cases}$$
$$k = 0, 1, 2, \ldots.$$

The subscripts are modulo $2n$; the partial sums increase periodically: $A(k+2n) = A(k) + A$, $B(k+2n) = B(k) + B$.

Furthermore, we turn the partial sums into (piecewise linear) continuous functions, that is, for nonnegative real number t, define

$$A(t) = A(\lfloor t \rfloor) + (t - \lfloor t \rfloor)[A(\lfloor t \rfloor + 1) - A(\lfloor t \rfloor)],$$
$$B(t) = B(\lfloor t \rfloor) + (t - \lfloor t \rfloor)[B(\lfloor t \rfloor + 1) - B(\lfloor t \rfloor)].$$

Then $A(\cdot)$ and $B(\cdot)$ are both non-decreasing continuous, periodic functions on $\mathbb{R}_{\geq 0}$.

As $A + B = 4$, there exists a positive integer L such that $\left\lfloor \frac{L}{A} \right\rfloor + \left\lfloor \frac{L}{B} \right\rfloor \geq L$. For $l = 0, 1, 2, \ldots, L$, by continuity of $A(\cdot)$, we may choose suitable $t_l \in \mathbb{R}_{\geq 0}$ such that $A(t_l) = l$ (let $t_0 = 0$). Since $L \geq A \left\lfloor \frac{L}{A} \right\rfloor = A\left(2n \cdot \left\lfloor \frac{L}{A} \right\rfloor\right)$, we may further require that $t_L \geq 2n \cdot \left\lfloor \frac{L}{A} \right\rfloor$. Now,

$$B(t_L) - B(t_0) = B(t_L) \geq B\left(2n \cdot \left\lfloor \frac{L}{A} \right\rfloor\right) \geq B\left\lfloor \frac{L}{A} \right\rfloor$$
$$\geq B\left(L - \left\lfloor \frac{L}{B} \right\rfloor\right) \geq (B-1) \cdot L.$$

Consequently, there exists some $l = 0, 1, \ldots, L-1$ such that $B(t_{l+1}) - B(t_l) \geq (B-1)$, or

$$B(t_l + 2n) - B(t_{l+1}) \leq 1.$$

Take nonnegative integers c and d satisfying $2c \le t_l \le 2c+2$, $2d-1 \le t_{l+1} \le 2d+1$. We have

$$\begin{aligned} A(t_1) &\le A(2c+1) = A(2c+2), \\ A(t_{l+1}) &\ge A(2d) = A(2d-1); \\ B(t_1) &\ge B(2c+1) = B(2c), \\ B(t_{l+1}) &\le B(2d) = B(2d+1). \end{aligned}$$

It turns out that

$$\begin{aligned} x_{2c+3} + x_{2c+5} + \cdots + x_{2d-1} &= A(2d) - A(2c+1) \le A(t_{l+1}) - A(t_l) = 1, \\ x_{2d+2} + x_{2d+4} + \cdots + x_{2c+2n} &= B(2c+1+2n) - B(2d) \\ &\le B(t_l + 2n) - B(t_{l+1}) \le 1, \end{aligned}$$

namely $p = 2c+2$, $q = d-c-1$ satisfy the problem conditions (if $q < 0$, let $q = 0$; if $q \ge n$, let $q = n-1$). Done. □

Solution 2 Let $x_1 + x_3 + \cdots + x_{2n-1} = A$, $x_2 + x_4 + \cdots + x_{2n} = B$.

If one of A and B is less than or equal to 1, the conclusion is trivial. Suppose $A > 1$, $B > 1$, and for $0 \le k \le n-1$, let $m(k) \in \{1, 2, \ldots, n-1\}$ satisfy

$$\sum_{i=0}^{m(k)} x_{2k+2i+1} > 1, \quad ①$$

$$\sum_{i=0}^{m(k)-1} x_{2k+2i+1} \le 1. \quad ②$$

Notice that if $x_{2k+2m(k)+2} + x_{2k+2m(k)+4} + \cdots + x_{2k+2n-2} \le 1$, then this inequality and ② together give the answer $p = 2k$, $q = m(k)$. In the following, assume

$$x_{2k+2m(k)+2} + x_{2k+2m(k)+4} + \cdots + x_{2k+2n-2} > 1. \quad ③$$

Treat $0, 1, 2, \ldots, n-1$ as vertices and draw directed edges from k to $k + m(k) + 1 \pmod n$. It is easy to see that the graph contains a cycle, say $k_1 \to k_2 \to \cdots \to k_t \to k_1$ is one with the shortest length.

If $t = 1$, $\sum_{i=0}^{n-2} x_{2k_1+2i+1} \leq 1$, then take $p = 2k_1$, $q = n-1$.

If $t > 1$, let

$$\left\{\frac{k_2 - k_1}{n}\right\} + \left\{\frac{k_3 - k_2}{n}\right\} + \cdots + \left\{\frac{k_t - k_{t-1}}{n}\right\} + \left\{\frac{k_1 - k_t}{n}\right\} = s.$$

For $k = k_1, k_2, \ldots, k_t$, there are totally $s \cdot n$ terms in ①; by definition of directed edges, each term of $x_1, x_3, \ldots, x_{2n-1}$ is added s times. Meanwhile, for $k = k_1, k_2, \ldots, k_t$, there are totally $(t-s) \cdot n$ terms in ③, and from directed edges we infer that each of $x_2, x_4, \ldots, x_{2n}$ is added $t - s$ times, because x_{2j} is added in ③ if and only if x_{2j+1} is not added in ①.

Now we sum all ①'s for $k = k_1, k_2, \ldots, k_t$ to obtain $s \cdot A > t$; likewise, sum all ③'s for $k = k_1, k_2, \ldots, k_t$ to obtain $(t-s) \cdot B > t$. Together, they lead to $A + B > t\left(\frac{1}{s} + \frac{1}{t-s}\right) \geq 4$, a contradiction. Therefore, the assumption is untrue and the conclusion follows. □

Solution 3 Let $x_1 + x_3 + \cdots + x_{2n-1} = A$, $x_2 + x_4 + \cdots + x_{2n} = B$. Similar to the second proof, assume $A > 1$, $B > 1$. We claim that there exists k such that

$$x_{2k} \geq x_{2k+1} \cdot \frac{B}{A},$$

$$x_{2k} + x_{2k+2} \geq (x_{2k+1} + x_{2k+3}) \cdot \frac{B}{A},$$

$$\cdots$$

$$x_{2k} + x_{2k+2} + \cdots + x_{2k+2n-2} \geq (x_{2k+1} + x_{2k+3} + \cdots + x_{2k+2n-1}) \cdot \frac{B}{A}.$$

Define $a_i = x_{2i} - x_{2i+1} \cdot \frac{B}{A}$. Then $a_0 + a_1 + \cdots + a_{n-1} = 0$. Evidently, there exists k such that

$$a_k \geq 0,$$

$$a_k + a_{k+1} \geq 0,$$

$$\cdots$$

$$a_k + a_{k+1} + \cdots + a_{k+n-1} \geq 0,$$

and the claim is justified.

Assume $k = 0$ (or else subtract $2k$ from each subscript). Let $m \in \{0, 1, \ldots, n-1\}$ satisfy

$$\sum_{i=0}^{m} x_{2i+1} > 1, \quad \sum_{i=0}^{m-1} x_{2i+1} \leq 1.$$

Take $p = 0$, $q = m$. Finally, it requires to show

$$x_{2m+2} + x_{2m+4} + \cdots + x_{2n-2} \leq 1.$$

Indeed,

$$\begin{aligned}
&x_{2m+2} + x_{2m+4} + \cdots + x_{2n-2} \\
&\quad = B - (x_0 + x_2 + \cdots + x_{2m}) \\
&\quad \leq B - (x_1 + x_3 + \cdots + x_{2m+1}) \cdot \frac{B}{A} \\
&\quad < B - \frac{B}{A} = 1 + \frac{AB - A - B}{A} \\
&\quad \leq 1 (AB \leq 4 = A + B)
\end{aligned}$$

and we are done. □

6 Let n be a fixed positive integer. Denote D as the set of all positive divisors of n. If A and B are subsets of D, satisfying that: for any $a \in A$, $b \in B$, neither a divides b, nor b divides a. Prove that

$$\sqrt{|A|} + \sqrt{|B|} \leq \sqrt{|D|}.$$

(Contributed by Ai Yinghua)

Solution 1 Given A and B, partition D into four disjoint subsets $D = X \cup Y \cup Z \cup W$, where

$$\begin{aligned}
X &= \{x \in D : \exists a|x, \exists b|x\}, \quad & Y &= \{x \in D : \exists a|x, \nexists b|x\}, \\
Z &= \{x \in D : \nexists a|x, \exists b|x\}, \quad & W &= \{x \in D : \nexists a|x, \nexists b|x\}.
\end{aligned}$$

The problem conditions indicate that $A \subseteq Y$, $B \subseteq Z$. We shall prove a stronger statement: for any two nonempty subsets A and B, the inequality

$\sqrt{|Y|}+\sqrt{|Z|}\le\sqrt{|D|}$ holds. Equivalently,

$$|Y|+|Z|+2\sqrt{|Y|\cdot|Z|}\le|D|=|X|+|Y|+|Z|+|W|,$$

or $2\sqrt{|Y|\cdot|Z|}\le|X|+|W|$, which can be derived from

$$|Y|\cdot|Z|\le|X|\cdot|W|,$$

which can be written as

$$\begin{aligned}(|X|+|Y|)(|X|+|Z|)&=|X|(|X|+|Y|+|Z|)+|Y|\cdot|Z|\\&\le|X|(|X|+|Y|+|Z|)+|X|\cdot|W|=|X|\cdot|D|.\end{aligned}$$

Define $U=X\cup Y$, $V=X\cup Z$. The above inequality becomes $|U|\cdot|V|\le|U\cap V|\cdot|D|$. Notice that $U=\{x\in D:\exists a|x\}$ satisfies: if $x\in U$ and $x|x'$, then $x'\in U$. We say U is upward closed in D; similarly, $V=\{x\in D:\exists b|x\}$ is also upward closed.

Claim For nonempty upward closed sets U and V of D, we have $|U|\cdot|V|\le|U\cap V|\cdot|D|$.

Proof of claim Let $n=p_1^{\alpha_1}\cdots p_k^{\alpha_k}$ be the prime factorization. Induct on k: denote $p=p_k$, $\alpha=\alpha_k$, and $n=p^\alpha n'$.

Define $D_k=\{x\in D|v_p(x)=k\}$,

$$U_k=U\cap D_k,\quad V_k=V\cap D_k.$$

For every $k=0,1,\ldots,\alpha-1$, and every $x\in U_k$, since U is upward closed, it follows that $px\in U_{k+1}$, $|U_k|\le|U_{k+1}|$, and the sequence $\{|U_k|\}_k$ is increasing; likewise, $\{|V_k|\}_k$ is also increasing. Notice that $\frac{1}{p^k}U_k$ and $\frac{1}{p^k}V_k$ are upward closed sets of $\frac{1}{p^k}D_k=D(n')$. By the induction hypothesis,

$$\left|\left(\frac{1}{p^k}U_k\right)\bigcap\left(\frac{1}{p^k}V_k\right)\right|\ge\frac{1}{|D(n')|}\cdot\left|\left(\frac{1}{p^k}U_k\right)\right|\cdot\left|\left(\frac{1}{p^k}V_k\right)\right|,$$

which means $|U_k \cap V_k| \geq \dfrac{1+\alpha}{|D|}|U_k| \cdot |V_k|$. By the rearrangement inequality, it follows that

$$\begin{aligned}
|U \cap V| &= \sum_{k=0}^{\alpha} |U_k \cap V_k| \geq \frac{1+\alpha}{|D|} \sum_{k=0}^{\alpha} |U_k| \cdot |V_k| \\
&\geq \frac{1+\alpha}{|D|} \cdot \frac{1}{1+\alpha} \left(\sum_{k=0}^{\alpha} |U_k| \right) \cdot \left(\sum_{k=0}^{\alpha} |V_k| \right) \\
&= \frac{1}{|D|} |U| \cdot |V|.
\end{aligned}$$

This completes the proof of the claim and also the desired inequality.

□

Solution 2 (organized from Zhang Zhicheng's proof).

Lemma *Let P be a finite set, $M_1, \ldots, M_t$ be distinct subsets of P. Then the sets $\{M_i \backslash M_j | 1 \leq i, j \leq t\}$ contain at least t elements in total.*

Proof of lemma Take induction on $|M_1 \cup \cdots \cup M_t| = k$. For $k = 0$, the conclusion is trivial; assume $k \geq 1$ and the lemma is true whenever $|M_1 \cup \cdots \cup M_t| < k$. Now, for $|M_1 \cup \cdots \cup M_t| = k$, take $a \in M_1 \cup \cdots \cup M_t$ and assume that a belongs to $M_1, \ldots, M_s$, but not to $M_{s+1}, \ldots, M_t$, $1 \leq s \leq t$. If $s = t$, replace $M_i \backslash \{a\}$ by M_i, and the conclusion follows from the induction hypothesis. In the following, assume $s < t$ and define four classes of subsets

$$\begin{aligned}
S_1 &= \{M_i, 1 \leq i \leq s | \exists s+1 \leq j \leq t, M_i \backslash \{a\} = M_j\}, \\
S_2 &= \{M_1, \ldots, M_s\} \backslash S_1, \\
S_3 &= \{M \backslash \{a\} | M \in S_1\}, \\
S_4 &= \{M_{s+1}, \ldots, M_t\} \backslash S_3.
\end{aligned}$$

Clearly, S_1, S_2, S_3, and S_4 form a partition of $\{M_1, \ldots, M_t\}$. Define $S_5 = \{M \backslash \{a\} | M \in S_2\}$. Then

$$|S_1| = |S_3|, \quad |S_2| = |S_5|.$$

(i) Apply the induction hypothesis to S_3 to deduce that the sets $\{M\backslash N|M, N \in S_3\}$ contain at least $|S_3|$ elements. Hence, the sets

$$\{(M \cup \{a\})\backslash N|M, N \in S_3\} = \{M\backslash N|M \in S_1, N \in S_3\}$$

contain at least $|S_3|$ elements.

(ii) Apply the induction hypothesis to $S_3 \cup S_4 \cup S_5$ to deduce that the sets $\{M\backslash N|M, N \in S_3 \cup S_4 \cup S_5\}$ contain at least $|S_3|+|S_4|+|S_5|$ elements. For M, $N \in S_3 \cup S_4 \cup S_5$,

- If neither M nor N belongs to S_5, then $M\backslash N$ is of the form $M_i\backslash M_j$, where M_i, $M_j \in S_3 \cup S_4$;
- If both M and N belong to S_5, then $M\backslash N = (M \cup \{a\})\backslash(N \cup \{a\})$ is of the form $M_i\backslash M_j$, where M_i, $M_j \in S_2$;
- If $M \notin S_5$, while $N \in S_5$, then $M\backslash N = M\backslash(N \cup \{a\})$ is of the form $M_i\backslash M_j$, where

$$M_i \in S_3 \cup S_4, \quad M_j \in S_2;$$

- If $M \in S_5$, $N \notin S_5$, and $M\backslash N$ is not of any preceding form, then

$$(M\backslash N) \cup \{a\} = (M \cup \{a\})\backslash N$$

is of the form $M_i\backslash M_j$, where $M_i \in S_2$, $M_j \in S_3 \cup S_4$, and it does not repeat any preceding set $(M' \cup \{a\})\backslash N'$, M', $N' \in S_3$ (otherwise, $M\backslash N = M'\backslash N'$, contradicting the assumption that $M\backslash N$ is not of any preceding form).

Combining (i) and (ii), the number of elements of $\{M_i\backslash M_j|1 \leq i, j \leq t\}$ is at least

$$|S_3| + |S_3| + |S_4| + |S_5| = |S_1| + |S_3| + |S_4| + |S_5| = t.$$

This verifies the lemma.

For the original problem, let $G = \{\gcd(a, b)|a \in A, b \in B\}$ and

$$L = \{\text{lcm}(a, b)|a \in A, b \in B\}.$$

It is known from the problem that the sets A, B, G, and L are pairwise disjoint and $A \cup B \cup G \cup L \subseteq D$. So, it suffices to prove

$$\sqrt{|A|} + \sqrt{|B|} \leq \sqrt{|A| + |B| + |G| + |L|},$$

or $2\sqrt{|A||B|} \leq |G| + |L|$, or equivalently $|A| \cdot |B| \leq |G| \cdot |L|$.

Define two mappings

$$\rho_1 : A \times B \to D \times D, (a, b) \mapsto (\gcd(a, b), \operatorname{lcm}(a, b)),$$

$$\rho_2 : G \times L \to D \times D, (x, y) \mapsto (\gcd(x, y), \operatorname{lcm}(x, y)).$$

We will prove:
For each

$$(g, l) \in D \times D, |\rho_1^{-1}(g, l)| \le |\rho_2^{-1}(g, l)|. \qquad ①$$

As ① implies $|A \times B| \le |G \times L|$, the conclusion follows once ① is verified.

Proof of ①. If $|\rho_1^{-1}(g, l)| = 0$, the conclusion is obvious. Assume $|\rho_1^{-1}(g, l)| = t > 0$, and let

$$\rho_1^{-1}(g, l) = \{(gm_i, gn_i) |\ i = 1, \ldots, t\},$$

in which $gm_i \in A$, $gn_i \in B$, $m_i n_i = l/g$, and $(m_i, n_i) = 1$. Define

$$P = \{p^\alpha | p \text{ is a prime, } \alpha \ge 1,\ p^\alpha \| l/g\},$$

$$P_i = \{p^\alpha | p \text{ is a prime, } \alpha \ge 1, p^\alpha \| m_i\}.$$

Then $P_i \subseteq P$, and $m_i = \prod\limits_{p^\alpha \in P_i} p^\alpha$, $n_i = \prod\limits_{p^\alpha \in P \backslash P_i} p^\alpha$. Now for any $1 \le i$, $j \le t$,

$$\gcd(m_i, n_j) = \prod_{p^\alpha \in P_i \backslash P_j} p^\alpha, \quad \operatorname{lcm}(m_j, n_i) = \prod_{p^\alpha \in P_j \cup (P \backslash P_i)} p^\alpha,$$

in which $P_j \cup (P \backslash P_i)$ is exactly the complement of $P_i \backslash P_j$ in P. As a result, $\gcd(m_i, n_j)$ and $\operatorname{lcm}(m_j, n_i)$ are coprime and their product is l/g. We use the notations

$$x_{ij} := g \cdot \gcd(m_i, n_j) = \gcd(gm_i, gn_j),$$

$$y_{ij} := g \cdot \operatorname{lcm}(m_j, n_i) = \operatorname{lcm}(gm_j, gn_i).$$

Then $x_{ij} \in G$, $y_{ij} \in L$, and $\rho_2(x_{ij}, y_{ij}) = (g, l)$. As $P_1, \ldots, P_t$ are t distinct subsets of P, according to the lemma, the sets $\{P_i \backslash P_j | 1 \le i, j \le t\}$ contain t or more elements, and so do the sets $\{(x_{ij}, y_{ij}) | 1 \le i, j \le t\}$. This gives $|\rho_2^{-1}(g, l)| \ge t$ and ①. □

Chinese Girls' Mathematical Olympiad

2020 (Yingtan, Jiangxi)

The 19th Chinese Girls' Mathematical Olympiad (CGMO) was held on August 8–11 at Yingtan No. 1 Middle School, Yingtan, Jiangxi. A total of 30 teams, 120 female students from all provinces, municipalities and autonomous regions across the country participated in the competition. Due to COVID-19, teams from Hong Kong and Macao participated in the competition online.

After two rounds of competition (4 hours, 4 problems for each round), Li Siyu and 34 other students were awarded gold medals (first prize), Lei Feiran and 54 other students were awarded silver medals (second prize), Wu Feitong and 29 other students were awarded bronze medals (third prize). Moreover, the top 15 scorers including Li Siyu were qualified to participate in the 2020 Chinese Mathematical Olympiad (CMO).

The 19th CGMO committee chairman: Xiong Bin (East China Normal University); committee members: Ai Yinghua (Tsinghua University), Fu Yunhao (Southern University of Science and Technology), He Yijie (East China Normal University), Ji Chungang (Nanjing Normal University), Lai Li (Fudan University), Li Ting (Sichuan University), Lin Tianqi (East China Normal University), Qu Zhenhua (East China Normal University), Wang Xinmao (University of Science and Technology of China), Wu Yuchi (East China Normal University) and Yu Hongbing (Soochow University).

First Day
(2:30 pm – 6:30 pm; August 9, 2020)

1. As shown in Fig. 1.1, in the quadrilateral $ABCD$, $AB = AD$, $CB = CD$, $\angle ABC = 90°$. Let E, F be points on the segments AB, AD, respectively, and P, Q be points on the segment EF (P is between E and Q) such that $\dfrac{AE}{EP} = \dfrac{AF}{FQ}$. Drop perpendiculars from B, D to CP, CQ, with feet X, Y, respectively. Prove that X, P, Q, Y are concyclic.

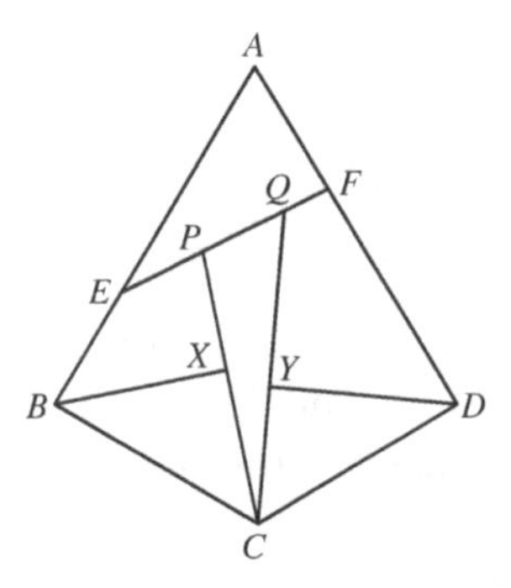

Fig. 1.1

Solution To begin, we have $\triangle ABC$ and $\triangle ADC$ symmetric about line AC, and $\angle ABC = \angle ADC = 90°$.

As shown in Fig. 1.2, let AC and EF meet at point K. As AK bisects $\angle EAF$, $\dfrac{AE}{AF} = \dfrac{KE}{KF}$.

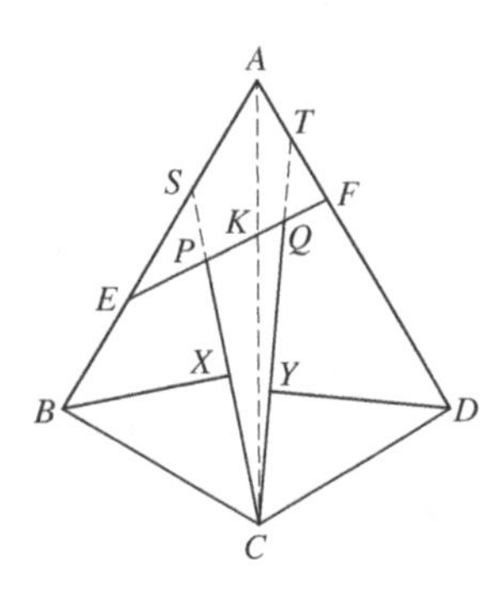

Fig. 1.2

Also, $\frac{AE}{AF} = \frac{EP}{FQ}$, and thus, $\frac{KE}{KF} = \frac{EP}{FQ}$ or $\frac{KE}{EP} = \frac{KF}{FQ}$.

Extend CP to meet AE at point S, and extend CQ to meet AF at point T. Apply Menelaus' theorem to $\triangle KPC$ and straight line ESA, obtaining

$$\frac{KE}{EP} \cdot \frac{PS}{CS} \cdot \frac{CA}{AK} = 1.$$

In the same manner, we find $\frac{KF}{FQ} \cdot \frac{QT}{CT} \cdot \frac{CA}{AK} = 1$. Then,

$$\frac{CS}{PS} = \frac{KE}{EP} \cdot \frac{CA}{AK} = \frac{KF}{FQ} \cdot \frac{CA}{AK} = \frac{CT}{QT},$$

which gives $\frac{CS}{CP} = \frac{CT}{CQ}$, or $\frac{CP}{CQ} = \frac{CS}{CT}$.

Note that $BX \perp CP$, $DY \perp CQ$ and $\angle ABC = \angle ADC = 90°$, and hence,

$$\frac{CX \cdot CP}{CY \cdot CQ} = \frac{CX \cdot CS}{CY \cdot CT} = \frac{CB^2}{CD^2} = 1.$$

We conclude that X, P, Q, Y are concyclic. ☐

2 Let $n \geq 2$ be a fixed integer, and $x_1, x_2, \ldots, x_n$ be any real numbers. Find the maximum value of $2 \sum_{1 \leq i < j \leq n} [x_i x_j] - (n-1) \sum_{i=1}^{n} [x_i^2]$, where $[x]$ represents the largest integer less than or equal to x.

Solution Notice that

$$2 \sum_{1 \leq i < j \leq n} [x_i x_j] - (n-1) \sum_{i=1}^{n} [x_i^2] = \sum_{1 \leq i < j \leq n} (2[x_i x_j] - [x_i^2] - [x_j^2]).$$

Since

$$2[x_i x_j] \leq 2x_i x_j \leq x_i^2 + x_j^2 < [x_i^2] + [x_j^2] + 2,$$

in which the leftmost and the rightmost quantities are both integers, it follows that

$$2[x_i x_j] \leq [x_i^2] + [x_j^2] + 1,$$

and equal sign is possible only when $[x_i^2]$ and $[x_j^2]$ have different parities (since $2[x_i x_j]$ is even).

As a result, when $[x_i^2]$ and $[x_j^2]$ have different parities,

$$2[x_ix_j] - [x_i^2] - [x_j^2] \le 1;$$

when $[x_i^2]$ and $[x_j^2]$ have the same parity,

$$2[x_ix_j] - [x_i^2] - [x_j^2] \le 0.$$

Suppose, among $[x_1^2]$, $[x_2^2]$, $[x_n^2]$, there are k odd numbers and $n-k$ even numbers. Then

$$\sum_{1\le i<j\le n} (2[x_ix_j] - [x_i^2] - [x_j^2]) \le k(n-k) \le \left[\frac{n^2}{4}\right].$$

On the other hand, we can let $m = \left[\frac{n}{2}\right]$ and

$$x_1 = x_2 = \cdots = x_m = 1.4, \quad x_{m+1} = x_{m+2} = \cdots = x_n = 1.5.$$

For $1 \le i < j \le m$, or $m+1 \le i < j \le n$,

$$2[x_ix_j] - [x_i^2] - [x_j^2] = 0;$$

while for $1 \le i \le m$ and $m+1 \le j \le n$, $2[x_ix_j] - [x_i^2] - [x_j^2] = 1$. Together, they give

$$\sum_{1\le i<j\le n} (2[x_ix_j] - [x_i^2] - [x_j^2]) = m(n-m) = \left[\frac{n^2}{4}\right].$$

Therefore, the desired maximum value is $\left[\frac{n^2}{4}\right]$. □

3 There are three classes each with n students, and all these $3n$ students have distinct heights. Divide them into n groups of three students, one from each class, and call the tallest person in each group a "leader". It is known that no matter how the students are divided, there always exist 10 leaders in each class. Prove that $n \ge 40$.

Solution First, we show that $n = 40$ is sufficient. Let the three classes be A, B, and C. Label the students from tallest to smallest as $1, 2, \ldots, 120$. Suppose that $1, 2, \ldots, 10$ and $71, 72, \ldots, 100$ are in class A; $11, 12, \ldots, 30$ and $101, 102, \ldots, 120$ are in class B; $31, 32, \ldots, 70$ are in class C. Clearly, the tallest 10 students in class A are all leaders. Among $11, 12, \ldots, 30$, at most 10 of them are group mates of $1, 2, \ldots, 10$ and they are not leaders, but the other 10 students must be leaders in their groups. Hence, class B also has 10 leaders. Finally, for class C, at most 30 of the students $31, 32, \ldots, 70$

are group mates of $1, 2, \ldots, 30$ and they are not leaders, but the other 10 students must be leaders. So, this example meets the conditions and $n = 40$ suffices.

For necessity, we give two solutions.

Solution 1

Lemma *Suppose the conditions are all satisfied. Then for each class i $(1 \leq i \leq 3)$, there is a positive integer k_i, such that among the tallest k_i students from all classes, the number of those from class i is at least* 10 *more than those from the other two classes.*

Proof of Lemma Pick any class, say class A, and rank their heights from tallest to smallest as $a_1 < a_2 < \cdots < a_n$. The other classes B and C have $x_1 < x_2 < \cdots < x_{2n}$. For every $1 \leq i \leq n-9$, let a_{i+9} (from class A) and x_i (from class B or C) be in a group. Then add a class B student to every group of class A and class C students; add a class C student to every group of class A and class B students. Now, other than $a_1, a_2, \ldots, a_9$, there must be another leader from class A, say a_m. We must have $a_m < x_{m-9}$, meaning that among all students taller or equal to a_m, at least m of them are from class A, and at most $m - 10$ from B or C. The lemma is verified.

Return to the original problem. Suppose the classes A, B, C correspond to integers k_1, k_2, k_3 as in the lemma, respectively, and $k_1 \leq k_2 \leq k_3$. Among $1, 2, \ldots, k_1$, at least 10 students are from class A; among $1, 2, \ldots, k_2$, at least $10 + 10 = 20$ students are from class B; among $1, 2, \ldots, k_3$, at least $10 + 20 + 10 = 40$ students are from class C. This implies that each class has at least $n = 40$ students.

Solution 2 We show that the conditions are not met if $n < 40$. First, there must be $10 * 3 = 30$ or more groups so as to have 10 leaders in each class, hence $n \geq 30$. Rank the students in each class from tallest to smallest as $a_1 < a_2 < \cdots < a_n$, $b_1 < b_2 < \cdots < b_n$ and $c_1 < c_2 < \cdots < c_n$. Consider a_{n-19}, b_{n-19}, c_{n-19}, and assume a_{n-19} is the tallest. Since $n \leq 39$, we infer that $a_1, a_2, \ldots, a_{n-19}$ are all taller than $b_{20}, b_{21}, \ldots, b_n$, $c_{20}, c_{21}, \ldots, c_n$. For $1 \leq i \leq n - 19$, make a_i, b_{i+19}, and c_{i+19} a group, each with a leader from class A; for the others, make groups in an arbitrary way. Then class B and C together have at most 19 leaders, a contradiction. Thus, $n \geq 40$. □

Remark (By Chen Haoran) In general, if there always exist k leaders in each class, then $n \geq 4k$. Furthermore, if there are m classes, then $n \geq 2^{m-1}k$. In this problem, $(m, k) = (3, 10)$.

4 Let p, q $(p > q)$ be prime numbers. Show that the greatest common divisor of $p! - 1$ and $q! - 1$ cannot exceed $p^{\frac{p}{3}}$.

Solution Define $D = \gcd(p! - 1, q! - 1)$. It is evident that $2! - 1$, $3! - 1$, $5! - 1$, $7! - 1$ are pairwise coprime, and hence, the statement is true for $p \leq 7$. Now, assume $p \geq 11$. Note that $p! - q!$ is divisible by D, but $q!$ and D are coprime, so D divides $\frac{p!}{q!} - 1$. It follows that $D \leq \frac{p!}{q!} \leq p^{p-q}$. If $q \geq \frac{2}{3}p$, the former inequality implies $D \leq p^{\frac{p}{3}}$ and the statement. Next, assume $p > \frac{3}{2}q$.

Observe that $D \mid (p! - q!^2)$, and $p! - q!^2 \neq 0$ (since $p! - q!^2$ is not divisible by prime p).

(i) If $p > 2q$, then $p!$ and $q!^2$ have the common divisor $q!^2$, which is coprime with D. This implies that D divides

$$\frac{p! - q!^2}{q!^2} = \frac{p!}{q!^2} - 1$$

and $D \leq \frac{p!}{q!^2}$; in addition, $D \mid (q! - 1)$ implies $D \leq q!$, yielding

$$D \cdot D^2 \leq \frac{p!}{q!^2} \cdot q!^2 = p! \leq p^p.$$

Hence, $D \leq p^{\frac{p}{3}}$.

(ii) If $\frac{3}{2}q < p \leq 2q$, then $p!$ and $q!^2$ have the common divisor $q!(p-q)!$, which is coprime with D. This indicates that D divides $\frac{p! - q!^2}{q!(p-q)!} \neq 0$, and thus,

$$D \leq \left| \frac{p!}{q!(p-q)!} - \frac{q!}{(p-q)!} \right| \leq \max\left\{ \frac{p!}{q!(p-q)!}, \frac{q!}{(p-q)!} \right\}.$$

Notice that $\frac{p!}{q!(p-q)!} < 2^p \leq 11^{\frac{p}{3}} \leq p^{\frac{p}{3}}$; in addition, $\frac{q!}{(p-q)!}$ can be written as a product of $2q - p$ consecutive integers less than or equal to p, and thus, $\frac{q!}{(p-q)!} \leq p^{2q-p} \leq p^{\frac{p}{3}}$ (the latter comes from $\frac{3}{2}q < p$). Now, $D \leq p^{\frac{p}{3}}$ follows. □

Second Day
(8:00 – 12:00 pm; August 10, 2020)

5 Find all real sequences $\{b_n\}_{n\geq 1}$ and $\{c_n\}_{n\geq 1}$ satisfying that, for each positive integer n, (1) $b_n \leq c_n$; (2) b_{n+1} and c_{n+1} are the two roots of the quadratic equation $x^2 + b_n x + c_n = 0$.

Solution By (1) and (2), it is clear that b_{n+1} and c_{n+1} are uniquely determined by b_n and c_n. By Viète's formulas,

$$b_n = -(b_{n+1} + c_{n+1}), \qquad ①$$

$$c_n = b_{n+1}c_{n+1}. \qquad ②$$

If $b_1 = c_1 = 0$, then $b_n = c_n = 0$ for all $n \geq 1$. The sequences $\{b_n\}_{n\geq 1} = \{c_n\}_{n\geq 1} = \{0, 0, \ldots\}$ satisfy the problem conditions.

We prove that all other sequences do not meet the requirements.

First, assume that one of b_1, c_1 is zero. If $b_1 = 0$, then $c_1 > 0$, $x^2 + b_1 x + c_1 = 0$ has no real roots; if $c_1 = 0$, then $b_1 < 0$, $x^2 + b_1 x + c_1 = 0$ has roots $b_2 = 0$, $c_2 = -b_1 > 0$, but $x^2 + b_2 x + c_2 = 0$ has no real roots.

Then, assume that both b_1, c_1 are non-zero. If $b_n > 0$, $c_n > 0$, then from ①, ②, we have $b_{n+1} < 0$, $c_{n+1} < 0$; if $b_n < 0$, $c_n < 0$, then $b_{n+1} < 0$, $c_{n+1} > 0$; if $b_n < 0$, $c_n > 0$, then $b_{n+1} > 0$, $c_{n+1} > 0$. Therefore, the signs of (b_n, c_n) are 3-periodic: positive and positive, negative and negative, negative and positive, ...

Suppose $b_n > 0$, $c_n > 0$. For the roots of $x^2 + b_n x + c_n = 0$ to be real, we must require the discriminant

$$b_n^2 - 4c_n \geq 0,$$

and thus, $b_n^2 \geq 4c_n \geq 4b_n$, $c_n \geq b_n \geq 4$. Due to the periodicity of the signs of (b_n, c_n), assume further that $n > 3$. By ①, ②, we have

$$b_{n-1} = -(b_n + c_n) < 0, \quad c_{n-1} = b_n c_n > 0,$$

$$b_{n-2} = -(b_{n-1} + c_{n-1}) = b_n + c_n - b_n c_n < 0,$$

$$c_{n-2} = b_{n-1}c_{n-1} = -(b_n + c_n)b_n c_n < 0,$$

$$b_{n-3} = -(b_{n-2} + c_{n-2}) \geq -c_{n-2} = (b_n + c_n)b_n c_n \geq 4b_n.$$

Let $i \in \{1, 2, 3\}$ be the first subscript such that $b_i > 0$, $c_i > 0$. Then $b_{i+3k} > 0$, $c_{i+3k} > 0$ for every positive integer k; moreover, repeatedly apply the above inequalities to get

$$b_i \geq 4^k b_{i+3k} \geq 4^{k+1}$$

for arbitrary k, but this is impossible.

Based on the above argument, $\{b_n\}_{n\geq 1} = \{c_n\}_{n\geq 1} = \{0, 0, \ldots\}$ are the only sequences satisfying ① and ②. □

6 Let $p, q > 1$ be integers and $(p, 6q) = 1$. Show that

$$\sum_{k=1}^{q-1}\left[\frac{pk}{q}\right]^2 \equiv 2p\sum_{k=1}^{q-1} k\left[\frac{pk}{q}\right] \pmod{(q-1)},$$

in which $[x]$ represents the largest integer less than or equal to x.

Solution For $\alpha \in \mathbb{R}$, denote $\{\alpha\} = \alpha - [\alpha]$. We have

$$\begin{aligned}
2p\sum_{k=1}^{q-1} k\left[\frac{pk}{q}\right] &= 2q\sum_{k=1}^{q-1}\frac{pk}{q}\left[\frac{pk}{q}\right] \\
&= q\sum_{k=1}^{q-1}\left(\left(\frac{pk}{q}\right)^2 + \left[\frac{pk}{q}\right]^2 - \left(\frac{pk}{q} - \left[\frac{pk}{q}\right]\right)^2\right) \\
&= q\sum_{k=1}^{q-1}\left(\frac{pk}{q}\right)^2 + q\sum_{k=1}^{q-1}\left[\frac{pk}{q}\right]^2 - q\sum_{k=1}^{q-1}\left\{\frac{pk}{q}\right\}^2. \quad ①
\end{aligned}$$

Since $(p, q) = 1$, the remainders of $p \cdot 1, p \cdot 2, \ldots, p \cdot (q-1)$ modulo q are exactly $1, 2, \ldots, q-1$. This implies

$$\sum_{k=1}^{q-1}\left\{\frac{pk}{q}\right\}^2 = \sum_{k=1}^{q-1}\left(\frac{k}{q}\right)^2,$$

plugged into ① to yield

$$\begin{aligned}
2p\sum_{k=1}^{q-1} k\left[\frac{pk}{q}\right] &= \frac{p^2-1}{q}\sum_{k=1}^{q-1} k^2 + q\sum_{k=1}^{q-1}\left[\frac{pk}{q}\right]^2 \\
&= \frac{(p^2-1)(q-1)(2q-1)}{6} + q\sum_{k=1}^{q-1}\left[\frac{pk}{q}\right]^2. \quad ②
\end{aligned}$$

As $(p, 6) = 1$, $6 \mid (p^2 - 1)$; additionally from ②, it follows that

$$2p\sum_{k=1}^{q-1} k\left[\frac{pk}{q}\right] \equiv \sum_{k=1}^{q-1}\left[\frac{pk}{q}\right]^2 \pmod{(q-1)}.$$

□

7 As illustrated in Fig. 7.1, the circumcircle of $\triangle ABC$ is centred at O, $AB < AC$, and $\angle BAC = 120^\circ$. Let M be the midpoint of $\overparen{BAC}$; P, Q be the points such that PA, PB, QA, QC are all tangent to the circumcircle; H and I be the orthocentre and the incentre of $\triangle POQ$, respectively. Let N be the midpoint of OI; the line MN meet $\odot O$ at another intersection D. Prove: $IH \perp AD$.

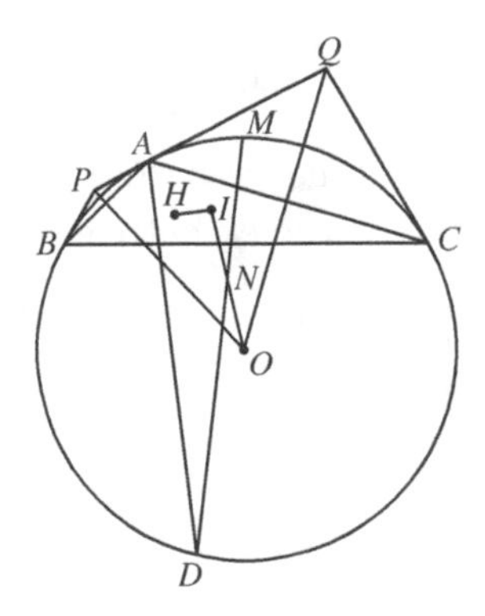

Fig. 7.1

Solution As shown in Fig. 7.2, extend BP, CQ beyond P, Q, respectively, to meet at point L.

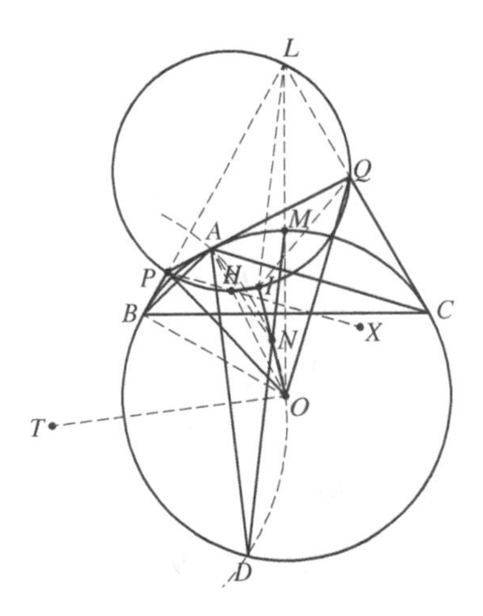

Fig. 7.2

To start, let

$$\angle ABC = 30^\circ + \vartheta, \quad \angle ACB = 30^\circ - \vartheta.$$

We have

$$\begin{aligned}\angle OPA &= 90^\circ - \angle POA \\ &= 90^\circ - \angle ACB \\ &= 60^\circ + \vartheta,\end{aligned}$$

and similarly, $\angle OQA = 60^\circ - \vartheta$, hence $\angle POQ = 60^\circ$, which gives $\angle PIQ = \angle PHQ = 120^\circ$. Meanwhile, O is the excentre of $\triangle LPQ$ relative to L,

$$\angle PLQ = 2(90^\circ - \angle POQ) = 60^\circ,$$

which implies that L, P, H, I, and Q are concyclic. Since OI bisects $\angle POQ$ and

$$\begin{aligned}\angle AOI &= \angle POI - \angle POA = 30^\circ - (30^\circ - \vartheta) = \vartheta, \\ \angle AOM &= \angle BOM - \angle BOA = 60^\circ - 2(30^\circ - \vartheta) = 2\vartheta,\end{aligned}$$

we infer that OI bisects $\angle AOM$, and

$$\angle AON = \angle AOI = \frac{1}{2}\angle AOM = \angle ADM = \angle ADN,$$

hence, A, N, O, D all lie on a circle, say centred at T.

Extend PH to point X. By properties of incentre and orthocentre,

$$\begin{aligned}\angle OHI &= \angle IHX + \angle OHX \\ &= \angle PQI + \angle OQP \\ &= \frac{3}{2}\angle OQP = \frac{3}{2}(60^\circ - \vartheta) \\ &= 90^\circ - \frac{3}{2}\vartheta.\end{aligned}$$

On the other hand, notice in the right $\triangle LOB$, $\angle BLO = 30^\circ$, $LO = 2OB = 2OM$, and thus $MN // LI$. We have

$$\begin{aligned}\angle LIO &= \angle LIQ + \angle QIO \\ &= \angle LPQ + 90^\circ + \frac{1}{2}\angle OPQ\end{aligned}$$

$$= 270^\circ - \frac{3}{2}\angle OPQ$$
$$= 270^\circ - \frac{3}{2}(60^\circ + \vartheta)$$
$$= 180^\circ - \frac{3}{2}\vartheta.$$

This implies $\angle MNI = \angle ANI = \frac{3}{2}\vartheta$, $\angle AOD = \angle AND = 180^\circ - 3\vartheta$. As $OA = OD$, OT bisects $\angle AOD$, $AD \perp OT$, and thus,

$$\angle AOT = \frac{1}{2}\angle AOD = 90^\circ - \frac{3}{2}\vartheta,$$

yielding $\angle OHI = \angle AOT$. Note that A, O, H are collinear, hence $IH // OT$. Finally, from $AD \perp OT$, we arrive at $IH \perp AD$. □

8 For a fixed positive integer n, we call a finite sequence of positive integers $(a_1, \ldots, a_m)$ an "n-even sequence", if $n = a_1 + \cdots + a_m$, and there is an even number of pairs (i, j) satisfying $1 \leq i < j \leq m$ and $a_i > a_j$. Find the number of n-even sequences $E(n)$. For example, $E(4) = 6$: the sequences are (4), $(1, 3)$, $(2, 2)$, $(1, 1, 2)$, $(2, 1, 1)$, and $(1, 1, 1, 1)$.

Solution Call $(a_1, \ldots, a_m)$ an "n-sequence", if $n = a_1 + \cdots + a_m$; if $(a_1, \ldots, a_m)$ is not an n-even sequence, then call it an "n-odd sequence". Let $O(n)$ be the number of n-odd sequences, and $S(n) = E(n) - O(n)$. Take $S(0) = 1$. We claim: for $n \geq 1$,

$$S(n) = 1 + \sum_{k=1}^{[\frac{n}{2}]} S(n - 2k). \qquad (*)$$

In fact, we may divide all n-sequences into the following three categories (when $n = 1$, only (i) occurs; when $n = 2$, only (i) and (iii) occur). Define $S_j(n)$ $(1 \leq j \leq 3)$ as the difference of the numbers of n-even and n-odd sequences in the respective categories, $S(n) = \sum_{j=1}^{3} S_j(n)$.

(i) $(a_1) = (n)$. Clearly, we have $S_1(n) = 1 - 0 = 1$.

(ii) $a_1 \neq a_2$. Notice that two n-sequences $(a_1, a_2, a_3, \ldots, a_m)$ and $(a_2, a_1, a_3, \ldots, a_m)$ have different parities, and this is a one-to-one correspondence, so we have $S_2(n) = 0$.

(iii) $a_1 = a_2$. Among all n-sequences with $a_1 = a_2 = k\left(1 \le k \le \left[\frac{n}{2}\right]\right)$, since $(a_1, a_1, a_3, \ldots, a_m)$ and $(a_3, \ldots, a_m)$ have the same parity, we infer that the difference is $S(n-2k)$ (note that $n = 2k$, $S(0) = 1$ corresponds to (k, k)).

Based on (i), (ii), (iii), the claim $(*)$ is verified.

From $(*)$, we see for $n \ge 3$,

$$\begin{aligned} S(n) - S(n-2) &= \sum_{k=1}^{\left[\frac{n}{2}\right]} S(n-2k) - \sum_{k=1}^{\left[\frac{n}{2}-1\right]} S(n-2k) \\ &= S(n-2), \end{aligned}$$

that is, $S(n) = 2S(n-2)$.

For $n = 2$, also, $S(2) = 2 = 2S(0)$.

Therefore, when n is odd,

$$S(n) = 1 + S(1) + S(3) + \cdots + S(n-2) = 2^{\frac{n-1}{2}};$$

when n is even,

$$S(n) = 1 + S(0) + S(2) + \cdots + S(n-2) = 2^{\frac{n}{2}}.$$

Finally, observe that each n-sequence $(a_1, \ldots, a_m)$ uniquely corresponds to an increasing sequence $(a_1, a_1+a_2, \ldots, a_1+a_2+\cdots+a_m)$, and each latter sequence corresponds to a subset of $\{1, 2, \ldots, n\}$ containing n. Hence, the number of n-sequences is 2^{n-1}, and

$$E(n) = \frac{2^{n-1} + S(n)}{2} = 2^{n-2} + 2^{\left[\frac{n-2}{2}\right]}. \qquad \square$$

Remark (By Chen Haoran) To find the number of n-sequences, place n stones in a row and insert sticks between the stones, at most one stick at each gap. Then a_i is the number of stones between the $(i-1)$th and the ith sticks. Since each gap contains 0 or 1 stick, and there are $n-1$ gaps, $E(n) + O(n) = 2^{n-1}$. Moreover, as indicated by (ii), one can further deduce that every n-sequence $(a_1, a_1, \ldots, a_k, a_k, a_i, a_j, \ldots)$ $(k \ge 0, a_i \ne a_j)$ can be paired with $(a_1, a_1, \ldots, a_k, a_k, a_j, a_i, \ldots)$ and they have different parities. The remaining unpaired n-sequences have the form $(a_1, a_1, \ldots, a_k, a_k)$ or $(a_1, a_1, \ldots, a_k, a_k, a_{k+1})$ and they are all n-even sequences. To find the number of such sequences, place n stones and insert sticks only at gaps between the $2i$ and the $2i+1$ stones $\left(1 \leq i \leq \left[\frac{n}{2}\right]\right)$ (if n is even, the last "gap" is at the very end). Then a_j $(j \leq k)$ is the number of stones between the $(j-1)$th and the jth sticks divided by 2; if there is no stick at the very end, a_{k+1} is the number of stones after the last stick. Since there are $\left[\frac{n}{2}\right]$ gaps, we have $E(n) - O(n) = 2^{\left[\frac{n}{2}\right]}$, and the conclusion follows.

Chinese Girls' Mathematical Olympiad

2021 (Changchun, Jilin)

The 20th Chinese Girls' Mathematical Olympiad (CGMO) was held on August 10–15 at The High School Attached to Northeast Normal University, Changchun, Jilin. The competition was sponsored by the Chinese Mathematical Society, organized by Changchun City Bureau of Education and Jilin Mathematical Society, and co-organized by The High School Attached to Northeast Normal University.

A total of 33 teams, 134 female students from all provinces, municipalities, autonomous regions, Hong Kong and Macao participated in the competition. Due to COVID-19, the competition was conducted online.

After two rounds of competition (4 hours, 4 problems for each round), Fu Jiarui and 35 other students were awarded gold medals (first prize), Chen Xi and 50 other students were awarded silver medals (second prize), Zhao Wenkuan and 45 other students were awarded bronze medals (third prize). Moreover, the top 17 scorers including Fu Jiarui were qualified to participate in the 2021 Chinese Mathematical Olympiad (CMO).

The 20th CGMO committee chairman: Qu Zhenhua (East China Normal University); committee members: Ai Yinghua (Tsinghua University), Han Jingjun (Fudan University), He Yijie (East China Normal University), Lai Li (Fudan University), Lin Tianqi (East China Normal University), Luo Zhenhua (East China Normal University), Wu Yuchi (East China Normal University), Xiong Bin (East China Normal University), and Yao Yijun (Fudan University).

First Day
(8 – 12 pm; August 12, 2021)

1 Let n be a positive integer, $x_1, x_2, \ldots, x_{n+1}$, p, q be positive real numbers, such that $p < q$, and

$$x_{n+1}^p > x_1^p + x_2^p + \cdots + x_n^p.$$

Prove: (1) $x_{n+1}^q > x_1^q + x_2^q + \cdots + x_n^q$; (2) $(x_{n+1}^p - x_1^p - x_2^p - \cdots - x_n^p)^{1/p} < (x_{n+1}^q - x_1^q - x_2^q - \cdots - x_n^q)^{1/q}$.

(Contributed by Wu Yuchi)

Solution Both inequalities are homogeneous, and we are led to assume $x_{n+1} = 1$ (otherwise, replace x_i by $\dfrac{x_i}{x_{n+1}}$, $i = 1, 2, \ldots, n+1$). Given the condition

$$x_1^p + x_2^p + \cdots + x_n^p < 1,$$

we have $x_1, x_2, \ldots, x_n \in (0, 1)$.

(1) As $p < q$, it follows that

$$x_1^q + x_2^q + \cdots + x_n^q < x_1^p + x_2^p + \cdots + x_n^p < 1.$$

(2) Denote $a = (1 - x_1^p - x_2^p - \cdots - x_n^p)^{1/p}$, $a \in (0, 1)$. Since $q > p$,

$$a^q + x_1^q + x_2^q + \cdots + x_n^q < a^p + x_1^p + x_2^p + \cdots + x_n^p = 1,$$

and thus, $1 - x_1^q - x_2^q - \cdots - x_n^q > a^q$, yielding

$$(1 - x_1^q - x_2^q - \cdots - x_n^q)^{1/q} > a = (1 - x_1^p - x_2^p - \cdots - x_n^p)^{1/p},$$

as desired. □

Remark The inequality (1) can also be derived from the norm inequality $\left(\sum_{i=1}^{n} x_i^p\right)^{1/p} \geq \left(\sum_{i=1}^{n} x_i^q\right)^{1/q}$.

2 As shown in Fig. 2.1, in the acute triangle ABC, $AB < AC$, I is the incentre, and J is the excentre relative to vertex A. Points X, Y lie on the minor arcs $\overparen{AB}$, $\overparen{AC}$ of the circumcircle of the triangle ABC, respectively, such that $\angle AXI = \angle AYJ = 90°$. Point K is on the

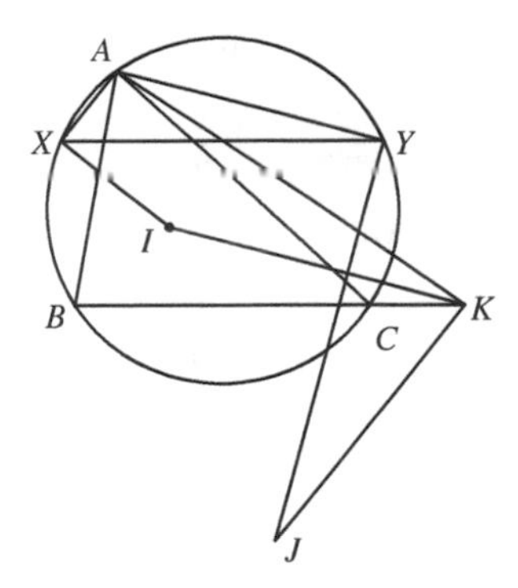

Fig. 2.1

extension of BC beyond C, such that $KI = KJ$. Prove: AK bisects the segment XY.

(Contributed by He Yijie)

We give three proofs as follows.

Solution 1 As illustrated in Fig. 2.2, let IJ cross the circumcircle of $\triangle ABC$ at M and A' be the antipode of A.

Since $\angle AXI = \angle AYJ = 90°$, the lines XI and YJ pass through A'. From properties of incentre and excentre, $MI = MJ = MB$, and together with $KI = KJ$, it follows that $KM \perp IJ$, and A' lies on KM. This implies $A'I = A'J$, $\angle AIX = \angle A'IJ = \angle A'JI = \angle AJY$. Since $\angle AXI = \angle AYJ$, we have $\triangle AIX \backsim \triangle AJY$, and $\angle XAM = \angle XAI = \angle YAJ = \angle YAM$, $\overset{\frown}{XM} = \overset{\frown}{YM}$. Also, $\overset{\frown}{BM} = \overset{\frown}{CM}$, and thus $\overset{\frown}{BX} = \overset{\frown}{CY}$, $XY // BC$. Moreover, from

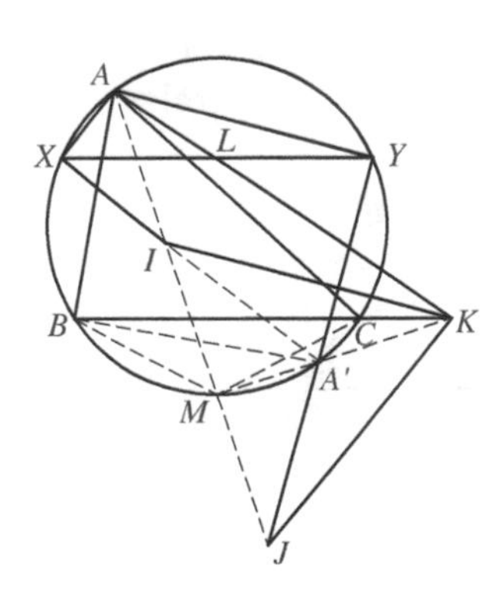

Fig. 2.2

properties of similar triangles, we obtain

$$\frac{AX}{AY} = \frac{AI}{AJ}. \quad ①$$

From $\overset{\frown}{BM} = \overset{\frown}{CM}$, it follows that $\angle MA'B = \angle MCB = \angle MBK$, $\triangle MA'B \backsim MBK$, and hence $MA' \cdot MK = MB^2 = MI \cdot MJ$. Now, $\triangle MIA' \backsim \triangle MKJ$, which gives $IA' \perp KJ$, or $XA' \perp KJ$. Notice $AX \perp XA'$, hence $AX // KJ$. In a similar way, we derive $AY // KI$. Together, they imply

$$\angle XAK = 180° - \angle AKJ, \quad \angle YAK = \angle AKI. \quad ②$$

Let AK and XY meet at L. From ① and ②, we deduce that

$$\frac{XL}{YL} = \frac{S_{\triangle AXL}}{S_{\triangle AYL}} = \frac{AX}{AY} \cdot \frac{\sin \angle XAK}{\sin \angle YAK} = \frac{AI}{AJ} \cdot \frac{\sin \angle AKJ}{\sin \angle AKI} = \frac{AI}{AJ} \cdot \frac{S_{\triangle AKJ}}{S_{\triangle AKI}} = 1,$$

and therefore, AK bisects the segment XY. □

Solution 2 Let $\odot \Omega$ be the circumcircle of $\triangle ABC$. It is evident that A, I, J lie on a line; suppose this line meets $\odot \Omega$ at M other than A. Then, it follows from simple properties of incentre and excentre that $MI = MJ = MB = MC$. Draw $\odot M$ centred at M of radius MB, and denote the circles with diameters AI, AJ by $\odot \omega_1$, $\odot \omega_2$, respectively, as illustrated in Fig. 2.3.

Extend AX, AY beyond X, Y to meet the line BC at points X', Y', respectively.

Analogous to the first proof, we have $KM \perp IJ$, $XY // BC$.

Since the line AX is the radical axis of $\odot \omega_1$ and $\odot \Omega$, and the line BC is the radical axis of $\odot \Omega$ and $\odot M$, we infer that X', the intersection of the lines AX and CB, is the radical centre of the circles $\odot \Omega$, $\odot \omega_1$ and $\odot M$. Therefore, X lies on the radical axis of $\odot \omega_1$ and $\odot M$, $X'I \perp AI$. In the same manner, Y' is the radical centre of the circles $\odot \Omega$, $\odot \omega_2$ and $\odot M$, $Y'J \perp AJ$.

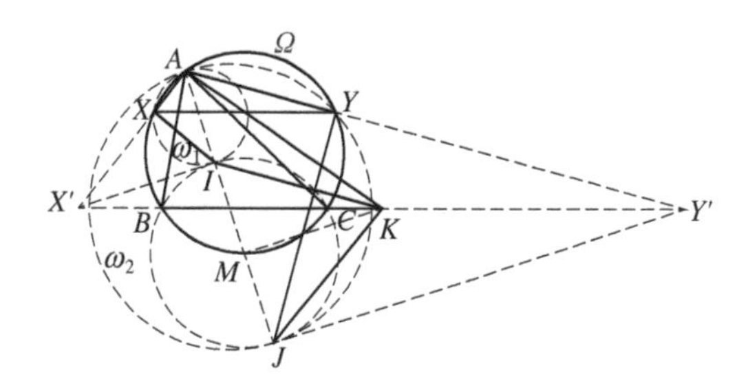

Fig. 2.3

Note that $KM \perp AM$, hence $X'I \,//\, KM // Y'J$.

Finally, we have $\dfrac{X'K}{Y'K} = \dfrac{IM}{JM} = 1$, or $X'K = Y'K$; together with $XY // X'Y'$, the statement that AK bisects the segment XY is verified.

□

Solution 3 Let $\odot\Omega$ be the circumcircle of $\triangle ABC$, and A' be the antipode of A. Suppose that the line through A and parallel to BC crosses $\odot\Omega$ at $T \neq A$; AK crosses $\odot\Omega$ at $P \neq A$. Further, suppose that the bisector of $\angle BAC$ meets the lines PA', TA' and $\odot\omega$ respectively at points Q, R and M(other than A), as shown in Fig. 2.4. Clearly, I and J also lie on this angle bisector.

Analogous to the first proof, we deduce that M, A', K are collinear, $XY // BC$, and thus, $AT // BC$.

Notice that M is the midpoint of $\widehat{BC}$, $MA = MT$; since $\angle ATR = 90°$, $MA = MR$. In addition, $AP \perp A'P$, that is, $AK \perp A'Q$. It follows that $\triangle AMK \backsim \triangle A'MQ$ and

$$MA \cdot MQ = MA' \cdot MK.$$

Now, from $MA = MR$ and $MA' \cdot MK = MI^2$ (as in the first proof), it follows that

$$MR \cdot MQ = MI^2.$$

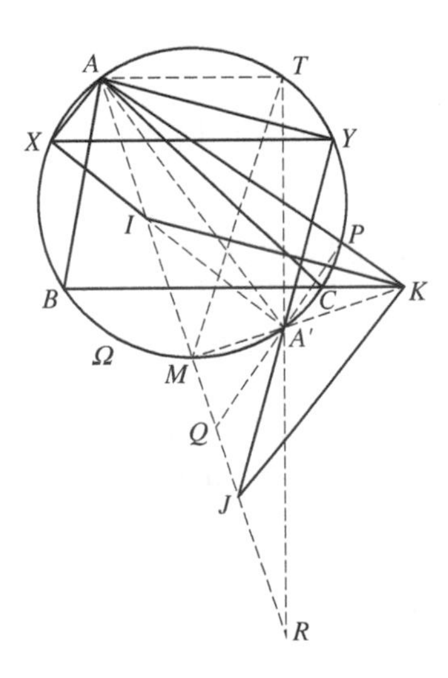

Fig. 2.4

Since M is the midpoint of IJ, it is clear that I, J, Q, and R form a harmonic range of points, and $A'I$, $A'J$, $A'Q$, $A'R$ are a harmonic pencil of lines. From

$$AX \perp A'I, \quad AY \perp A'J, \quad AK \perp A'Q, \quad AT \perp A'R,$$

we deduce that AX, AY, AK, and AT are also a harmonic pencil of lines. As $AT // XY$, and according to properties of harmonic pencil, AK bisects the segment XY. The argument is complete. □

3 Find the smallest positive integer n, such that one can colour each square of an $n \times n$ grid red, yellow, or blue, satisfying the following three properties:

(i) the number of squares coloured in each colour is the same;
(ii) if a row contains red square(s), then it also contains blue square(s), but no yellow squares;
(iii) if a column contains blue square(s), then it also contains red square(s), but no yellow squares.

(Contributed by He Yijie)

Solution Assume a colouring algorithm exists on an $n \times n$ grid with all the desired properties. Evidently, we have $3 \mid n^2$, or $3 \mid n$. Let $n = 3k$. There are $3k^2$ squares of each colour on the grid.

Notice that if two rows or two columns are interchanged, it still maintains all required properties. Due to this, we may assume that rows $1, 2, \ldots, u$ and columns $1, 2, \ldots, v$ contain yellow squares.

Use a vertical line and a horizontal line to divide the grid into four rectangular regions A, B, C, and D, as shown in Fig. 3.1: A, B each has u rows; A, C each has v columns. (If $u = 3k$, then C, D do not exist; if $v = 3k$, then B, D do not exist.)

Since rows $1, 2, \ldots, u$ contain yellow squares, by property (i), A and B do not contain red squares (and C, D must exist). Since columns $1, 2, \ldots, v$ contain yellow squares, by property (ii), A and C do not contain blue squares (and B, D exist). It follows that all squares in A are yellow. On the other hand, all yellow squares are in A, $uv = 3k^2$. Furthermore, B only contains blue squares and C only contains red squares.

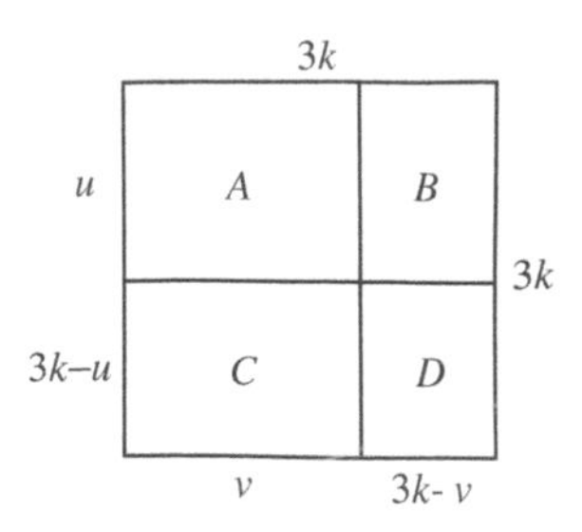

Fig. 3.1

For D, according to properties (ii) and (iii), every row and every column contain red and blue squares. So, the number of blue squares in B is smaller than the total number of blue squares, that is, $u(3k - v) < 3k^2 = uv$; similarly, $(3k - u)v < uv$. After simplification, they become

$$u > \frac{3}{2}k, \quad v > \frac{3}{2}k. \qquad ①$$

As $3 \mid uv$, by symmetry, we may assume $3 \mid u$ (otherwise, transpose the rows to columns and swap the red and blue squares). Denote $\frac{u}{3} = rs$ and $v = rt$, in which $r = \left(\frac{u}{3}, v\right)$. Now, $st = \frac{uv}{3r^2} = \left(\frac{k}{r}\right)^2$ is a perfect square, $(s, t) = 1$, hence s, t are both perfect squares. Let $s = p^2$, $t = q^2$, and we have

$$u = 3p^2r, \quad v = q^2r, \quad k = pqr.$$

Plug them into ① to get $3p^2r > \frac{3}{2}pqr$, $q^2r > \frac{3}{2}pqr$, which reduce to $\frac{3}{2}p < 2p - 1$, or $p \geq 3$. Since $q > \frac{3}{2}p$, $q \geq 5$. It follows that $n = 3k = 3pqr \geq 45$.

Consider a 45×45 grid. From $uv = 675$, $u, v > \frac{45}{2}$ (by ①) and $3 \mid u$, we find $u = 27$, $v = 25$. Colour the upper left 27×25 squares in yellow, the upper right 27×20 squares in blue, the lower left 18×25 squares in red, and for the lower right region, colour 225 squares in red, 135 squares in blue, such that every row contains blue squares and every column contains red squares. Fig. 3.2 shows an algorithm that meets all the requirements.

In conclusion, the minimum of n is 45. □

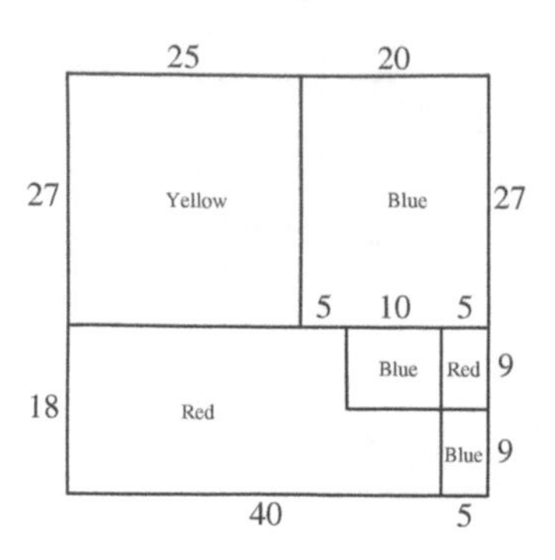

Fig. 3.2

4 A sequence of positive integers $\{a_n\}_{n\geq 1}$ is called a "CGMO sequence" if: (i) $\{a_n\}_{n\geq 1}$ is strictly increasing; (ii) for each integer $n \geq 2022$, a_n is the smallest integer greater than a_{n-1}, such that for some non-empty subset $A_n \subseteq \{a_1, a_2, \ldots, a_{n-1}\}$, the product $a_n \cdot \prod\limits_{a\in A_n} a$ is a perfect square.

Prove that there exist constants c_1, $c_2 > 0$, such that for each CGMO sequence $\{a_n\}_{n\geq 1}$, there exists a positive integer N (depending on the sequence), for every $n \geq N$,

$$c_1 \cdot n^2 \leq a_n \leq c_2 \cdot n^2.$$

(Contributed by Qu Zhenhua)

Solution Take $c_1 = 2^{-4042}$, $c_2 = 2$. We show $c_1 \cdot n^2 \leq a_n \leq c_2 \cdot n^2$ for sufficiently large n.

(1) *For the upper bound*: Suppose $\{a_n\}_{n\geq 1}$ is any CGMO sequence. By definition, there exists $A_{2022} \subset \{a_1, a_2, \ldots, a_{2011}\}$ such that $a_{2022} \cdot \prod\limits_{a\in A_{2022}} a = L^2$, $L \in \mathbb{N}^+$. We claim: for $k \geq 0$,

$$a_{2022+k} \leq (L+k)^2.$$

Induct on k. When $k = 0$, obviously $a_{2022} \leq a_{2022} \cdot \prod\limits_{a\in A_{2022}} a = L^2$. Assume it is true for k, namely $a_{2022+k} \leq (L+k)^2$. Then for $k+1$, note that

$(L+k+1)^2 > (L+k)^2 \geq a_{2022+k}$ and

$$(L+k+1)^2 \cdot a_{2022} \cdot \prod_{a \in A_{2022}} a = (L+k+1)^2 L^2$$

is a perfect square. By the choice of $a_{2022+k+1}$, we obtain $a_{2022+k+1} \leq (L+k+1)^2$ and the validity of the claim.

Now, $a_n \leq (L+n-2022)^2$ holds for $n \geq 2022$. Take $N > |L-2022|/(\sqrt{2}-1)$, N only dependent on $\{a_n\}_{n\geq 1}$. Then for $n \geq N$, $(L+n-2022)^2 \leq 2n^2$, and the upper bound is verified.

(2) *For the lower bound*: Note that every positive integer m can be uniquely written as $m = ab^2$, a, b as positive integers, and a has no square factor other than 1. Let $a = f(m)$ denote the square-free part of m. Then f has simple properties: $f(xy^2) = f(x)$, $f(xy) = f(f(x)f(y))$.

Let $S = \{a_1, a_2, \ldots, a_{2021}\}$, and $F = \left\{ f\left(\prod_{a\in B} a\right) | B \subseteq S \right\}$ (if $B = \varnothing$, the product is 1). The set F satisfies: for any $x_1, x_2, \ldots, x_t \in F$,

$$f(x_1 x_2 \cdots x_t) \in F.$$

Indeed, let $x_i = f\left(\prod_{a\in B_i} a\right)$, $B_i \subseteq S$, $1 \leq i \leq t$, and take $B = B_1 \triangle B_2 \triangle \cdots \triangle B_t$ (here, $X \triangle Y$ is the symmetric difference of two sets X, Y: $X \triangle Y = (X \backslash Y) \cup (Y \backslash X)$). Then

$$f(x_1 x_2 \cdots x_t) = f\left(\prod_{a\in B} a\right) \in F.$$

We prove by induction that for every positive integer n, the square-free part of a_n belongs to F.

Obviously, the claim is true for $1 \leq n \leq 2021$; assume it is true for $1 \leq n \leq m$, $m \geq 2021$. By definition of a_{m+1}, there exists $A_{m+1} \subseteq \{a_1, a_2, \ldots, a_m\}$, such that $a_{m+1} \cdot \prod_{a\in A_{m+1}} a$ is a perfect square. Also, by the induction hypothesis, for any $a \in A_{m+1}$, $f(a) \in F$. Using the above property of F, we find

$$f(a_{m+1}) = f\left(\prod_{a\in A_{m+1}} a\right) = f\left(\prod_{a\in A_{m+1}} f(a)\right) \in F,$$

and the induction is complete.

Let $|F| = m \le 2^{2021}$. For any positive integer n, consider the square-free parts of $a_1, a_2, \ldots, a_n$. By the Pigeonhole principle, among them there are $k \ge \left\lceil \frac{n}{m} \right\rceil \ge \left\lceil \frac{n}{2^{2021}} \right\rceil$ identical parts, say $f(a_{i_1}) = f(a_{i_2}) = \cdots = f(a_{i_k}) = u$, where

$$1 \le i_1 < i_2 < \cdots < i_k \le n.$$

Since $a_{i_1} < a_{i_2} < \cdots < a_{i_k}$ and their square-free parts are all equal to u,

$$a_{i_k} \ge uk^2 \ge k^2 \ge \left(\frac{n}{2^{2021}}\right)^2,$$

and thereby, $a_n \ge a_{i_k} \ge 2^{-4042} \cdot n^2$. □

Second Day
(8 – 12 pm; August 13, 2021)

5 Prove: for any 4 numbers selected from the set $\{1, 2, \ldots, 20\}$ (with repetition allowed), one can always take 3 of them as a, b, c, such that the congruence equation $ax \equiv b \pmod{c}$ has integer solutions.

(Contributed by Qu Zhenhua)

Solution It is well known that $ax \equiv b \pmod{c}$ has integer solutions if and only if $\gcd(a, c) \mid b$.

Assume on the contrary that for $1 \le a_1, a_2, a_3, a_4 \le 20$, the statement is not true. There are three cases to consider:

(1) If one of the 4 numbers is a prime power, say $a_1 = p^\alpha$, p prime, $\alpha \ge 1$. Suppose $v_p(a_2) = \beta \le v_p(a_3) = \gamma$ (here $v_p(x)$ is the highest power of p in the factorization of x). Take $a = a_1$, $b = a_3$, $c = a_2$. Since $\gcd(a, c) = p^{\min(\alpha, \beta)} \mid p^\gamma$, it follows that $\gcd(a, c) \mid b$ and the assumption is untrue.

(2) If one number divides another, say $a_1 \mid a_2$. Take $a = a_1$, $b = a_2$, $c = a_3$. From $\gcd(a, c) \mid a_1$, we have $\gcd(a, c) \mid b$ and the assumption is untrue.

(3) a_1, a_2, a_3, a_4 contain no prime powers, and $a_i \nmid a_j$ for $1 \le i, j \le 4$. Each of the four numbers must have distinct prime factors, and no two of them are equal. Consequently, they must be four of the following seven numbers:

$$6,\ 12,\ 18,\ 10,\ 15,\ 14,\ 20.$$

Note that the prime factor 7 of 14 does not appear in the other numbers. This indicates that when checking $\gcd(a,c) \mid b$, 14 is equivalent to 2 and cannot be selected (or else it has integer solutions). For the remaining six numbers, $6 \mid 12$, $6 \mid 18$, $10 \mid 20$, and hence, only $\{12, 18, 15, 10 \text{ or } 20\}$ has no divisibility relation. Take $a = 12$, $b = 18$, $c = 15$, $\gcd(a,c) = 3 \mid b$. Again, the assumption is untrue.

Therefore, the original statement is true. □

Remark The number 20 in the problem can be improved to 104, and 104 is optimal. On one hand, if 20 is replaced by 105 or larger, then take the products of any three numbers from 2, 3, 5, 7 to get 30, 42, 70, 105: since the greatest common divisor of any two of them does not divide another, the statement is untrue for these numbers. On the other hand, we show for any four numbers selected from $\{1, 2, \ldots, 104\}$, there are two of them whose greatest common divisor can divide another. Suppose the assertion is not true and a, b, c, d is a counterexample. If $\gcd(a,b,c,d) = x > 1$, use $\frac{a}{x}, \frac{b}{x}, \frac{c}{x}, \frac{d}{x}$ instead of a, b, c, d and it is still a counterexample. Now, assume $\gcd(a,b,c,d) = 1$, and one of them must be odd, say a is odd.

Consider $d_1 = \gcd(a,b)$, $d_2 = \gcd(a,c)$, $d_3 = \gcd(a,d)$. Assume d_1, d_2, d_3 have no divisibility relation (otherwise, say $d_2 \mid d_1$, then $(a,c) \mid b$, and the congruence equation $ax \equiv b \pmod{c}$ has integer solutions). This assumption indicates that a has three pairwise non-divisible factors d_1, d_2, d_3, and a is not a prime power. If a has 3 distinct prime factors, then $a \geq 3 \times 5 \times 7 > 104$ which is impossible. Hence, a has two distinct prime factors, say $a = p^{\alpha}q^{\beta}$.

Let $d_i = p^{\alpha_i}q^{\beta_i}$, $i = 1, 2, 3$. Since d_1, d_2, d_3 have no divisibility relation, α_1, α_2, α_3 must be distinct, and β_1, β_2, β_3 are distinct as well. It follows that $\alpha \geq 2$, $\beta \geq 2$, but $a \geq 3^2 \times 5^2 > 104$, impossible. The assertion is now verified.

6 Let S be a finite set and $P(S)$ be the power set of S (the set of all subsets of S). For any function $f : P(S) \to \mathbb{R}$, prove

$$\sum_{A \in P(S)} \sum_{B \in P(S)} f(A)f(B)2^{|A \cap B|} \geq 0,$$

where $|X|$ represents the number of elements of set X.

(Contributed by Fu Yunhao and Zhang Ruixiang)

Solution First, notice that

$$2^{|A\cap B|} \sum_{X\subset A\cap B} 1 = \sum_{\substack{X\subset A \\ X\subset B}} = 1,$$

and the left-hand side of the desired inequality can be written as

$$\sum_{A\in P(S)} \sum_{B\in P(S)} \sum_{\substack{X\subset A \\ X\subset B}} f(A)f(B).$$

Change the order of summation, and we find

$$\begin{aligned}
&\sum_{X\in P(S)} \sum_{\substack{A\in P(S) \\ A\supset X}} \sum_{\substack{B\in P(S) \\ B\supset X}} f(A)f(B) \\
&= \sum_{X\in P(S)} \left(\sum_{\substack{A\in P(S) \\ A\supset X}} f(A) \right) \left(\sum_{\substack{B\in P(S) \\ B\supset X}} f(B) \right) \\
&= \sum_{X\in P(S)} \left(\sum_{\substack{A\in P(S) \\ A\supset X}} f(A) \right)^2 \geq 0,
\end{aligned}$$

where the last inequality holds since the sum of squares of real numbers is always nonnegative. Done. □

7 As shown in Fig. 7.1, in acute $\triangle ABC$, $AB > AC$, and O is the circumcentre. Let K be the symmetric point of B with respect to AC, and L be the symmetric point of C with respect to AB. Let X be a point inside $\triangle ABC$, such that $AX \perp BC$, $XK = XL$. Y, Z are points on the segments BK, CL, respectively, such that $XY \perp CK$, $XZ \perp BL$. Prove that B, C, Y, O, and Z all lie on a circle.

(Contributed by Lin Tianqi)

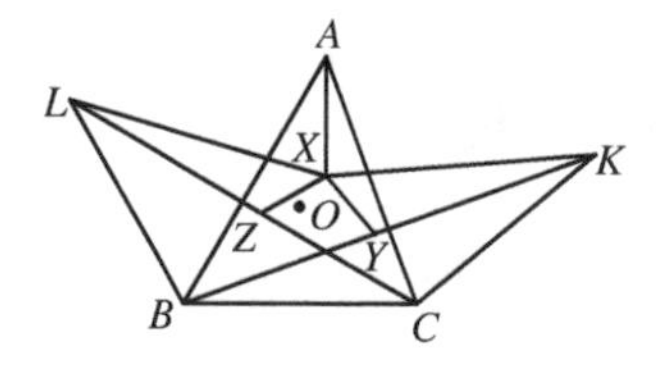

Fig. 7.1

Solution Let the lines BL, CK meet at P; $AD \perp BC$ with foot D, as shown in Fig. 7.2.

Assume that T is the circumcentre of $\triangle PKL$, $TL = TP = TK$. In the isosceles triangle TLP, by Stewart's Theorem, we have $TB^2 = TP^2 - BL \cdot BP$. In the same manner, $TC^2 = TP^2 - CK \cdot CP$.

It is known that $BL = BC = CK$, and hence,

$$TB^2 - TC^2 = (TP^2 - BL \cdot BP) - (TP^2 - CK \cdot CP) = BC(CP - BP).$$

Notice that A is the excentre of $\triangle PBC$ relative to the vertex P, and D is the point of tangency, hence

$$CP - BP = BD - CD,$$

which implies $TB^2 - TC^2 = (BD + CD)(BD - CD) = BD^2 - CD^2$ and further $XD \perp BC$.

Moreover, $TK = TL, X, T$ both lie on the perpendicular bisectors of AD and KL. Since $AB > AC$, KL and BC are not parallel; AD and the perpendicular bisector of KL are not parallel either, and they have a unique intersection. It follows that $T = X$, and X is the circumcentre of $\triangle PKL$.

Since $XY \perp CK$, or $XY \perp PK$, we infer that Y is on the perpendicular bisector of PK, and $YP = YK$. Together with $BC = CK$, we find $\angle CPY = \angle CKY = \angle CBY$, and B, Y, C, P are concyclic. Similarly, B, Z, C, P are concyclic. Finally, since

$$\begin{aligned}\angle BPC &= \angle CBL + \angle BCK - 180^\circ \\ &= 2(\angle ABC + \angle ACB) - 180^\circ \\ &= 180^\circ - 2\angle BAC \\ &= 180^\circ - \angle BOC,\end{aligned}$$

B, P, O, C are concyclic.

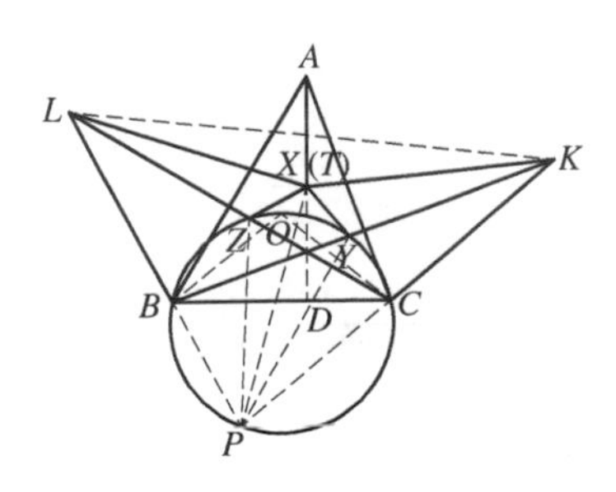

Fig. 7.2

As we have shown that B, P, C, Y, O, and Z are concyclic, the conclusion follows immediately. □

8 For positive integers m and n, define

$$f(x) = (x-1)(x^2-1)\cdots(x^m-1),$$
$$g(x) = (x^{n+1}-1)(x^{n+2}-1)\cdots(x^{n+m}-1).$$

Show that there exists an integral polynomial $h(x)$ of degree mn, such that $f(x)h(x) = g(x)$, and all the $mn+1$ coefficients of $h(x)$ are positive integers.

(Contributed by Wang Bin)

Solution More generally, for nonnegative integers m and n,

$$f_{m,n}(x) = \frac{(x^{m+1}-1)(x^{m+2}-1)\cdots(x^{m+n}-1)}{(x-1)(x^2-1)\cdots(x^n-1)}$$

is a polynomial of degree mn, and all the $mn+1$ coefficients of $f_{m,n}(x)$ are positive integers. For $n=0$, take the denominator as 1.

We prove the above assertion by induction on $m+n$. First, $f_{m,n}(x) = f_{n,m}(x)$, as indicated by

$$f_{m,n}(x) = \frac{\prod_{i=1}^{m+n}(x^i-1)}{\prod_{i=1}^{m}(x^i-1)\prod_{i=1}^{n}(x^i-1)} = f_{n,m}(x).$$

For $m+n \le 3$,

$$f_{m,0}(x) = f_{0,n}(x) = 1, \quad f_{1,1}(x) = 1+x, \quad f_{1,2}(x) = f_{2,1}(x) = 1+x+x^2,$$

and the assertion is true. Now, suppose $m,\ n > 0$, $m+n \ge 4$, and the assertion is true for all smaller $m+n$ values. Then

$$\frac{f_{m-1,n}(x)}{f_{m,n}(x)} = \frac{(x^m-1)(x^{m+1}-1)\cdots(x^{m+n-1}-1)}{(x^{m+1}-1)(x^{m+2}-1)\cdots(x^{m+n}-1)} = \frac{x^m-1}{x^{m+n}-1},$$

$$\frac{f_{m,n-1}(x)}{f_{m,n}(x)} = \frac{f_{n-1,m}(x)}{f_{n,m}(x)} = \frac{(x^n-1)(x^{n+1}-1)\cdots(x^{m+n-1}-1)}{(x^{n+1}-1)(x^{n+2}-1)\cdots(x^{m+n}-1)}$$
$$= \frac{x^n-1}{x^{m+n}-1},$$

and hence,

$$x^n \cdot \frac{f_{m-1,n}(x)}{f_{m,n}(x)} + \frac{f_{m,n-1}(x)}{f_{m,n}(x)} = \frac{x^n(x^m - 1)}{x^{m+n} - 1} + \frac{x^n - 1}{x^{m+n} - 1} = 1$$

By the induction hypothesis, $f_{m,n}(x) = x^n f_{m-1,n}(x) + f_{m,n-1}(x)$ is a polynomial of degree mn with nonnegative integer coefficients. Since all coefficients of $f_{m-1,n}(x)$ are positive, it is clear that in $f_{m,n}(x)$, the coefficients of $x^n, x^{n+1}, \ldots, x^{mn}$ are positive; in addition, all coefficients of $f_{m,n-1}(x)$ are positive, and hence, the coefficients of $x^0, x^1, \ldots, x^{mn-m}$ in $f_{m,n}(x)$ are positive. As $m, n > 0$, $mn - m \geq n - 1$, all terms $x^0, x^1, \ldots, x^{mn}$ of $f_{m,n}(x)$ have positive coefficients. The induction is completed and the assertion is secured.

China Southeastern Mathematical Olympiad

2020 (Zhuji, Zhejiang)

The 17th China Southeastern Mathematical Olympiad (CSMO) was held on August 5th–8th, 2020 at Hailiang Senior High School, Zhuji, Zhejiang Province. Due to COVID-19, except for a few local onsite students, most contestants attended the competition online.

The 17th CSMO committee members were: Tao Pingsheng (Jiangxi Science and Technology Normal University), Li Shenghong, Yang Xiaoming (Zhejiang University), Liu Bin (Peking University), Hu Zhiming, Zheng Wenxun (Tsinghua University), Wu Quanshui (Fudan University), Luo Ye (University of Hong Kong), Zhang Pengcheng (Fujian Normal University), Dong Qiuxian (Nanchang University), Wu Genxiu (Jiangxi Normal University), Xiong Bin, He Yijie (East China Normal University).

Lasted for two days, the competition consisted of two levels: 10th grade and 11th grade, and they were conducted at the same time on each day. The following are the problems and suggested solutions.

10th Grade
First Day
(1:30 – 5:30 pm; August 5, 2020)

1 Let $f(x) = a(3a+2c)x^2 - 2b(2a+c)x + b^2 + (c+a)^2$ $(a, b, c \in \mathbb{R})$ be a quadratic function such that $f(x) \leq 1$ for all $x \in \mathbb{R}$. Find the maximum of $|ab|$.

(Contributed by Yang Xiaoming, Zhao Bin)

Solution Rewrite $f(x) = a(3a+2c)x^2 - 2b(2a+c)x + b^2 + (c+a)^2$ as a quadratic polynomial in c:

$$f(x) = c^2 + 2(ax^2 - bx + a)c + 3(ax)^2 - 4abx + b^2 + a^2 \leq 1.$$

Complete the square and simplify, obtaining

$$(c + ax^2 - bx + a)^2 + (1 - x^2)(ax - b)^2 \leq 1 \Leftarrow (1 - x^2)(ax - b)^2 \leq 1.$$

Let $x = \pm\frac{\sqrt{3}}{3} \Rightarrow \frac{4\sqrt{3}}{3}|ab| \leq \frac{3}{2} \Rightarrow |ab| \leq \frac{3\sqrt{3}}{8}$, and the equality holds when $a = \frac{3\sqrt{2}}{4}, b = -\frac{\sqrt{6}}{4}, c = -\frac{5\sqrt{2}}{4}$. We have

$$f(x) = -\frac{3}{8}x^2 + \frac{\sqrt{3}}{4}x + \frac{7}{8} = -\frac{3}{8}\left(x - \frac{\sqrt{3}}{3}\right)^2 + 1 \leq 1. \qquad \square$$

2 As illustrated in Fig. 2.1, in $\triangle ABC$, $AB < AC$, PB, PC are tangent to the circumcircle O of $\triangle ABC$. Let R be a point on $\overparen{AC}$, $AR // BC$, and Q be the other intersection of PR and circle O. Let I be the incentre of $\triangle ABC$, $ID \perp BC$ with foot D, and G be the other intersection of QD and circle O. Suppose that the line through I and perpendicular to AI intersects AB, AC at points M, N, respectively. Prove: A, G, M, N are concyclic.

(Contributed by Zhang Pengcheng)

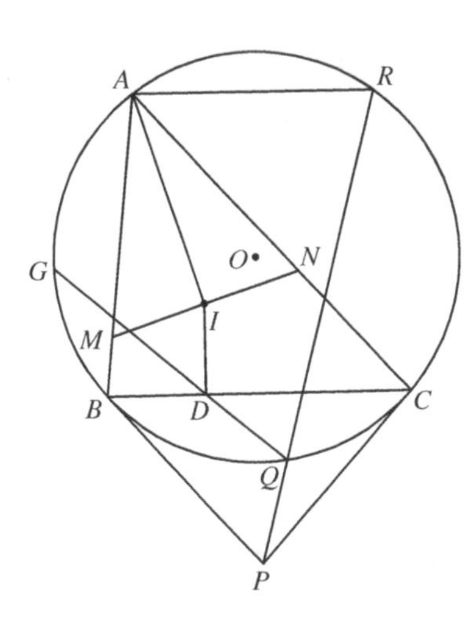

Fig. 2.1

Solution Connect $GB, GC, GM, GN, RB, RC, QB, QC, BI$, and CI. See Fig. 2.2.

Since PB, PC are tangent to circle O,

$$\frac{RB}{BQ} = \frac{PR}{PB} = \frac{PR}{PC} = \frac{RC}{CQ},$$

and hence $\dfrac{BQ}{CQ} = \dfrac{RB}{RC}$.

Moreover, $\dfrac{BD}{CD} = \dfrac{S_{\triangle BGQ}}{S_{\triangle CGQ}} = \dfrac{BG \cdot BQ}{CG \cdot CQ} = \dfrac{BG \cdot RB}{CG \cdot RC}$;

from $AR//BC$, we have $RB = AC, RC = AB$. Together, they imply that

$$\frac{BD}{CD} = \frac{BG \cdot AC}{CG \cdot AB},$$

and

$$\frac{BG}{CG} = \frac{BD \cdot AB}{CD \cdot AC}$$

$$= \frac{BI \cdot \cos\dfrac{B}{2}\sin C}{CI \cdot \cos\dfrac{C}{2}\sin B} = \frac{BI \cdot \sin\dfrac{C}{2}}{CI \cdot \sin\dfrac{B}{2}} = \frac{BI^2}{CI^2}. \quad ①$$

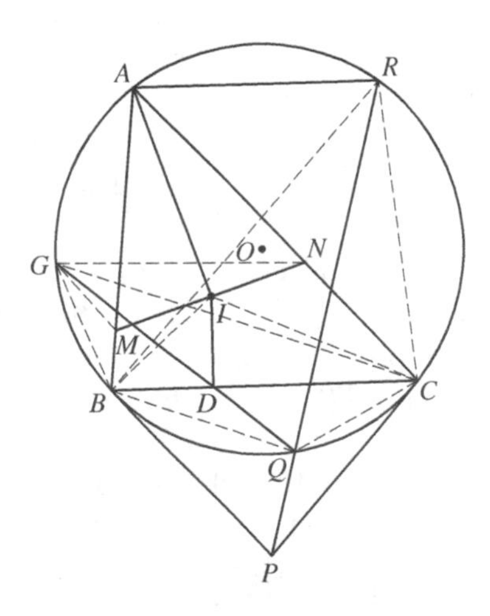

Fig. 2.2

As $AI \perp MN$, it follows that

$$\angle AMN = \angle ANM = \frac{B+C}{2}, \angle BMI = \angle INC = 180^\circ - \frac{B+C}{2},$$

$$\angle MBI = \frac{B}{2} = \frac{B+C}{2} - \frac{C}{2} = \angle ANI - \angle NCI = \angle NIC,$$

and $\triangle MBI \backsim \triangle NIC$. Hence, $\dfrac{BM}{CN} = \dfrac{BM}{IN} \cdot \dfrac{IM}{CN} = \dfrac{BI^2}{CI^2}$. ②

By ① and ②, we find $\dfrac{BG}{CG} = \dfrac{BM}{CN}$; in addition, $\angle GBM = \angle GBA = \angle GCA = \angle GCN$. Therefore, $\triangle GBM \backsim \triangle GCN$, giving

$$\angle GMA = 180^\circ - \angle GMB = 180^\circ - \angle GNC = \angle GNA.$$

It follows that A, G, M, N lie on a circle. □

3 Let $f(x) = x^{2020} + \sum_{i=0}^{2019} c_i x^i$, $c_i \in \{-1, 0, 1\}$ be a polynomial, and N be the number of positive integer roots of $f(x)$ (counted with their multiplicities). Given that $f(x)$ has no negative integer roots, find the maximum value of N.

(Contributed by Luo Ye)

Solution The maximum of $N = 10$. Notice that $c_i \in \{-1, 0, 1\}$, and hence any real root of $f(x)$ must have absolute value less than 2. It follows that $-1, 0$, and 1 are the only possible integer roots of $f(x)$.

Consider the monic polynomial $f(x) = (x-1)(x^3-1)(x^5-1)(x^{11}-1)(x^{21}-1)(x^{43}-1)(x^{85}-1)(x^{171}-1)(x^{341}-1)(x^{683}-1)x^{656}$, which has degree 2020 and integer roots $x = 0, 1$ only. The multiplicity of $x = 1$ is $N = 10$.

Denote $f_i = x^{q_i} - 1$, for which $q_1 = 1, q_2 = 3, q_3 = 5, q_4 = 11, q_5 = 21, q_6 = 43, q_7 = 85, q_8 = 171, q_9 = 341, q_{10} = 683$. Since every term is larger than the sum of all previous terms, it is easy to verify by induction that in the expanded form of $\prod_{i=1}^{J} f_i$, each term $c_k x^k$ has coefficient $c_k \in \{-1, 0, 1\}$, $J = 1, 2, \ldots 10$.

It remains to prove that $N = 10$ is the maximum.

Suppose $N \geq 11$. Then $(x-1)^{11} | x^{2020} + \sum_{i=0}^{2019} c_i x^i$. Let $x = -1$, and we have $2^{11} | f(-1)$. However, $|f(-1)| \neq 0$, and this implies $|f(-1)| \geq 2^{11}$. On the other hand, $|f(-1)| \leq 1 + \sum_{i=0}^{2019} c_i \leq 2021 < 2^{11}$, which leads to a contradiction. Therefore, the maximum value of N is 10. □

4 Let $a_1, a_2, \ldots, a_{17}$ be a permutation of $1, 2, \ldots, 17$, satisfying that

$$(a_1 - a_2)(a_2 - a_3) \cdots (a_{16} - a_{17})(a_{17} - a_1) = n^{17}.$$

How large can the positive integer n possibly be?

(Contributed by He Yijie)

Solution Denote $S = (a_1-a_2)(a_2-a_3)\cdots(a_{16}-a_{17})(a_{17}-a_1)$, and $a_{18} = a_1$. Our argument consists of four steps:

(1) n is even;
(2) $n < 9$;
(3) $n \neq 8$;
(4) we give a permutation of $1, 2, \ldots, 17$: $a_1, a_2, \ldots, a_{17}$, such that $S = 6^{17}$.

(1) Since $a_1, a_2, \ldots, a_{17}$ are 9 odd integers and 8 even integers, there exists $i \in \{1, 2, \ldots, 17\}$, such that both a_i and a_{i+1} are odd. It follows that $a_i - a_{i+1}$ is even and so is n.

(2) Write

$$S = \prod_{1\le i\le 17} |a_i - a_{i+1}| = \prod_{1\le i\le 17} (\max\{a_i, a_{i+1}\} - \min\{a_i, a_{i+1}\})$$

and let

$$U = \sum_{1\le i\le 17} \max\{a_i, a_{i+1}\} - \sum_{1\le i\le 17} \min\{a_i, a_{i+1}\}.$$

Since $a_1, a_2, \ldots, a_{17}$ is a permutation of $1, 2, \ldots, 17$, every integer appears in the two sums exactly twice, and clearly we have $U \le (17+\cdots+10)\times 2 - (8+\cdots+1)\times 2 = 144$. By the AM-GM inequality,

$$S \le \left(\frac{U}{17}\right)^{17} \le \left(\frac{144}{17}\right)^{17} < 9^{17}.$$

Hence, $n < 9$.

(3) Let $t \in 1, 2, 3, 4\}$.

Clearly, the remainders of $a_1, a_2, \ldots, a_{17}$ modulo 2^t traverse all $0, 1, \ldots, 2^t - 1$.

For $i = 1, 2, \ldots, 17$, when the remainders of a_i and a_{i+1} modulo 2^t are distinct (there are at least 2^t of such subscripts i), $2^t \nmid (a_i - a_{i+1})$.

Therefore, for at most $17 - 2^t$ subscripts $i \in \{1, 2, \ldots, 17\}$, $2^t \,|(a_i - a_{i+1})$ holds.

Specifically, in all factors $a_i - a_{i+1} (i = 1, 2, \ldots, 17)$, at most 15 of them are even, at most 13 of them are multiples of 2^2, 9 of them are multiples of 2^3, and 1 of them is a multiple of 2^4.

Consequently, the multiplicity of prime factor 2 in the product $S = \prod_{i=1}^{17} (a_i - a_{i+1})$ cannot exceed $15 + 13 + 9 + 1 = 38$.

If $n = 8$, then $S = 8^{17} = 2^{51}$, which is impossible. So $n \neq 8$.

(4) Define $a_1, a_2, \ldots, a_{17}$ as follows:

$$1, 9, 17, 8, 7, 16, 14, 5, 13, 4, 6, 15, 12, 3, 11, 2, 10.$$

We have $S = (-8) \cdot (-8) \cdot 9 \cdot 1 \cdot (-9) \cdot 2 \cdot 9 \cdot (-8) \cdot 9 \cdot (-2) \cdot (-9) \cdot 3 \cdot 9 \cdot (-8) \cdot 9 \cdot (-8) \cdot 9 = (-1)^8 \cdot 2^{3+3+1+3+1+3+3} \cdot 3^{2+2+2+2+2+1+2+2+2} = 6^{17}$.

In summary, n can be as large as 6. □

Second Day
(8:00 am – 12:00 pm; August 6, 2020)

1 Given $I = \{1, 2, \ldots, 2020\}$. Define "Wu" set $W = \{w(a, b) = (a + b) + ab | a, b \in I\} \cap I$, "Yue" set $Y = \{y(a, b) = (a + b) \cdot ab | a, b \in I\} \cap I$, and "Xizi" set $X = W \cap Y$. Find the sum of the largest and the smallest elements of "Xizi" set X.

(Contributed by Tao Pingsheng)

Solution If $n \in W$, then there exist integers a, b in I, such that $n = a + b + ab$, or $n + 1 = (a + 1)(b + 1)$. So, $n \in W$ if and only if $n + 1 \in I$ is composite. We have the following observations.

(1) Every integer in the set $\{1, 2, 4, 6, 10, 12, \ldots\} = \{p - 1 | p \text{ is prime}\}$ is not in W (we call such number a "pre-prime"), while all others are in W;
(2) For set Y: $2 = y(1, 1) \in Y$; when $(a, b) \neq (1, 1), y(a, b)$ is composite, and this implies that all odd primes are not in Y. For each composite number in Y, either it is a product of two consecutive integers $y(1, n) = (n + 1) \cdot n$; or it is a product of three integers, the largest one being the sum of the two smaller ones, $y(a, b) = (a + b) \cdot a \cdot b$.

We claim that the smallest element of X is 20. On one hand, $20 = w(2, 6) = y(1, 4)$; on the other hand, among all integers less than 20, excluding the primes and pre-primes, there are four composite numbers 8, 9, 14, 15, and they all belong to W, but not to Y.

Now we search for the largest element of X. Since $2000 = w(2, 666) = y(10, 10), 2000 \in Y$. For all integers in I larger than 2000, excluding the three prime numbers 2003, 2011, 2017 and their corresponding pre-primes, there are 14 numbers left, and they have the factorizations:

$$2001 = 3 \cdot 23 \cdot 29, 2004 = 2^2 \cdot 3 \cdot 167, 2005 = 5 \cdot 401,$$

$$2006 = 2 \cdot 17 \cdot 59, 2007 = 3^2 \cdot 223, 2008 = 2^3 \cdot 251, 2009 = 7^2 \cdot 41,$$

$$2012 = 2^2 \cdot 503, 2013 = 3 \cdot 11 \cdot 61, 2014 = 2 \cdot 19 \cdot 53,$$

$$2015 = 5 \cdot 13 \cdot 31, 2018 = 2 \cdot 1009, 2019 = 3 \cdot 673, 2020 = 2^2 \cdot 5 \cdot 101.$$

None of them meets the criterion in (2), indicating that they are not in Y, and thus the largest element of X is 2000.

Therefore, the sum of the largest and the smallest elements of X is $2000 + 20 = 2020$. □

2 In a quadrilateral $ABCD, \angle ABC = \angle ADC < 90°$. The circle with diameter AC is centred at O and intersects BC, CD at points E, F (other than C), respectively. Let M be the midpoint of $BD, AN \perp BD$ with foot N, as shown in Fig. 2.1.

Show that M, N, E, F are concyclic.

(Contributed by Zhang Pengcheng)

Solution Let G be the intersection of circle O and AB. Connect AE, GC, GD, GF, EN, EF, and BF, as shown in Fig. 2.2. Evidently, we have

$$\angle BGF = 90° + \angle CGF = 90° + \angle CAF,$$

$$S_{\triangle BGF} = GB \cdot GF \sin \angle BG = GB \cdot GF \cos \angle CAF = \frac{GB \cdot GF \cdot AF}{AC}.$$

Similarly, $S_{\triangle DGF} = \dfrac{DF \cdot FG \cdot CG}{AC}$.

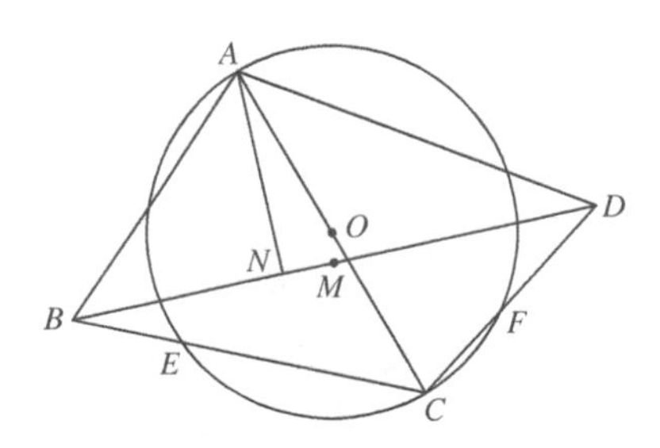

Fig. 2.1

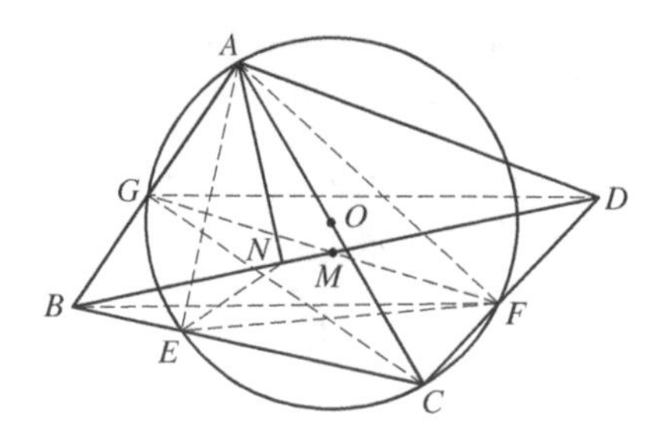

Fig. 2.2

Note that $\angle ABC = \angle ADC$, and hence

$$\text{Rt}\triangle GBC \backsim \text{Rt}\triangle FDA,$$

which implies $\dfrac{GB}{FD} = \dfrac{CG}{AF}$, or

$$GB \cdot AF = FD \cdot CG.$$

It follows that $S_{\triangle BGF} = S_{\triangle DGF}$,

and G, M, F are collinear.

Moreover, from $AE \perp EC, AN \perp BN$,

we infer that A, B, E, N are concyclic,

$$\angle BNE = \angle BAE = \angle GAE = \angle GFE = \angle MFE,$$

and thus M, N, E, F are concyclic. □

3 For any prime number $p \geq 3$, show that for all $x \in \mathbb{N}$ sufficiently large, one of the integers $x+1, x+2, \ldots, x+\dfrac{p+3}{2}$ has a prime factor larger than p.

(Contributed by Luo Ye)

Solution Let $q = \dfrac{p+3}{2}$.

Assume that the statement is untrue, and none of the integers $x+1, x+2, \ldots, x+\dfrac{p+3}{2}$ has a prime factor larger than p.

Let $p_i^{\alpha_i}$ be the highest prime power of $x+i$ (that is, $p_i^{\alpha_i+1} \nmid x+i$). As the number of primes $\leq p$ is less than or equal to $\dfrac{p+1}{2} = q-1$, we deduce that whenever $x \geq (q-1)^{q-1}, p_i^{\alpha_i} > x^{\frac{1}{q-1}} \geq (q-1)$ holds for $i = 1, 2, \ldots, q$.

Since $p_i \le p$, and the number of primes $\le p$ does not exceed $\dfrac{p+1}{2} = q-1$, there must exist distinct $i_1, i_2 \in \{1, 2, \ldots, q\}$ such that $p_{i_1} = p_{i_2}$. Then we have $(x+i_1, x+i_2) \ge \min(p_{i_1}^{\alpha_1}, p_{i_2}^{\alpha_2}) > q-1$. Meanwhile, $(x+i_1, x+i_2) = (x+i_1, i_2-i_1) \le |i_2-i_1| \le q-1$, which is contradictory. Therefore, the statement is true for all $x \ge (q-1)^{q-1}$. □

4 An inkjet printer is used to print on a paper strip with $1 \times n$ grids. When the nozzle prints on the ith $(1 \le i \le n)$ grid, it becomes black; in addition, each of the adjacent grids, the $(i-1)$th and the $(i+1)$th grids (if they exist), independently has a probability of $\frac{1}{2}$ becoming black. Let $T(n)$ be the expected number of prints needed to make all grids black, provided that the optimal strategy is adopted (make as few prints as possible). Find the formula of $T(n)$.

(Contributed by Luo Ye)

Solution Use a subset $S \subseteq \{1, \ldots, n\}$ to represent all uncoloured grids at the current state: $i \in S$ means that the ith grid is not coloured yet; $i \notin S$ means it is already black. Let $f(S)$ be the expected number of prints needed to make all grids in S black. If $S \subseteq S'$, then obviously the optimal strategy for S' can also be implemented on S, and hence $f(S) \le f(S')$. Let $g(S, i)$ be the expected number of prints after the nozzle prints on i. We have

$$f(S) = 1 + \min_{\{1 \le i \le n\}} g(S, i).$$

$$g(S,i) = \frac{1}{4}f(S - \{i\}) + \frac{1}{4}f(S - \{i, i-1\}) + \frac{1}{4}f(S - \{i, i+1\}) + \frac{1}{4}f(S - \{i, i-1, i+1\}).$$

Thus, for every $1 \le i \le n$, $\quad f(S) - 1 \le g(S, i) \le f(S - \{i\})$.

Lemma *In all circumstances, if the ith grid is already black, $1 \le i \le n$, then there exists an optimal strategy such that the nozzle never prints on it thereafter.*

Proof of Lemma Assume to the contrary that printing on i at a later time is an optimal strategy.

Consider $i-1$ and $i+1$ (if they exist). If they are both black, clearly, printing on i does no good.

If one grid is black, say $i-1$, then printing on i has a probability of $\frac{1}{2}$ making $i+1$ black, while printing on $i+1$ has a probability 1 and definitely does a better job.

If both $i-1$ and $i+1$ are uncoloured, consider two strategies: one is to print on $i-1$, and the other on $i+1$. We claim that

$$g(S,i-1)+g(S,i+1)\le 2\,g(S,i).$$

It is easy to see that

$$\begin{aligned}g(S,i)=&\frac{1}{4}f(S)+\frac{1}{4}f(S-\{i-1\})+\frac{1}{4}f(S-\{i+1\})\\&+\frac{1}{4}f(S-\{i-1,i+1\}),\end{aligned}$$

and $f(S)=1+g(S,i)$. Therefore,

$$g(S,i)=\frac{1}{3}(f(S-\{i-1\})+f(S-\{i+1\})+f(S-\{i-1,i+1\})+1).$$

On the other hand, we have

$$\begin{aligned}g(S,i-1)\le f(S-\{i-1\})\le 1+f(S-\{i-1,i+1\}),g(S,i+1)\\ \le f(S-\{i+1\})\le 1+f(S-\{i-1,i+1\}).\end{aligned}$$

This implies that

$$\begin{aligned}&g(S,i-1)+g(S,i+1)\\&\le\left(\frac{1}{3}(1+f(S-\{i-1,i+1\}))+\frac{2}{3}f(S-\{i-1\})\right)\\&\quad+\left(\frac{1}{3}(1+f(S-\{i-1,i+1\}))+\frac{2}{3}f(S-\{i+1\})\right)=2g(S,i).\end{aligned}$$

Consequently, we have either $g(S,i-1)\le g(S,i)$ or $g(S,i+1)\le g(S,i)$, meaning that one of the two strategies, printing on either $i-1$ or $i+1$, is superior to printing on i. When the nozzle prints on one side of i and continues to work on that side until all those grids become black, based on the previous argument, it does not need to print on i thereafter. The lemma is verified.

For the original problem, let $T(k)$ be the expected number of prints to make $1\times k$ grids black. Define $T(-1)=T(0)=0,T(1)=1,T(2)=\frac{3}{2}$. If the nozzle prints on $i\in\{1,2,\ldots,k\}$, then according to the lemma, there is an optimal strategy that does not print on i anymore, and thus the problem reduces to printing on the two sides separately. We have

$$T(k) = 1 + \min_{1\le i\le k} \frac{1}{2}(T(i-1) + T(i-2) + T(k-i-1) + T(k-i)).$$

Let $S(k) = T(k)(k = -1, 0, 1, 2)$. When $k \ge 3, S(k) = 1 + \frac{1}{2}(S(0) + S(1) + S(k-2) + S(k-3)) = \frac{3}{2} + \frac{1}{2}S(k-2) + \frac{1}{2}S(k-3)$. It is straightforward to check that $T(3) = S(3) = 2, T(4) = S(4) = \frac{11}{4}, T(5) = S(5) = \frac{13}{4}, T(6) = S(6) = \frac{31}{8}$.

Let $h(k) = S(k) - \frac{3}{5}k$. Then we have the homogeneous recursive relation $h(k) = \frac{1}{2}h(k-2) + \frac{1}{2}h(k-3)$. Some routine calculation gives the formula for $h(k)$, then for $S(k)$, as:

$$S(k) = h(k) + \frac{3}{5}k = \frac{3}{5}k + \frac{7}{25} + b\lambda^k + \bar{b}\bar{\lambda}^k. \quad (k \ge 0)$$

Here, $\lambda = \dfrac{-1+i}{2}$ is a root of the characteristic equation $x^3 - \frac{1}{2}x - \frac{1}{2} = 0$, and $b = \dfrac{-7+i}{50}$. We assert that for $k \ge 6, S(0) + S(1) + S(k-2) + S(k-3) = \min_{1\le i\le k}(S(i-1) + S(i-2) + S(k-i-1) + S(k-i))$. With this, induction on k gives $T(k) = S(k)$.($k \le 5$ is already verified.)

It suffices to show, for $1 \le i \le k$,

$$\begin{aligned} &S(0) + S(1) + S(k-2) + S(k-3) \\ &\quad \le S(i-1) + S(i-2) + S(k-i-1) + S(k-i). \qquad ① \end{aligned}$$

By symmetry, we only need to consider $1 \le i \le \frac{k+1}{2}$.

If $i = 1$, (1) can be simplified to $1 + S(k-3) \le S(k-1)$. When $k \ge 6$,

$$\begin{aligned} S(k-1) - S(k-3) &= \frac{6}{5} - (b\lambda^{k-1} + \bar{b}\bar{\lambda}^{k-1} - b\lambda^{k-3} - \bar{b}\bar{\lambda}^{k-3}) \\ &\ge \frac{6}{5} - 4||b|| \cdot ||\lambda||^3 = 1. \end{aligned}$$

If $i = 2$, the equality in (1) holds.

If $i \geq 3$, substitute $S(k)$ in (1) by the formula and simplify, obtaining

$$\mathrm{Re}(b(\lambda+1)(1-\lambda^{i-2})(1-\lambda^{k-i-1})) \leq 0. \quad ②$$

To prove ②, first, notice that $\mathrm{Arg}(b(\lambda+1)) = \mathrm{Arg}\left(\frac{-4-3i}{50}\right) = -\pi + \sin^{-1}\frac{3}{5} > -\pi$.

Claim for $t \geq 1, \mathrm{Arg}(1-\lambda^t) \in \left(-\frac{\pi}{6}, \sin^{-1}\frac{1}{\sqrt{5}}\right]$.

Proof of Claim Indeed, for $1 \leq t \leq 3$, this is evident; for $t \geq 4$, we have $|\mathrm{Arg}(1-\lambda^t)| \leq \sin^{-1}|\lambda^t| \leq \sin^{-1}\frac{1}{4} \leq \frac{\pi}{12}$.

Let $X = \mathrm{Arg}(b(\lambda+1)) + \mathrm{Arg}(1-\lambda^{k-i-1}) + \mathrm{Arg}(1-\lambda^{i-2})$. Then,

$$-\frac{4\pi}{3} = -\pi - 2\cdot\frac{\pi}{6} < X \leq -\pi + \sin^{-1}\frac{3}{5} + 2\sin^{-1}\frac{1}{\sqrt{5}} = -\frac{\pi}{2}.$$

It follows that $X \in \left(-\frac{4\pi}{3}, -\frac{\pi}{2}\right]$, which validates ②, and further ①.

Therefore, the general formula for $T(k)$ is

$$T(k) = \frac{3}{5}k + \frac{7}{25} + b\lambda^k + \bar{b}\bar{\lambda}^k, \quad \lambda = \frac{-1+i}{2}, \quad b = \frac{-7+i}{50}.$$

The optimal strategy is this: for $n = 1, 2$, print on the first grid; for $n \geq 3$, print on the second grid from the left, and (if the first grid is not black) then treat the two sides separately, using the strategy for $n = 1$ on the single grid. □

11th Grade
First Day
(1:30 – 5:30 pm; August 5, 2020)

1 Let $a_1, a_2, \ldots, a_{17}$ be a permutation of $1, 2, \ldots, 17$, satisfying

$$(a_1 - a_2)(a_2 - a_3)\cdots(a_{16} - a_{17})(a_{17} - a_1) = 2^n.$$

Find the maximum value of the integer n.

(Contributed by He Yijie)

Solution Denote $S = (a_1 - a_2)(a_2 - a_3)\cdots(a_{16} - a_{17})(a_{17} - a_1)$, and $a_{18} = a_1$. Let $t \in \{1, 2, 3, 4\}$.

Clearly, the remainders of $a_1, a_2, \ldots, a_{17}$ modulo 2^t traverse all $0, 1, \ldots, 2^t - 1$.

For $i = 1, 2, \ldots, 17$, when the remainders of a_i and a_{i+1} modulo 2^t are different (there are at least 2^t of such subscripts i), $2^t \nmid (a_i - a_{i+1})$.

Therefore, for at most $17 - 2^t$ subscripts $i \in \{1, 2, \ldots, 17\}$, $2^t | (a_i - a_{i+1})$ holds.

In specific, among all factors $a_i - a_{i+1}$ $(i = 1, 2, \ldots, 17)$, at most 15 of them are even, 13 of them are multiples of 2^2, 9 of them are multiples of 2^3, and 1 of them is a multiple of 2^4.

Consequently, the multiplicity of prime factor 2 in the product $S = \prod_{i=1}^{17} (a_i - a_{i+1})$ cannot exceed $15 + 13 + 9 + 1 = 38$.

On the other hand, we may define a permutation of $1, 2, \ldots, 17$ as follows:

$$1, 17, 9, 5, 13, 11, 3, 7, 15, 16, 8, 12, 4, 6, 14, 10, 2,$$

Then,

$$S = (-16) \cdot 8 \cdot 4 \cdot (-8) \cdot 2 \cdot 8 \cdot (-4) \cdot (-8) \cdot (-1) \cdot 8 \cdot (-4) \cdot 8 \cdot (-2) \cdot (-8) \cdot 4 \cdot 8 \cdot 1 = 2^{38}.$$

In conclusion, the maximum value of n is 38. □

2 As illustrated in Fig. 2.1, in $\triangle ABC$, $AB < AC$, PB, PC are tangent to the circumcircle O of $\triangle ABC$. Let R be a point on $\overset{\frown}{AC}$, $AR // BC$,

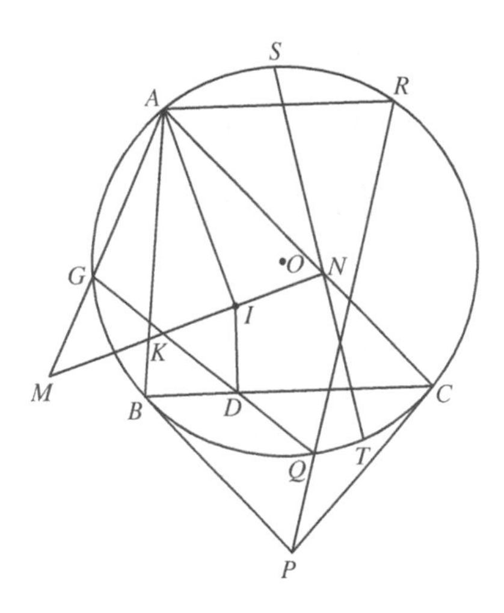

Fig. 2.1

and Q be the other intersection of PR and circle O. Let I be the incentre of $\triangle ABC, ID \perp BC$ with foot D and G be the other intersection of QD and circle O. Suppose that the line through I and perpendicular to AI intersects the lines AG, AC at the points M, N, respectively. Let S be the midpoint of $\overparen{AR}, T$ be the other intersection of the line SN and circle O. Prove that M, B, T are collinear.

(Contributed by Zhang Pengcheng)

Solution Let AB meet MN at K. Connect $GB, GC, GK, GN, RB, RC, QB, QC, BI, CI$. See Fig. 2.2.

Since PB, PC are tangent to circle O,

$$\frac{RB}{BQ} = \frac{PR}{PB} = \frac{PR}{PC} = \frac{RC}{CQ},$$

and hence $$\frac{BQ}{CQ} = \frac{RB}{RC}.$$

Moreover, $$\frac{BD}{CD} = \frac{S_{\triangle BGQ}}{S_{\triangle CGQ}} = \frac{BG \cdot BQ}{CG \cdot CQ} = \frac{BG \cdot RB}{CG \cdot RC};$$

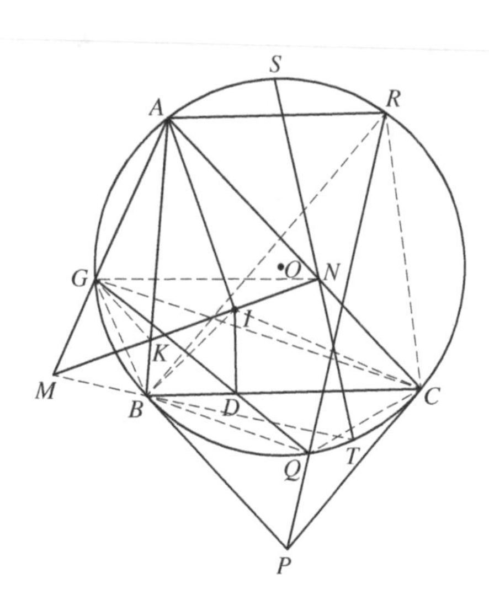

Fig. 2.2

from $AR//BC$, we have $RB = AC, RC = AB$. Together, they imply

$$\frac{BD}{CD} = \frac{BG \cdot AC}{CG \cdot AB},$$

$$\text{and} \quad \frac{BG}{CG} = \frac{BD \cdot AB}{CD \cdot AC}$$

$$= \frac{BI \cdot \cos\frac{B}{2} \sin C}{CI \cdot \cos\frac{C}{2} \sin B} = \frac{BI \cdot \sin\frac{C}{2}}{CI \cdot \sin\frac{B}{2}} = \frac{BI^2}{CI^2}. \quad ①$$

As $AI \perp KN$, we find

$$\angle AKN = \angle ANK = \frac{B+C}{2}, \angle BKI = \angle INC = 180° - \frac{B+C}{2},$$

$$\angle KBI = \frac{B}{2} = \frac{B+C}{2} - \frac{C}{2} = \angle ANI - \angle NCI = \angle NIC,$$

$$\text{and} \quad \triangle KBI \backsim \triangle NIC. \quad \text{Hence,} \ \frac{BK}{CN} = \frac{BK}{IN} \cdot \frac{IK}{CN} = \frac{BI^2}{CI^2}. \quad ②$$

By ① and ②, $\frac{BG}{CG} = \frac{BK}{CN}$; in addition, $\angle GBK = \angle GBA = \angle GCA = \angle GCN$. Therefore, $\triangle GBK \backsim \triangle GCN$, giving

$$\angle GKA = 180° - \angle GKB = 180° - \angle GNC = \angle GNA.$$

It follows that A, G, K, N lie on a circle. Connect BT. Since S is the midpoint of $\overset{\frown}{AR}$ and $AR//BC$, we have

$$\angle BTN = \angle BTS = \frac{B+C}{2} = \angle AKN,$$

and hence K, B, T, N are concyclic.

Finally, by the radical axis theorem we conclude that the lines AG, KN, BT are concurrent, and M, B, T are collinear. □

3 The same as that of Problem 3, 10th grade, First Day.

4 Given that $a_n \geq a_{n-1} \geq \cdots \geq a_1 \geq 0, \sum_{i=1}^n a_i = 1$.

Prove: for any nonnegative reals $x_i, y_i, 1 \leq i \leq n$,

$$\left(\sum_{i=1}^n a_i x_i - \prod_{i=1}^n x_i^{a_i}\right)\left(\sum_{i=1}^n a_i y_i - \prod_{i=1}^n y_i^{a_i}\right)$$
$$\leq a_n^2\left(n\sqrt{\sum_{i=1}^n x_i \sum_{i=1}^n y_i} - \sum_{i=1}^n \sqrt{x_i}\sum_{i=1}^n \sqrt{y_i}\right)^2.$$

(Contributed by Li Shenghong)

Solution First, we claim

$$\sum_{i=1}^n a_i x_i - \prod_{i=1}^n x_i^{a_i} \leq a_n\left\{n\sum_{i=1}^n x_i - \left(\sum_{i=1}^n \sqrt{x_i}\right)^2\right\}. \qquad ①$$

In fact, by the reversed Cauchy inequality,

$$\left(\sum_{i=1}^n a_i x_i - \prod_{i=1}^n x_i^{a_i}\right)\left(\sum_{i=1}^n a_i y_i - \prod_{i=1}^n y_i^{a_i}\right)$$
$$\leq a_n^2\left\{n\sum_{i=1}^n x_i - \left(\sum_{i=1}^n \sqrt{x_i}\right)^2\right\}\left\{n\sum_{i=1}^n y_i - \left(\sum_{i=1}^n \sqrt{y_i}\right)^2\right\}$$
$$\leq a_n^2\left\{n\sqrt{\sum_{i=1}^n x_i \sum_{i=1}^n y_i} - \sum_{i=1}^n \sqrt{x_i}\sum_{i=1}^n \sqrt{y_i}\right\}^2.$$

Then, ① follows from the identities

$$n\sum_{i=1}^n x_i - \left(\sum_{i=1}^n \sqrt{x_i}\right)^2 = \sum_{i<j}\left((\sqrt{x_i} - \sqrt{x_j}\right)^2$$

and

$$\sum_{i<j}\left(\sqrt{x_i} - \sqrt{x_j}\right)^2 = \sum_{i<j}(x_i + x_j) - 2\sum_{i<j}\sqrt{x_i x_j}$$
$$= \sum_{i=1}^n x_i - n\left(\prod_{i=1}^n x_i\right)^{\frac{1}{n}} + (n-2)\sum_{i=1}^n x_i$$
$$+ n\left(\prod_{i=1}^n x_i\right)^{\frac{1}{n}} - 2\sum_{i<j}\sqrt{x_i x_j}.$$

Define

$$f_n(x_1,\ldots,x_n)=(n-2)\sum_{i=1}^{n}x_i+n\left(\prod_{i=1}^{n}x_i\right)^{\frac{1}{n}}-2\sum_{i<j}\sqrt{x_ix_j},$$

$$g_n(x_1,\ldots,x_n)=\frac{1}{n}\sum_{i=1}^{n}x_i-\left(\prod_{i=1}^{n}x_i\right)^{\frac{1}{n}},$$

$$h_n(x_1,\ldots,x_n)=\sum_{i=1}^{n}a_ix_i-\prod_{i=1}^{n}x_i^{a_i}.$$

Next, we prove

(i) $f_n(x_1,\ldots,x_n)\geq 0$;
(ii) $g_n(x_1,\ldots,x_n)\geq \dfrac{1}{na_n}h_n(x_1,\ldots,x_n)$.

Proof of (ii). If $a_i=\dfrac{1}{n}$, $i\geq 1$, then $g_n(x_1,\ldots,x_n)=h_n(x_1,\ldots,x_n)$, and (ii) is true.

If $\{a_i\}$ are not all equal, assume $a_n-a_i\geq 0$ for $i\geq 1$, and they are not all zeros. Then

$$\begin{aligned}g_n(x_1,\ldots,x_n)-\frac{1}{na_n}h_n(x_1,\ldots,x_n)&=\frac{1}{na_n}\sum_{i=1}^{n}(a_n-a_i)\,x_i\\&\quad+\frac{1}{na_n}\left(\prod_{i=1}^{n}x_i^{a_i}\right)-\left(\prod_{i=1}^{n}x_i\right)^{\frac{1}{n}}.\end{aligned}$$

Since $\sum\limits_{i=1}^{n-1}\dfrac{(a_n-a_i)}{na_n}+\dfrac{1}{na_n}=1$, $\dfrac{a_n-a_i}{na_n}+\dfrac{a_i}{na_n}=\dfrac{1}{n}$, by the AM–GM inequality, the conclusion (ii) follows, and furthermore, when $a_2=\cdots=a_n>a_1$, $x_2=\cdots=x_n, x_1=0$, we have

$$g_n(x_1,\ldots,x_n)=\frac{1}{na_n}\mathrm{h}_n(x_1,\ldots,x_n).$$

Proof of (i). By induction.

When $n=2$, $f_n(x_1,\ldots,x_n)=0$. Suppose for $n>2$, $f_n(x_1,\ldots,x_n)\geq 0$. Now for $n+1$, without loss of generality, assume $x_1\leq x_2\leq\cdots\leq x_n\leq x_{n+1}$.

Let $x_1' = x_1,\ x_i' = (x_2 \cdots x_{n+1})^{\frac{1}{n}} = a,\ i = 2, \ldots, n+1$. We have

$$f_{n+1}(x') = (n-1)x_1 + (n+1)x_1^{\frac{1}{n+1}} a^{\frac{n}{n+1}} - 2n(x_1 a)^{\frac{1}{2}}.$$

When $x_1 = 0$, $f_n(x_1, \ldots, x_n) = 0$.

When $x_1 > 0$, $\dfrac{f_{n+1}(x')}{2nx_1^{\frac{1}{1+n}}} = \dfrac{(n-1)}{2n}x_1 + \dfrac{(n+1)}{2n}a^{\frac{n}{n+1}} - x_1^{\frac{n-1}{2(n+1)}} a^{\frac{1}{2}} \geq 0$,
and $f_n = 0 \Leftrightarrow x_1 = 0$. We find that

$$f_{n+1}(x) - f_{n+1}(x') = (n-1)\sum_{i=2}^{n+1} x_i - 2x_1^{\frac{1}{2}}\left(\sum_{i=2}^{n+1}\sqrt{x_i} - n\sqrt{a}\right)$$

$$-2\sum_{2 \leq i < j \leq n}\sqrt{x_i x_j} = f_n(x_2, \ldots, x_{n+1}) + \sum_{i=2}^{n+1} x_i$$

$$-2x_1^{\frac{1}{2}}\left(\sum_{i=2}^{n+1}\sqrt{x_i} - n\sqrt{a}\right) - n\prod_{i=2}^{n+1} x_i^{\frac{1}{n}}.$$

Since $x_1 \leq a$, $a = (x_2 \cdots x_{n+1})^{\frac{1}{n}}$, and $f_n(x_2, \ldots, x_{n+1}) \geq 0$, it follows that

$$f_{n+1}(x) - f_{n+1}(x') \geq \sum_{i=2}^{n+1} x_i$$

$$-2\sqrt{a}\left(\sum_{i=2}^{n+1}\sqrt{x_i}\right) + na = \sum_{i=2}^{n+1}(\sqrt{x_i} - \sqrt{a})^2 \geq 0$$

and thereby $f_{n+1}(x) \geq f_{n+1}(x') \geq 0$, yielding the desired inequality. □

Second Day
(8:00 am – 12:00 pm; August 6, 2020)

1 Given $I = \{1, 2, \ldots, 2020\}$. Define "Wu" set $W = \{w(a,b) = (a+b) + ab | a, b \in I\} \cap I$, "Yue" set $Y = \{y(a,b) = (a+b) \cdot ab | a, b \in I\} \cap I$, and "Xizi" set $X = W \cap Y$. Elements of X are called "Xizi" numbers: for example, $54 = W(4, 10) = Y(3, 3)$, $56 = W(2, 18) = Y(1, 7)$ are both Xizi numbers.

(i) Find the sum of the largest and the smallest Xizi numbers.

(ii) If $n \in Y$ and the representation of n as $Y(a, b)$ is not unique, then it is called an "elite number". For example, 30 is elite as $30 = Y(1,5) = Y(2,3)$. How many numbers in Y are elite?

(Contributed by Tao Pingsheng)

Solution

(i) The same as that of Problem 1, 10th grade, Second Day.

(ii) If m is elite, there exist positive integers a, b, c, d, such that $Y(a,b) = Y(c,d) = m$, or $m = ab(a+b) = cd(c+d)$. Assume $a \le b,\ c \le d$, and ordered pairs $\langle a, b\rangle \neq \langle c, d\rangle$. Further, assume $a < c$, and we obtain $a < c \le d < b$ or else $Y(a,b) < Y(c,d)$. Let $p = (a,b),\ \ q = (c,d)$. Without loss of generality, first, assume $(p,q) = 1$ (otherwise, divide a, b, c, d by (p,q)). Clearly, $c \le 10$, or else $Y(c,d) = cd(c+d) \ge 11 \cdot 11 \cdot 22 > 2020$. There are three situations on p, q.

Case 1: $p = q = 1$. Then a, b, c, d are pairwise coprime. As $m \le 2020$, m has at most four distinct prime factors, or else $m \ge 2 \cdot 3 \cdot 5 \cdot 7 \cdot 11 = 2310 > 2020$. Since $a < c < d < b$, it must be $a + b > c + d$; in addition, $c, d, c+d$ are pairwise coprime, by the pigeonhole principle, at least two of the three numbers $c, d, c+d$ are prime powers, and the third has at most two prime factors.

It is easy to see $a \le 8, c \le 9$; otherwise, $c \ge 10$, $d \ge 11$, and $Y(c,d) = cd(c+d) \ge 10 \times 11 \times 21 > 2020$. As $c, d, c+d$ are pairwise coprime, $m = cd(c+d)$ has either three or four prime factors. We discuss as follows.

Case 1A: m has three prime factors, $m = p_1^{\alpha_1} p_2^{\alpha_2} p_3^{\alpha_3}$. As $q = (c,d) = 1$, we must have $c, d, c+d$ in the form of $f_1 = p_1^{\alpha_1},\ f_2 = p_2^{\alpha_2},\ f_3 = p_3^{\alpha_3}$, respectively, where $f_1 < f_2,\ f_3 = f_1 + f_2$.

Since $a, b, a+b$ are pairwise coprime, $ab(a+b) = m = p_1^{\alpha_1} p_2^{\alpha_2} p_3^{\alpha_3}$, it follows that all prime factors of $a, b, a+b$ are p_1, p_2, p_3, $a < c = f_1 = \min\{p_1^{\alpha_1}, p_2^{\alpha_2}, p_3^{\alpha_3}\}$, and it cannot be $a = p_j^r\ \ (1 \le r < \alpha_j)$ (otherwise, $p_j^{\alpha_j - r}$ divides b or $a+b$, contradicting that $a, b, a+b$ are pairwise coprime!). So $a = 1$, and $(b, a+b) = (f_i f_j, f_k)$ or $(f_k, f_i f_j)$. Since $f_3 = f_1 + f_2$, $a+b > c+d = f_3$, it must be $a+b = f_1 f_2$ (otherwise, $a+b = f_1 f_3$ or $f_2 f_3$, then $a+b \ge 2f_3 > \max\{f_1, f_2\} + 1 \ge b + a$ is contradictory), and thus $b = f_3$. Solving $f_1 f_2 = a + b = 1 + f_3 = 1 + f_1 + f_2$, we find $f_1 = 2,\ f_2 = 3$ and one solution $(a, b, c, d) = (1, 5, 2, 3)$.

Case 1B: m has four prime factors, $m = p_1^{\alpha_1} p_2^{\alpha_2} p_3^{\alpha_3} p_4^{\alpha_4}$. Now denote $f_j = p_j^{\alpha_j}, j = 1, 2, 3, 4$ and assume f_4 is the largest. Then, among the pairwise coprime numbers $c, d, c + d$, two of them are distinct f_i, while the third is the product of the remaining two f_i, f_j, and it cannot contain f_4 as otherwise $f_i f_j > 2f_4$ becomes larger than the sum of the other two. Thus,

$$c + d = f_4 = f_1 + f_2 f_3 \qquad ①$$

or

$$c + d = f_1 f_2 = f_3 + f_4. \qquad ②$$

Recall that $p = (a, b) = 1$, and $a, b, a + b$ are pairwise coprime.

(1^0) If $a > 1$, then among $a, b, a + b$, two of them are distinct f_i, while the third is the product of the two f_i, f_j excluding f_4. In specific, $a + b$ is either f_4, or non-f_4 product of two f_i's. As $a + b > c + d$, the two sums cannot be both f_4, nor can be both non-f_4 products of two f_is, as otherwise $c + d = f_1 f_2 = f_3 + f_4$, $a + b = f_1 f_3 = f_2 + f_4$, subtracting $f_1 f_2 = f_3 + f_4$ from $f_1 f_2 = f_3 + f_4$, we get $(f_1 + 1)(f_2 - f_3) = 0$, a contradiction.

Consequently, $c + d = f_4 = f_1 + f_2 f_3$, $a + b = f_1 f_2 = f_3 + f_4$, or vice versa (swap $a + b$ and $c + d$). Plug $f_4 = f_1 + f_2 f_3$ into the other equation to get

$$f_3(f_2 + 1) = f_1(f_2 - 1). \qquad ③$$

As $(f_1, f_3) = 1$, and $(f_2 + 1,\ f_2 - 1) = 1$ or 2:

If $(f_2 + 1,\ f_2 - 1) = 1$, then f_2 is even, $f_2 = 2^u$, and by ③, we have $f_1 = f_2 + 1, f_3 = f_2 - 1$. When $f_2 = 2^2 = 4$, $(a, b, c, d) = (3, 17, 5, 12)$ is a solution. When $f_2 \geq 2^3$, $m > 2020$, no solutions exist.

If $(f_2 + 1,\ f_2 - 1) = 2$, then f_2 is an odd prime power, $\left(\dfrac{f_2 + 1}{2}, \dfrac{f_2 - 1}{2}\right) = 1$, and by ③, we find

$$f_1 = \frac{f_2 + 1}{2}, f_3 = \frac{f_2 - 1}{2}.$$

When $f_2 = 5$, $(a, b, c, d) = (2, 13, 3, 10)$ is a solution. When $f_2 \geq 7$, it leads to $m > 2020$ and no solutions exist.

(2^0) If $a = 1$, then $(a, b, a + b) = (1, b, b + 1)$, and neither of $b, b + 1$ can be the product of three f_i's: because $f_1 f_2 f_3 f_4 = ab(a + b) = b(b + 1)$, if $b = f_1 f_2 f_3$, then $f_4 = b + 1 = f_1 f_2 f_3 + 1$; or else $b + 1 = f_1 f_2 f_3$, then

$f_4 = b = f_1 f_2 f_3 - 1$, nevertheless it must be $f_4 > f_1 + f_2 f_3$, and $f_1 f_2 < f_3 + f_4$, indicating the nonexistence of $c, d, c + d$.

Hence, one of $b, b + 1$ is the product of f_4 and the minimum of $\{f_1, f_2, f_3\}$, and the other one is the product of the other two f_i's. As we require $ab(a + b) \leq 2020$, or $b(b + 1) \leq 2020$, $b \leq 44$.

For $b \in \{1, 2, \ldots, 44\}$, there are nine b values such that $b, b + 1$ are both products of two prime powers, which are

$$14, 20, 21, 33, 34, 35, 38, 39, 44.$$

Accordingly, we find seven solutions

$$(1, 14, 3, 7), (1, 20, 5, 7), (1, 21, 3, 11), (1, 33, 6, 11), (1, 34, 7, 10),$$
$$(1, 38, 6, 13), (1, 44, 9, 11).$$

There are totally 10 solutions in Case (1).

Case 2: $pq > 1$. We only assume $a \leq b$, $c \leq d$ for now. The relation between a and c, b and d will be discussed later.

First, let $q > p \geq 1$, $q^3 | cd(c + d)$. As $(p, q) = 1$, q^3 divides one of the integers $a, b, a + b$. If $q \geq 4$, then $a + b \geq 4^3 = 64$, and m must be larger than 2020. Hence, $q = 2$ or 3.

Let $c = qc_1, d = qd_1, (c_1, d_1) = 1$. For q:

(1^0) If $q = 3$, then $27 | ab(a + b)$. As $a \leq 10$, we must have $27 | b$ or $27 | a + b$. From $m \leq 2020$, it follows that $b = 27$ or $a + b = 27$.

(A) If $b = 27$, then $a(a + 27) = c_1 d_1 (c_1 + d_1) < \dfrac{2020}{27} < 75$, $a < 3$, and hence a is 1 or 2. If $a = 1$, $28 = c_1 d_1 (c_1 + d_1)$, since $28 = 2^2 \times 7$ has only two prime factors, $c_1 = 1$, and it becomes $28 = d_1(1 + d_1)$, which has no integer solutions.

If $a = 2$, the above equation becomes $58 = c_1 d_1 (c_1 + d_1)$, and $c_1 = 1$, giving $58 = d_1(1 + d_1)$, again, no integer solutions.

(B) If $a + b = 27$, then $a(27 - a) = c_1 d_1 (c_1 + d_1) < \dfrac{2020}{27} < 75$, and $a < 4$; as $(a, a + b) = 1$, $(a, 27) = 1$, a can only be 1 or 2.

For $a = 1$, we have $26 = c_1 d_1 (c_1 + d_1)$, $c_1 = 1$, and $26 = d_1(1 + d_1)$ has no integer solutions.

For $a = 2$, we have $50 = c_1 d_1 (c_1 + d_1)$, $c_1 = 1$, and $50 = d_1(1 + d_1)$ has no integer solutions, either.

(2^0) If $q = 2$, then $p = 1,\ 8|ab(a+b)$.

(A) If $8|a$, then $a = 8,\ b \geq 8$ is odd,

$$b(8+b) = c_1 d_1 (c_1 + d_1) < \frac{2020}{8} = \frac{505}{2},$$

but $c_1 d_1 (c_1 + d_1)$ is even, b must be even, a contradiction.

(B) If $8\,|b$, let $b = 8k$, and we have

$$m_1 = \frac{m}{8} = ak(a+8k) = c_1 d_1 (c_1 + d_1) < \frac{2020}{8} = \frac{505}{2}.$$

As $c_1 d_1 (c_1 + d_1)$ is even, a is odd, k must be even: from $8k = b \leq 44$, we get $k = 2$ or $k = 4$.

If $k = 2$, then $b = 16$, $m_1 = 2a(a+16) \leq 252$, and so $a \leq 5$, $a+16 > 2a$. We assert that $T = a+16$ is not a prime. Indeed, if $c_1 > 1$, then the number in $c_1,\ d_1,\ c_1+d_1$ that is divisible by T must be greater than the sum of the other two, which is impossible. Hence $c_1 = 1$, which gives $a + 16 = 2a + 1$ or $2a - 1$, and it follows that $a > 10$ either way, a contradiction.

Now, as T is not a prime, $a \geq 5$: when $a = 5$, we find two solutions $(a, b, c, d) = (5, 16, 2, 28)$ or $(5, 16, 6, 14)$.

If $k = 4$, then $b = 8k = 32$, $m_1 = 4a(a+32) \leq 252$, and $a = 1$, giving one solution $(a, b, c, d) = (1, 32, 2, 22)$.

(C) If $8\,|(a+b)$, let $a + b = 8k$, and we have

$$m_1 = \frac{m}{8} = abk = c_1 d_1 (c_1 + d_1) < \frac{2020}{8} = \frac{505}{2}.$$

As $c_1 d_1 (c_1 + d_1)$ is even, while a, b are both odd, it must be $2\,|k$, $a + b \leq 45$, and thus $k = 2$ or 4.

If $k = 4$, then $a + b = 32$,

$$m_1 = \frac{m}{8} = 4a(32-a) = c_1 d_1 (c_1 + d_1) < \frac{2020}{8} = \frac{505}{2},$$

hence $a < 3$; as a is odd, $a = 1$. Now $c_1 d_1 (c_1 + d_1) = 4 \times 31$ has only two prime factors, $c_1 = 1$, but d_1 has no integer solutions.

If $k = 2$, then $a + b = 16$,

$$m_1 = \frac{m}{8} = 2a(16-a) = c_1 d_1 (c_1 + d_1) < \frac{2020}{8} = \frac{505}{2}$$

always holds. On the other hand, a must be odd, $a \leq b$, and thus a is in $\{1, 3, 5, 7\}$.

When $a = 1$, $2a(16-a)=2\times3\times5=1\times5\times6$, giving $(a,b,c,d) = (1,15,2,10)$ or $(1,15,4,6)$;

when $a = 3$, $2a(16-a)=2\times3\times13$, as $13 > 2+3$, $1+6$, no solutions exist;

when $a = 5$, $2a(16-a)=2\times5\times11$, giving $(a,b,c,d) = (5,11,2,20)$;

when $a = 7$, $2a(16-a)=2\times7\times9$, giving $(a,b,c,d) = (7,9,4,14)$.

Altogether, there are 7 solutions in Case (2):

$$(5,16,2,28),(5,16,6,14),(1,32,2,22),(1,15,2,10),(1,15,4,6),$$
$$(5,11,2,20),(7,9,4,14).$$

Case 3: $(p,q)\neq1$. Reduce it to $(p,q) = 1$ by taking $(a,b,c,d) = k(\bar{a},\bar{b},\bar{c},\bar{d})$, where $k\geq2$ is an integer. Then $(\bar{a},\bar{b},\bar{c},\bar{d})$ must be one of the 17 solutions derived from Case (1) and (2). We call such (a,b,c,d) a derivative solution.

The elite number $m = ab(a+b) = k^3\bar{m}$, where $\bar{m} = (\bar{a}+\bar{b})\bar{a}\bar{b}$, it must be $\bar{m}\leq\dfrac{2020}{8}=\dfrac{505}{2}$, or $\bar{m}\leq252$.

In Case (1), only two solutions $(1,5,2,3)$, $\bar{m} = 30$ and $(1,14,3,7)$, $\bar{m}=210$, meet the criterion $\bar{m}\leq252$.

From $(1,5,2,3)$, we obtain derivative solutions $(2,10,4,6)$, $(3,15,6,9)$, and $(4,20,8,12)$;

From $(1,14,3,7)$, we obtain $(2,28,6,14)$.

In Case (2), there are two solutions $(1,15,2,10),(1,15,4,6)$ both with $\bar{m}=240$, and they give $(2,30,4,20)$ and $(2,30,8,12)$.

In all, there are 6 derivative solutions. Finally, note that there are 3 elite numbers each having three representations as $Y(a,b)$:

$$m = 240,\quad \text{corresponds to}\quad (1,15,2,10),\ (1,15,4,6),\ (2,10,4,6);$$
$$m = 1680,\quad \text{corresponds to}\quad (2,28,5,16),\ (2,28,6,14),\ (5,16,6,14);$$
$$m = 1920,\quad \text{corresponds to}\quad (2,30,4,20),\ (2,30,8,12),\ (4,20,8,12).$$

As each of them is counted two additional times, there are $10+7+6-3\times2=17$ elite numbers in total. □

2 In a quadrilateral $ABCD$, $\angle ABC = \angle ADC < 90^\circ$. The circle with diameter AC is centred at O and intersects BC, CD at points E, F (other than C), respectively. Let M be the midpoint of BD, $AN\perp BD$ with foot N, as shown in Fig. 2.1.

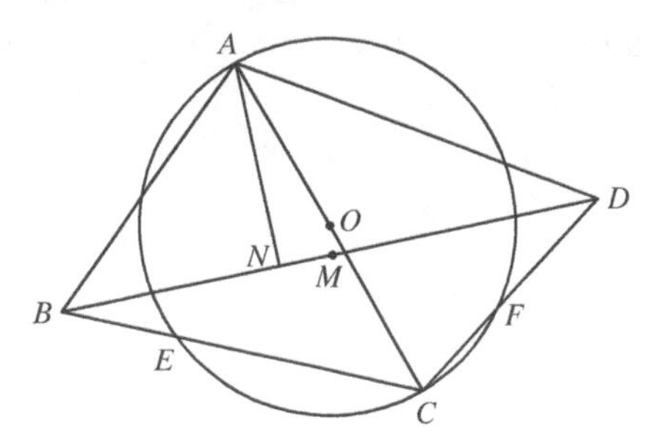

Fig. 2.1

Show that M, N, E, F are concyclic.

(Contributed by Zhang Pengcheng)

Solution The same as that of Problem 2, 10th grade, Second Day. □

3 Let $a_1, a_2, a_3, \ldots, a_n, \ldots$ be the sequence of all square-free positive integers in the ascending order. Show that there are infinitely many indices n satisfying $a_{n+1} - a_n = 2020$.

(Contributed by Yang Xiaoming and Wu Yupei)

Solution First, we need a lemma.

Lemma $\sum\limits_{p\,\text{prime}} \dfrac{1}{p^2} < 0.48.$

Proof of Lemma

$$\begin{aligned}\sum_{p\,\text{prime}} \frac{1}{p^2} &< \frac{1}{4} + \frac{1}{9} + \frac{1}{25} + \frac{1}{49} + \left(\frac{1}{9 \cdot 11} + \frac{1}{11 \cdot 13} + \cdots\right) \\ &= \frac{1}{4} + \frac{1}{9} + \frac{1}{25} + \frac{1}{49} + \frac{1}{18} < 0.48.\end{aligned}$$

For the original problem, we need to find a positive integer x such that x and $x + 2020$ are both square-free, but every $x + i$ $(1 \leq i \leq 2019)$ in between contains a square factor.

Take 2019 distinct prime numbers $p_1, p_2, \ldots, p_{2019}$, $p_i > 4039 (1 \leq i \leq 2019)$. Let $x_k = k\left(\prod\limits_{i=1}^{2019} p_i\right)^2 + x$, where k and x are positive integers,

$1 \le x \le \left(\prod_{i=1}^{2019} p_i\right)^2$ satisfies the system of congruence equations $x + i \equiv 0 \pmod{p_i^2}$, $1 \le i \le 2019$. The existence of x is guaranteed by the Chinese remainder theorem.

We must find a positive integer k such that x_k and $x_k + 2020$ are square-free. Let $N > 2 + 40000\left(\prod_{i=1}^{2019} p_i\right)^2$ be a fixed, large integer. Then $0.01N > 2\prod_{i=1}^{2019} p_i\sqrt{N+2}$. Since $p_i > 4039$, $p_i\,(1 \le i \le 2019)$ is not a prime factor of x_k or $x_k + 2020$.

For convenience, denote $M = \prod_{i=1}^{2019} p_i$.

For any prime p, the congruence equation $x_k = kM^2 + x \equiv 0 \pmod{p^2}$ has at most $\left\lceil \frac{N}{p^2} \right\rceil$ solutions in $1 \le k \le N$.

Similarly, $x_k + 2020 = kM^2 + x + 2020 \equiv 0 \pmod{p^2}$ also has at most $\left\lceil \frac{N}{p^2} \right\rceil$ solutions in $1 \le k \le N$.

It is worth noting that when $p > M\sqrt{N+2} > \sqrt{kM^2 + 2M^2} > \sqrt{kM^2 + x + 2020}$, the divisibility in the above two congruence equations cannot occur. Therefore, in $[1, N]$, the number of integers k such that x_k and $x_k + 2020$ are both square-free, is at least

$$N - 2\sum_{\substack{p\,\text{prime},\\ p \le M\sqrt{N+2}}} \left\lceil \frac{N}{p^2} \right\rceil \ge N - 2N\left(\sum_{\substack{p\,\text{prime},\\ p \le M\sqrt{N+2}}} \frac{1}{p^2}\right) - 2\left(\sum_{\substack{p\,\text{prime},\\ p \le M\sqrt{N+2}}} 1\right)$$

$$> 0.04N - 2M\sqrt{N+2} > 0.03N.$$

Here, the estimate on the second line comes from the lemma and the condition $N > 2 + 40000\left(\prod_{i=1}^{2019} p_i\right)^2$. Letting $N \to +\infty$, the conclusion follows.

□

4 An inkjet printer is used to print on a paper strip with $1 \times n$ grids. When the nozzle prints on the ith $(1 \le i \le n)$ grid, it becomes black; and each of the adjacent grids, the $(i-1)$th and the $(i+1)$th grids (if they exist), independently has a probability of $\frac{1}{2}$ becoming black.

Let $T(n)$ be the expected number of prints needed to make all grids black, provided that the optimal strategy is adopted (make as few prints as possible). Find the formula of $T(n)$.

(Contributed by Luo Ye)

Solution The same as that of Problem 4, 10th grade, Second Day. □

China Southeastern Mathematical Olympiad

2021 (Nanchang, Jiangxi)

The 18th China Southeastern Mathematics Summer Camp and China Southeast Mathematical Olympiad (Hongcheng Cup) was held on July 26th–31st, 2021 at Nanchang No. 2 Middle School, Nanchang, Jiangxi Province. The contest was carried out both onsite (about 1100 participants) and online (about 400 participants). Academician Tian Gang, chairman of the Chinese Mathematical Society, attended the opening ceremony and gave a one-hour inspirational lecture to the students. The contest committee consisted of 15 experts and scholars from Peking University, Tsinghua University, Fudan University, Zhejiang University, East China Normal University, Fujian Normal University, Nanchang University, Jiangxi Normal University and Hong Kong University. There were first year and second year (of senior high school) levels, and each lasted for 2 days. The following are the problems and suggested solutions.

10th Grade
First Day
(8 am – 12 pm; July 28, 2021)

1. Let $\{a_n\}$ be a decreasing sequence of positive real numbers, such that $a_1 = \frac{1}{2}$, and

$$a_n^2(a_{n-1}+1) + a_{n-1}^2(a_n+1) - 2a_na_{n-1}(a_na_{n-1}+a_n+1) = 0$$

for $n \geq 2$. (1) Find $a_n, n \geq 1$; (2) let $S_n = \sum_{i=1}^{n} a_i$, show that $\ln\left(\frac{n}{2}+1\right) < S_n < \ln(n+1)$.

Solution (1) The recursive formula can be written as

$$(a_n - a_{n-1})^2 - a_n a_{n-1}(a_n - a_{n-1}) - 2a_n^2 a_{n-1}^2 = 0, \quad \text{or}$$

$$(a_n - a_{n-1} + a_n a_{n-1})(a_n - a_{n-1} - 2a_n a_{n-1}) = 0.$$

Since $0 < a_n < a_{n-1}$, it follows that $a_n - a_{n-1} - 2a_n a_{n-1} < 0$, leaving $a_n - a_{n-1} + a_n a_{n-1} = 0$. Hence $\frac{1}{a_n} - \frac{1}{a_{n-1}} = 1, a_n = \frac{1}{n+1}$.

(2)(By Chen Haoran) Consider $f_1(x) = \frac{1}{1-x}$, $f_2(x) = e^x$, $f_3(x) = 1 + x$. They are touching each other at $(0,1)$ and $f_1 > f_2 > f_3$ for $0 < x < 1$. Let $x = 1/n$, take the logarithm of the functions, and find

$$\ln\left(1 + \frac{1}{n}\right) < \frac{1}{n} < \ln\left(1 + \frac{1}{n-1}\right).$$

Consequently,

$$S_n = \frac{1}{2} + \frac{1}{3} + \cdots + \frac{1}{n+1} < \ln 2 + \ln\frac{3}{2} + \cdots + \ln\frac{n+1}{n} = \ln(n+1),$$

$$S_n = \frac{1}{2} + \frac{1}{3} + \cdots + \frac{1}{n+1} > \ln\frac{3}{2} + \ln\frac{4}{3} + \cdots + \ln\frac{n+2}{n+1} = \ln\frac{n+2}{2},$$

as desired. □

2 In $\triangle ABC, AB = AC > BC, O$ and H are the circumcentre and orthocentre of $\triangle ABC$, respectively. Let G be the midpoint of AH, BE be the altitude on AC, as illustrated in Fig. 2.1. Prove that if $OE \| BC$, then H is the incentre of $\triangle GBC$.

Solution To begin, notice that GH bisects $\angle BGC$, and it suffices to prove $\angle GBH = \angle CBH$.

Let the extension of BG beyond G meet AC at point F; $\odot O$ and $\odot O'$ be the circumcircles of ABC and AHC, respectively, as shown in Fig. 2.2. By properties of orthocentre, it follows that $\odot O'$ and $\odot O$ are symmetric with respect to AC. Let the extension of BE beyond E meet $\odot O'$ at point D. Then

$$BE = ED. \qquad ①$$

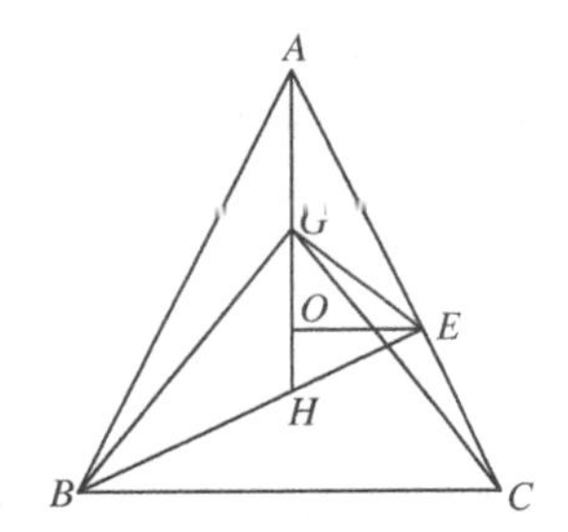

Fig. 2.1

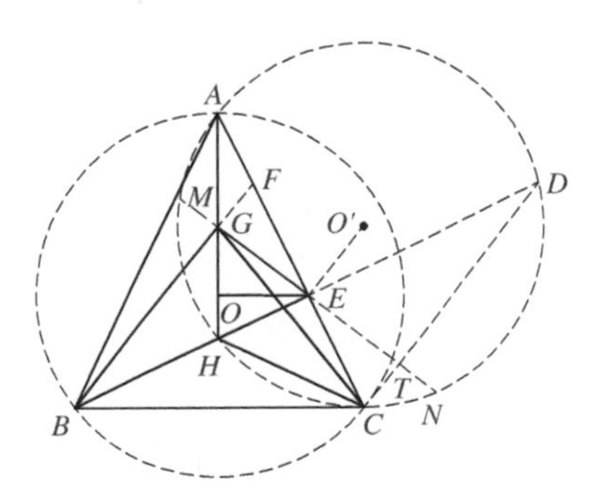

Fig. 2.2

Connect $CD, O'E$. Let the line GE meet CD at point T and meet $\odot O'$ at points M, N(M is on the extension of EG beyond G). Since $OE//BC$, we have $\angle AEO' = \angle AEO = \angle ACB = \angle AHE$.

Moreover, $AE \perp BE$, and hence $O'E \perp MN$, indicating that E is the midpoint of the chord MN. Apply the butterfly theorem to $\odot O'$ and obtain

$$EG = ET. \qquad ②$$

From ① and ②, it follows that $BG//TD, EF = EC$. Hence, $\angle GBH = \angle FBH = \angle CBH$, and H is the incentre of $\triangle GBC$. □

Remark (By Chen Haoran) Alternatively, drop perpendiculars from O, E to BC with feet K, L, respectively. Let $\angle CAK = \angle EBC = \alpha$, $BK = 1$. We find

$$OK = \frac{BK}{\tan \angle BOK} = \frac{1}{\tan \angle BAC} = \frac{1}{\tan(2\alpha)};$$

$$EL = BE \sin(\alpha) = BC \cos(\alpha) \sin(\alpha) = \sin(2\alpha).$$

Use $OE//BC$ to find $OK = EL$, and the above quantities are equal. To show BH bisects $\angle GBK$, it suffices to verify $\angle GBK = 2\angle HBK$. This is secured by

$$GH = \frac{1}{2}AH = \frac{1}{2}(AK - HK) = \frac{1}{2}\left(\frac{1}{\tan(\alpha)} - \tan(\alpha)\right)$$

and

$$\begin{aligned}\tan\angle GBK = GK &= \frac{1}{2}\left(\frac{1}{\tan(\alpha)} + \tan(\alpha)\right)\\ &= \frac{1}{\sin(2\alpha)} = \frac{1}{EL} = \frac{1}{OK} = \tan(2\alpha).\end{aligned}$$

3 Given an odd prime p and a sequence of integers $\{u_n\}, n \geq 0$. Define the sequence

$$v_n = \sum_{i=0}^{n} \binom{n}{i} p^i u_i, \quad n \geq 1.$$

Prove: if there are infinitely many values of n such that $v_n = 0$, then for all $n \geq 0, v_n = 0$.

Solution Define rational polynomial (polynomial with rational coefficients)

$$R_N(x) = \sum_{i=0}^{N} \frac{p^i}{i!} u_i x(x-1)\cdots(x-i+1),$$

where p is an odd prime, u_i is an integer, $0 \leq i \leq N$. Clearly, for $m \leq n$, we have $R_n(m) = v_m$. Suppose there exists a sequence of nonnegative integers $m_j, j = 1, 2, \ldots$ such that $v_{m_j} = 0$. We are led to prove $v_r = 0$ for every nonnegative integer r, that is, $p^l | v_r$ for all $l \in \mathbb{N}^+$.

For a rational number s, denote $v_p(s)$ as the highest power of p that divides s. It suffices to show $v_p(v_r) \geq l$. Take k sufficiently large, such that $k \cdot \frac{p-2}{p-1} \geq l, N \geq r$, and $N \geq m_k$. Then $v_r = R_N(r)$, and $R_N(m_i) = 0$, $i = 1, 2, \ldots, k$. This implies that

$$R_N(x) = (x - m_1)\cdots(x - m_k)h(x) \qquad ①$$

for some $h(x) \in \mathbb{Q}[x]$, and $v_p(v_r) = v_p(R_N(r)) \geq v_p(h(r))$. Now for $h(x) \in \mathbb{Q}[x]$, let

$$f(x) = \sum_{i=0}^{n} a_i x^i, v_p^{(i)}(f) = \min_{j \geq i} v_p(a_j).$$

Lemma *Let $f(x), g(x)$ be rational polynomials such that*

$$g(x) = (x - m)f(x)(^*)$$

for some integer m. Then, for every $i \leq \deg f, v_p^{(i)}(f) \geq v_p^{(i+1)}(g)$.

Proof of lemma Let $f(x) = \sum_{i=0}^{n} a_i x^i, g(x) = \sum_{i=0}^{n+1} b_i x^i$. From $(^*)$, it follows that $\forall j, 1 \leq j \leq n$,

$$a_j = b_{j+1} + mb_{j+2} + \cdots + m^{n-j} b_{n+1}.$$

Thus, for every $j \geq i, v_p(a_j) \geq \min_{j \geq i+1} v_p(b_j) = v_p^{(i+1)}(g)$. We conclude $v_p^{(i)}(f) \geq v_p^{(i+1)}(g)$.

Return to the original problem. Clearly,

$$v_p(v_r) = v_p(R_N(r)) \geq v_p(h(r)) \geq v_p^{(0)}(h(x)), \qquad ②$$

By ① and the lemma, it follows that $v_p^{(0)}(h(x)) \geq v_p^{(k)}(R_N(x))$.

Since $R_N(x) = \sum_{i=1}^{N} \frac{p^i}{i!} u_i x(x-1) \cdots (x-i+1)$, and $v_p\left(\frac{p^i}{i!}\right) \geq i\frac{p-2}{p-1}$, we infer that $v_p^{(k)}(R_N(x)) \geq k\frac{p-2}{p-1} \geq l$. Combine ② and the above inequalities to derive $v_p(v_r) \geq l$. The argument is complete. □

4 There are $n \geq 5$ points, labeled $1, 2, \ldots, n$, arbitrarily arranged on a circle. We call such a permutation S. For a permutation, a "descending chain" is a clockwise sequence of consecutive points (at least two) with descending labels, and it is not a sub chain of any longer descending chains; the "pivot" of a descending chain is the point with the largest label, and all other points are "non-pivots". For example, the clockwise permutation 5, 2, 4, 1, 3 contains two descending chains 5, 2 and 4, 1, where 5, 4 are pivots, 2, 1 are non-pivots. Apply the following operations on S: first, find all the descending chains of S and delete all the non-pivots; then, for the remaining points, find all the descending chains and delete all the non-pivots; and so on, until no descending chains can be found. Let $G(S)$ be the number of descending chains that have appeared in the whole process; $A(n)$ be the average of $G(S)$ among all permutations S of $1, 2, \ldots, n$.

(1) Find $A(5)$; (2) For $n \geq 6$, prove: $\frac{83}{120}n - \frac{1}{2} \leq A(n) \leq \frac{101}{120}n - \frac{1}{2}$.

Solution Evidently, $A(1) = 0, A(2) = 1$. Define $S(n) = A(1) + A(2) + \cdots + A(n)$. In the context, all the permutations, sequences and so on are in the clockwise direction.

Lemma For $n \geq 3, A(n)$ satisfies

$$\frac{2}{n-1}S(n-1) + \frac{1}{2} \leq A(n) \leq \frac{2}{n-1}S(n-1) + \frac{2n-3}{2(n-1)}. \qquad ①$$

Proof of Lemma Suppose S' is an arbitrary permutation of $1, 2, \ldots, n$ on a circle. Fix n as the starting number and consider $n-1$: they divide the circle into two sequences $S_{1'}$, $S_{2'}$, the former starts with n and the latter starts with $n-1$. For a sequence whose starting number is the largest, after a few operations it will reduce to a single number which is the starting number itself. We say this sequence is "completed". A key observation is that the number of descending chains that will appear in $S_{1'}$ and $S_{2'}$ are $G(S_{1'}), G(S_{2'})$, respectively, and there is an extra part when they combine.

During the process, if $S_{1'}$ is completed earlier, then n will catch up with $n-1$ and take the role of $n-1$ thereafter, generating no additional descending chains in the process. Symmetric to S', we interchange the numbers n and $n-1$ and denote this permutation as S''. In S'', $n-1$ will catch up with n first, and it must wait until the sequence that starts with n is completed, then n and $n-1$ takes one operation at last. In this case, the two symmetric permutations S' and S'' will generate $\frac{1}{2}$ descending chain on average beyond $G(S_{1'}) + G(S_{2'})$.

The only exception is when $S_{1'}$ and $S_{2'}$ are completed at the same time. Then S' and S'' (obtained from interchanging n and $n-1$ in S') both take one additional operation to eliminate $n-1$. In this case, they will generate another $\frac{1}{2}$ descending chain on average beyond $G(S_{1'}) + G(S_{2'}) + \frac{1}{2}$. By assumption $n \geq 3$: if one of $S_{1'}$ and $S_{2'}$ has length 1, of which the probability is $\frac{1}{n-1}$, the exceptional situation cannot occur, and $\frac{G(S') + G(S'')}{2} = G(S_{1'}) + G(S_{2'}) + \frac{1}{2}$; for all others, we have

$$G(S_{1'}) + G(S_{2'}) + \frac{1}{2} \leq \frac{G(S') + G(S'')}{2} \leq G(S_{1'}) + G(S_{2'}) + 1.$$

Putting them together and taking the average, it follows that

$$\frac{1}{n-1}\sum_{i=1}^{n-1}(A(i)+A(n-i))+\frac{1}{2}$$

$$\leq A(n) \leq \frac{1}{n-1}\sum_{i=1}^{n-1}(A(i)+A(n-i))+\left(1-\frac{1}{2(n-1)}\right),$$

which is exactly ①. The lemma is verified.

By ①, it follows that

$$\frac{n+1}{n-1}S(n-1)+\frac{1}{2} \leq S(n) \leq \frac{n+1}{n-1}S(n-1)+\frac{2n-3}{2(n-1)}.$$

Now we find

$$S(n)+\frac{n}{2} \geq \frac{n+1}{n-1}\left(S(n-1)+\frac{n-1}{2}\right), \qquad ②$$

and

$$S(n)+n-\frac{1}{4} \leq \frac{n+1}{n-1}\left(S(n-1)+(n-1)-\frac{1}{4}\right). \qquad ③$$

Use ② and ③ repeatedly, to derive

$$S(n) \geq \frac{(n+1)n}{(k+1)k}\left(S(k)+\frac{k}{2}\right)-\frac{n}{2}, \qquad ④$$

$$S(n) \leq \frac{(n+1)n}{(k+1)k}\left(S(k)+k-\frac{1}{4}\right)-\left(n-\frac{1}{4}\right) \qquad ⑤$$

for $k \geq 2$. Some direct calculation yields $A(3)=\frac{3}{2}$, $A(4)=\frac{7}{3}$. For $n=5$, it is necessary to count the number of exceptional situations, that is, the permutations for which $S_{1'}$ and $S_{2'}$ are completed at the same time. According to the lemma, it must satisfy

(a) that 5 and 4 are not adjacent;
(b) that one of the numbers 1, 2, 3 belongs to $S_{1'}$, while the other two belong to $S_{2'}$ and are descending.

Fix $n=5$. There are 2 ways to fill the number 4 and 3 ways to pick the one number for $S_{1'}$ in (b). So, there are $2 \times 3 = 6$ permutations each with 1 additional descending chain; all others have $\frac{1}{2}$ additional descending chain on average. Hence,

$$A(5) = \frac{2}{4}(A(1) + A(2) + A(3) + A(4)) + \frac{1}{2} + \frac{1}{2} \times \frac{2 \times 3}{4!} = \frac{73}{24},$$

and $S(5) = A(1) + \cdots + A(5) = \dfrac{63}{8}$.

Plug $k = 5$ into ④ and ⑤. For $n \geq 5$,

$$S(n) \geq n\left(\frac{n+1}{30} \times \frac{83}{8} - \frac{1}{2}\right),$$

$$S(n) \leq n\left(\frac{n+1}{30} \times \frac{101}{8} - 1\right) + \frac{1}{4}.$$

Substituting them into ①, we find

$$A(n) \geq \frac{2}{n-1} S(n-1) + \frac{1}{2} \geq \frac{83}{120} n - \frac{1}{2},$$

$$A(n) \leq \frac{101}{120} n - 1 \leq \frac{101}{120} n - \frac{1}{2},$$

that is, $\dfrac{83}{120} n - \dfrac{1}{2} \leq A(n) \leq \dfrac{101}{120} n - \dfrac{1}{2}$ as required. □

Second Day
(8 am – 12 pm; July 29, 2021)

5 To celebrate the 43rd anniversary of the restoration of mathematical competitions, a math lover arranges the first 2021 positive integers $1, 2, \ldots, 2021$ into a sequence $\{a_n\}$, such that any 43 consecutive numbers add up to a multiple of 43.

(1) Prove if the two ends of $\{a_n\}$ are joined to form a circle, then any 43 consecutive numbers on the circle also add up to a multiple of 43.

(2) Determine the number of all such sequences $\{a_n\}$.

Solution

(1) The sequence $\{a_n\}$ has sum $S = \dfrac{2021 \times 2022}{2} = 2021 \times 1011$. Since $2021 = 43 \times 47, 43 | S$. Every 43 consecutive numbers are called a "good segment", as their sum is divisible by 43. Join the two ends of $\{a_n\}$ and

consider 43 consecutive numbers on the circle including a_1 and a_{2021}. Let their sum be S_0. Remove those 43 numbers, leaving 1978 numbers that form a subsequence of $\{a_n\}$: let their sum be S_1. Since $1978 = 43 \times 46$, those 1978 numbers are exactly 46 good segments, and thus $43 | S_1$. As $S_0 = S - S_1$, we have $43 | S_0$.

(2) Divide $1, 2, \ldots, 2021$ into 43 residue classes modulo 43: $R_1, R_2, \ldots, R_{43}$, each with 47 numbers. We claim that in $\{a_n\}$, every good segment (as defined in (1)) forms a complete system of residues, and moreover

$$a_r \equiv a_{43k+r} \pmod{43}, \quad r = 1, 2, \ldots, 43, \quad k = 0, 1, \ldots, 46. \qquad ①$$

Indeed, for good segments $a_r, a_{r+1}, \ldots, a_{r+42}$ and $a_{r+1}, a_{r+2}, \ldots, a_{r+42}, a_{r+43}$, since $43 | (a_r + a_{r+1} + \cdots + a_{r+42}), 43 | (a_{r+1} + \cdots + a_{r+42} + a_{r+43})$, it follows that

$$43 | [(a_{r+1} + a_{r+2} \cdots + a_{r+42} + a_{r+43}) - (a_r + a_{r+1} + \cdots + a_{r+42})],$$

which is $43 | (a_{r+43} - a_r)$, hence $a_r \equiv a_{43+r} \pmod{43}$. In the same manner, $a_{43+r} \equiv a_{2\times 43+r} \pmod{43}$, and so on. So, $a_r, a_{43+r}, a_{2\times 43+r}, \ldots, a_{46\times 43+r}$ belong to R_r and exhaust all the numbers of R_r (here the subscript r refers to $a_r \in R_r$, not remainder r). Since this is true for every $r = 1, 2, \ldots, 43$, the claim is justified.

Note that a complete system of residues modulo 43 forms a good segment, as $\sum_{i=1}^{43} i \equiv 0 \pmod{43}$. This indicates that as long as ① is satisfied, the corresponding sequence $\{a_n\}$ meets the problem conditions. There are 43! ways to choose the residue classes of $a_1, a_2, \ldots, a_{43}$; once they are all determined, there are 47! ways for each residue class R_r to settle the 47 numbers in the positions $a_{43k+r}, k = 0, 1, \ldots, 46$. Altogether, there are $43! \cdot (47!)^{43}$ arrangements of $1, 2, \ldots, 2021$ for making $\{a_n\}$. The answer is therefore $43! \cdot (47!)^{43}$. □

6 In the cyclic quadrilateral $ABCD$, let E be an interior point on BC, F be a point on AE, and G be a point on the exterior bisector of $\angle BCD$, such that $EG = FG, \angle EAG = \frac{1}{2}\angle BAD$, as shown in Fig. 6.1.

Prove: $AB \cdot AF = AD \cdot AE$.

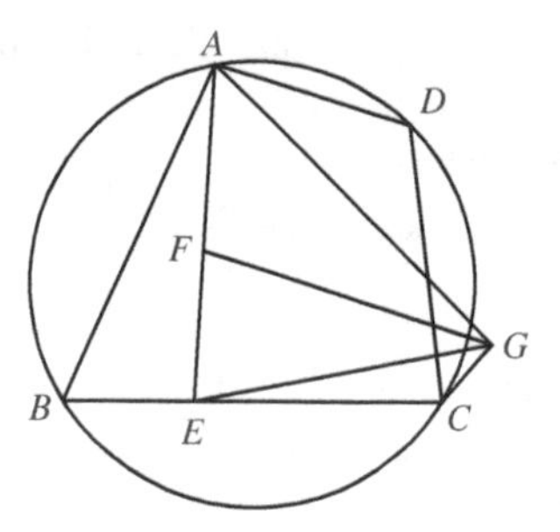

Fig. 6.1

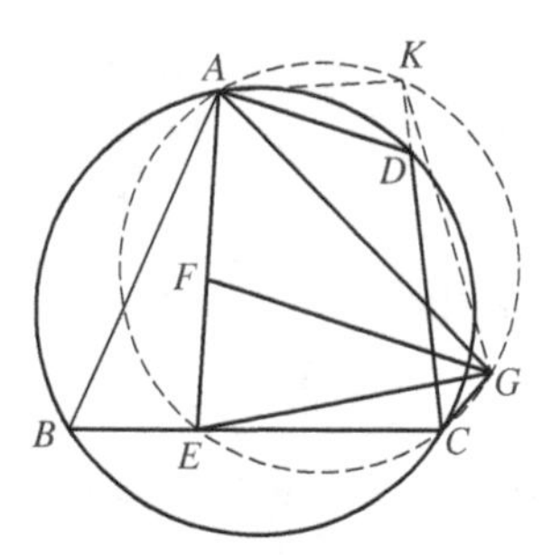

Fig. 6.2

Solution It is known that $\angle EAG = \frac{1}{2}\angle BAD, \angle ECG = 90^\circ + \frac{1}{2}\angle BCD.$ Hence,

$$\angle EAG + \angle ECG = 90^\circ + \frac{1}{2}(\angle BAD + \angle BCD) = 180^\circ,$$

and A, E, C, G all lie on a circle, say ω. Let the extension of CD beyond D cross circle ω at point K, as shown in Fig. 6.2. Notice that CG is the exterior bisector of $\angle ECK$, and thus G is the midpoint of $\overparen{ECK}$ of circle ω, AG bisects $\angle EAK$, that is, $\angle FAG = \angle KAG$. Meanwhile, from $GE = GF$ we find

$$\angle AKG = 180^\circ - \angle AEG = 180^\circ - \angle FEG = 180^\circ - \angle GFE = \angle AFG.$$

Together with $\angle FAG = \angle KAG$, they imply that $\triangle AKG \cong \triangle AFG, AF = AK$ ①.

Since A, B, C, D are concyclic, A, E, C, K are also concyclic, we infer $\angle ADK = \angle ABE$ and $\angle AKD = \angle AEB$, hence $\triangle ADK \backsim \triangle ABE$. By ①, it follows that

$$\frac{AF}{AE} = \frac{AK}{AF} = \frac{AD}{AB},$$

or $AB \cdot AF = AD \cdot AE$, as desired. □

7 Let a, b, c be distinct positive reals. Prove

$$\frac{ab+bc+ca}{(a+b)(b+c)(c+a)} < \frac{1}{7}\left(\frac{1}{|a-b|} + \frac{1}{|b-c|} + \frac{1}{|c-a|}\right).$$

Solution (By Chen Haoran) By symmetry, assume $a > b > c > 0$. On one hand,

$$\frac{ab+bc+ca}{(a+b)(b+c)(c+a)} = \frac{ab+bc+ca}{(a+b)(ab+bc+ca+c^2)} < \frac{1}{a+b}.$$

On the other hand,

$$\begin{aligned}\frac{1}{7}\left(\frac{1}{|a-b|} + \frac{1}{|b-c|} + \frac{1}{|c-a|}\right) &= \frac{1}{7}\left(\frac{1}{a-b} + \frac{1}{b-c} + \frac{1}{a-c}\right) \\ &> \frac{1}{7}\left(\frac{1}{a-b} + \frac{1}{b} + \frac{1}{a}\right).\end{aligned}$$

Based on the above inequalities, it suffices to prove

$$\frac{1}{a+b} < \frac{1}{7}\left(\frac{1}{a-b} + \frac{1}{b} + \frac{1}{a}\right), \quad \text{or} \quad \frac{a}{a-b} + \frac{a}{b} + 1 - \frac{7a}{a+b} > 0$$

by multiplying $7a$ on both sides and rearranging the terms. To justify it, let $x = a/b > 1$. We have

$$\begin{aligned}\frac{a}{a-b} + \frac{a}{b} + 1 - \frac{7a}{a+b} &= \frac{x}{x-1} + x + 1 - \frac{7x}{x+1} \\ &= \frac{x^3 - 5x^2 + 7x - 1}{(x-1)(x+1)} = \frac{(x-1)^2(x-3)+2}{(x-1)(x+1)}\end{aligned}$$

Define $f(x) = (x-1)^2(x-3)+2$. If $x \geq 3$, clearly $f(x) > 0$; if $1 < x < 3$, then

$$(x-1)^2(x-3) = -\frac{1}{2}(x-1)(x-1)(6-2x) \geq -\frac{1}{2}\left(\frac{4}{3}\right)^3 > -2$$

by the AM-GM inequality, and $f(x) > 0$, too. This completes the proof. □

8 Find all positive integer pairs $\{a, b\}$, such that

$$14\varphi^2(a) - \varphi(ab) + 22\varphi^2(b) = a^2 + b^2, (^*)$$

where $\varphi(n)$ represents the number of positive integers less than n and relatively prime to n.

Solution We denote by $p^\alpha \| b$ that the highest power of p that divides b is p^α, that is, $p^\alpha | b$, but $p^{\alpha+1} \nmid b$.

To begin, notice that if $\{a, b\}$ is a solution of $(^*)$, p is a prime and $p^2 | a, p^2 | b$, then $\left\{\frac{a}{p}, \frac{b}{p}\right\}$ is another solution. Henceforth, we assume for each prime p, at least one of a, b is relatively prime to p or divisible exactly by p.

The following properties of Euler's function can be easily verified:

(a) when $x > 2, \varphi(x)$ is even;
(b) if $2^\alpha \| \varphi(x)$, then x has at most α distinct odd prime factors.

If $a, b \le 3$, then $(^*)$ cannot hold as the left-hand side is larger. Thus, one of a and b must exceed 3, and $2 | \varphi(ab)$. This implies that the left-hand side of $(^*)$ is even, and a, b are of the same parity.

If a, b are both odd, then the right-hand side of $(^*)$ is congruent to 2 modulo 4. If ab has two or more distinct prime factors, then $4 | \varphi(ab)$, and one of $\varphi(a), \varphi(b)$ must be 1. Hence, one of a and b, say b, equals 1. Now, $(^*)$ becomes

$$14\varphi^2(a) - \varphi(a) + 22 = a^2 + 1. \quad ①$$

Suppose $8 | \varphi(a)$. Then ① leads to $22 \equiv 2 \pmod 8$, a contradiction. Hence, a has at most two distinct prime factors say p and q, but this means

$$13\varphi^2(a) = 13\left(\frac{p-1}{p} \cdot \frac{q-1}{q}\right)^2 a^2 \ge 3a^2 > a^2, \quad ②$$

negating ① since the left-hand side is always larger.

We are led to assume that a, b are both even: let $a = 2c, b = 2d$. If one of c and d equals 1, analogous to ①, the equation has no integer solutions $\{2c, 2d\}$. Suppose c and d are both larger than 1. By the method of descent, assume that one or both of c and d are odd. There are two situations, the first of which can be quickly ruled out:

(i) if c and d have different parities, then $4\|(a^2+b^2)$.

Since $2|\varphi(a), 2|\varphi(b)$, it follows that $8|14\varphi^2(a)+22\varphi^2(b)$, and $4\|\varphi(ab)$. Since c or d is even, $8|ab$. Let $ab=2^\delta m$, where $\delta \geq 3$ is an integer, m is odd. We see that $\varphi(ab)=2^{\delta-1}\varphi(m)$ is divisible exactly by 4, that is, $\varphi(m)$ is odd, $m=1, \delta=3$, and $ab=8$. However, this contradicts $ab=4cd \geq 16$ and no solutions exist.

In the following, we focus on the second situation:

(ii) if c and d have the same parity. Clearly, they must be odd integers, and we have

$$7\varphi^2(c)-\varphi(cd)+11\varphi^2(d)=2(c^2+d^2), \quad ③$$

where $c, d>2$ are odd. If $2\|\varphi(c), 2\|\varphi(d)$, it is easy to verify

$$7\varphi^2(c) \geq 7\times\left(\frac{2}{3}\right)^2 c^2 > 2.5c^2, \quad ④$$

$$11\varphi^2(d) \geq 11\times\left(\frac{2}{3}\right)^2 d^2 > 2.5d^2. \quad ⑤$$

Since $\varphi(cd)<cd \leq 0.5c^2+0.5d^2$, the left-hand side of ③ must be larger, and hence $4|\varphi(c)$ or $4|\varphi(d)$. First, assume $4|\varphi(c)$. If $16|\varphi(cd)$, take ③ modulo 16, to obtain

$$4 \equiv 2(c^2+d^2) \equiv 11\varphi(d)^2 \pmod{16},$$

which implies $2\|\varphi(d)$, but then $11\varphi^2(d) \equiv 12 \pmod{16}$ is contradictory. Therefore, $16 \nmid \varphi(cd)$. The above argument can also be applied to $4|\varphi(d)$.

As a result, there are two situations to consider: either $4\|\varphi(cd)$, or $8\|\varphi(cd)$. In any case, cd has at most three distinct prime factors; moreover, if indeed three factors, then they are all of the form $4k+3$.

Case 1, cd is a prime power, say the prime factor is p. According to $4|\varphi(cd), p$ must be of the form $4k+1$, and $p|c, p|d$. Then

$$7\varphi^2(c)>2.5c^2, \quad 11\varphi^2(d)>2.5d^2, \quad ⑥$$

which opposes ③.

Case 2, cd contains two distinct prime factors, say p, q. We have the following observations. First, notice that p or q must be 3, otherwise ⑥ holds, and in ③ the left hand side is larger, contradiction. Second, c and d cannot be relatively prime, as otherwise c and d each has only one prime factor, and ⑥ holds, leading to the same contradiction. Third, if $pq|c$ and $pq|d$, assuming $q > p$, then $p = 3$. This implies $16|\varphi(c), 16|\varphi(d)$, and $4\|\varphi(cd)$, hence p, q are both primes of the form $4k+3$. If $q > 7$, then ⑥ holds again; thus $q = 7$, and $\{p, q\} = \{3, 7\}$.

If $q^2|c$ or $q^2|d$, assume the former, then $q^2|\varphi^2(c), q^2|\varphi(cd)$, and so $q^2|11\varphi^2(d), q|\varphi(d)$. Since $q > p$, we have $q \nmid (p-1), q^2|d$, then q^2 divides both c and d, yet this violates the assumption at the very beginning of the proof. Therefore, $q\|c, q\|d$. Let $c = 3^\alpha \times 7, d = 3^\beta \times 7$. i) $\alpha = \beta$, then ③ leads to $17\varphi^2(c) = 4c^2, 17|c$, a contradiction. ii) $\alpha > \beta$, then β =1, and in ③ the left-hand side

$$\begin{aligned}
&28 \cdot 6^2 \cdot 3^{2\alpha-2} - 12 \cdot 7 \cdot 3^{\alpha+\beta-1} + 44 \cdot 6^2 \cdot 3^{2\beta-2} \\
&> (28 \cdot 36 - 12 \cdot 7) \cdot 3^{2\alpha-2} + 44 \cdot 6^2 \cdot 3^{2\beta-2} \\
&> 49(18 \cdot 3^{2\alpha-2} + 18 \cdot 3^{2\beta-2}) \\
&= 2(a^2 + b^2),
\end{aligned}$$

again, contradiction. iii) $\alpha < \beta$, no solutions for the same reason. In all, there are no solutions when $pq|c$ and $pq|d$.

Now, we are left with the last situation: c and d are not relatively prime, pq does not divide both c and d. Suppose c and d have the common factor p.

If $pq|d$, then c has only one prime factor, causing

$$11\varphi^2(d) \geq 11 \times \left(\frac{2}{3} \times \frac{4}{5}\right)^2 > 2.5, \quad 7\varphi^2(c) \geq \left(\frac{2}{3}\right)^2 > 2.5,$$

and ⑥ holds, contradiction. Thus $pq|c, p|d, (q, c) = 1$. Similar to the above argument, it must be $p = 3$. If $q > 7$, ⑥ holds again, and hence $q = 5$ or 7.

(i) $q = 5$, then $3 \nmid (q-1)$. If $3^2|a$ or $3^2|b$, assume $3^2|a$. Take ③ modulo 3^2 to obtain $3|11$, impossible. The situation is similar when $3^2|b$. So it must be $3\|a, 3\|b$, and $d = 3$. If $5^2|a$, take ③ modulo 5 to get $11\varphi^2(b) = 2b^2 \pmod 5$. This implies that 2 is a quadratic residue modulo 5, which is absurd. Hence, $5\|a, a = 15$. It is straightforward to check $(a, b) = (15, 3)$ is a solution of ③, and hence (30, 6) is a solution of (*).

(ii) $q = 7$, if $3^2|c$, then $3\|d, d = 3, 3|11\varphi(d)$, a contradiction. Hence $3\|c$. Let $c = 3 \times 7^\beta, d = 3^\alpha$. If $\alpha \geq 3$, take ③ modulo 27 to obtain

$$7 \times 2^2 \times 6^2 \times 7^{2\beta-2} \equiv 2 \times 7^{2\alpha} \times 9 \pmod{27},$$

which can be simplified to $16 \equiv 14 \pmod 3$, absurd. So $d = 3$ or 9, indicating that

$$\begin{aligned} 7\varphi^2(c) - \varphi(cd) &= 1008 \times 7^{2\alpha-2} - 18 \times 6 \times 7^{\alpha-1} \\ &\geq 900 \times 7^{2\alpha-2} > 882 \times 7^{2\alpha-2} = 2c^2. \end{aligned}$$

Meanwhile, $11\varphi^2(d) > 2d^2$. Now ③ becomes unbalanced, and no solutions exist.

Case 3, cd contains three distinct prime factors, each of which has the form $4k+3$. Then $8|\varphi(cd)$, which gives $4\|\varphi(c)$ or $4\|\varphi(d)$, that is, c or d has only one prime factor. If it is c, then

$$7\varphi^2(c) \geq 7 \times \left(\frac{2}{3}\right)^2 c^2 > 2.5c^2,$$

$$11\varphi^2(d) \geq 11 \times \left(\frac{2}{3} \times \frac{6}{7} \times \frac{10}{11}\right)^2 c^2 > 2.5d^2,$$

and ⑥ holds. Thus, d has only one prime factor, say p, and the other two prime factors of cd are q and r. First, $3|c$, or else ③ holds. If c and d are relatively prime, then $3 \nmid d, p \geq 7$, and

$$11\varphi^2(d) \geq 11 \cdot \frac{36}{49} d^2 > 8d^2.$$

In the meantime, c has only two prime factors, thus $7\varphi^2(c) \geq 7 \times \left(\frac{2}{3} \times \frac{36}{49}\right)^2 c^2 > 2.25$, yielding

$$\begin{aligned} 7\varphi^2(c) + 11\varphi^2(d) &> 2.25c^2 + 8d^2 > 2c^2 + 2d^2 + (0.25c^2 + 4d^2) \\ &> 2c^2 + 2d^2 + \varphi(cd), \end{aligned}$$

a contradiction. Hence, $(c, d) \neq 1, p|c$. If $p^2|c$, by the method of descent, $p = d$. Based on the previous argument, $p^2\|11\varphi(d)$, but this is impossible, so we conclude $p\|c$. Take ③ modulo p twice, to derive $p|7\varphi^2(c)$, and hence $p = 7$ or p divides one of $q - 1$ and $r - 1$, say $q - 1$.

If $p \geq 11$, then $q \geq 19$. A similar argument can show in ③ the left-hand side must be larger. Therefore, $p = 3$ or $p = 7$.

If $p = 7$, then $7|c, 7|d, 7^2|2(c^2+d^2)$. If $7^2|d$, then $7\|c$, and we must have $7^2|\varphi(cd), 7^2|\varphi^2(d)$, implying $7^2|7\varphi^2(c), 7|r-1$. The smallest prime r of the form $4k+3$ satisfying $7|r-1$ is $r = 43$, which gives

$$7\varphi^2(c) \geq \left(\frac{2}{3} \times \frac{6}{7} \times \frac{42}{43}\right)^2 > 2.18.$$

In addition, $11\varphi^2(d) \geq 11 \times \left(\frac{6}{7}\right)^2 > 8$, and thus

$$\begin{aligned} 7\varphi^2(c) + 11\varphi^2(d) &\geq 2(c^2+d^2) + (0.18c^2 + 6d^2) \\ &> 2(c^2+d^2) + \varphi(cd), \end{aligned}$$

a contradiction.

If $p = 3$, let $d = p^\alpha$. Notice that

$$\begin{aligned} \varphi^2(c) &= \left(\frac{2}{3} \times \frac{q-1}{q} \times \frac{p-1}{p}\right)^2 c^2, \\ \varphi^2(d) &= \frac{4}{9}d^2, \\ \varphi(cd) &= \frac{2}{3} \times \frac{q-1}{q} \times \frac{p-1}{p}cd. \end{aligned}$$

By the AM-GM inequality, we have $\frac{1}{2}\varphi^2(c) + 2d^2 \geq 2\varphi(cd) > \varphi(cd)$. Since $11\varphi^2(d) = \frac{44}{9}d^2 > 2d^2 + 2d^2$, in case $6.5\varphi^2(c) > 2c^2$, then the left-hand side of ③ is larger, which is contradictory.

If $q, r \neq 7$, we know

$$6.5 \times \left(\frac{2}{3} \times \frac{q-1}{q} \times \frac{p-1}{p}\right)^2 \geq \frac{4}{9} \times \frac{100}{121} \times \frac{324}{361} > 2,$$

and again the left hand side of ③ is larger. Hence, q or r must be 7. Assume $q = 7, r \geq 11$. Let $c = 3 \times 7^\beta r^\gamma$, and $3|p-1$. If $\alpha \geq 3$, then $27|\varphi(cd), 27|\varphi^2(d)$. Take ③ modulo 27 to find

$$7 \times 36 \times 4 \times (r-1)^2 \times 7^{2\beta-2} \times r^{2\gamma-2} \equiv 2 \times 3^2 \times 7^{2\beta} \times r^{2\gamma} \pmod{27},$$

which can be simplified to

$$8(r-1)^2 \equiv 7r^2 \pmod{3}.$$

Since $3 \nmid r$, it follows that $3 \nmid (r-1)$, and 2 is a quadratic residue modulo 3, a contradiction.

Therefore, $\alpha = 1$ or 2, and $d = 3$ or 9. Since $3|\varphi(c), \varphi(cd)$, taking ③ modulo 3 yields $3|11\varphi^2(d)$. Hence $3^2|d$, and $d = 9$. Let $e = 7^\alpha r^\gamma$. Plug $c = 3 \times 7^\alpha \times r^\gamma$ and $d = 9$ into ③, and we obtain

$$14\varphi^2(e) - 9\varphi(e) + 117 = 9e^2. \quad ⑦$$

If $\gamma \geq 2$, then $7|\varphi(e), 7|117$, which is untrue.

We deduce that $\beta \geq 2$ is impossible, and thus $\beta = 1$. Let $f = r^\gamma$ and simplify ⑦ to

$$56\varphi^2(f) - 6\varphi(f) + 13 = 49f^2.$$

In the same manner, if $\gamma > 1$, then $r|13, r = 13$. But 13 is a prime of the form $4k+1$, a contradiction.

If $\gamma = 1$, then $\varphi(f) = r - 1$, and ⑦ reduces to

$$7r^2 - 118r + 85 = 0,$$

which does not have integer solutions.

The final conclusion: with the assumption that for each prime p, at least one of a, b is relatively prime to p or divisible exactly by p, there is only one solution

$$\{a, b\} = \{2c, 2d\} = \{30, 6\}.$$

By the method of descent, all solutions of (*) are

$$\{a, b\} = \{30.2^\alpha.3^\beta, 6.2^\alpha.3^\beta\},$$

where α, β are arbitrary nonnegative integers. □

11th Grade
First Day
(8 am – 12 pm; July 28, 2021)

1 In $\triangle ABC, AB = AC > BC, O$ and H are the circumcentre and orthocentre of $\triangle ABC$, respectively. Let G be the midpoint of AH, BE be the altitude on AC, as shown in Fig. 1.1. Prove that if $OE//BC$, then H is the incentre of $\triangle GBC$.

Solution See 10th grade, First Day, Problem 2. □

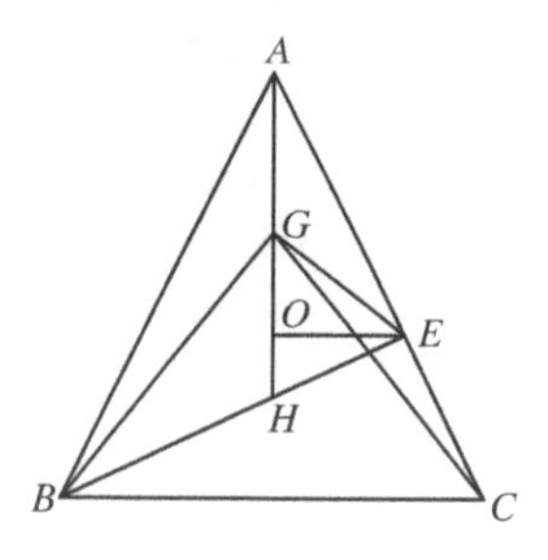

Fig. 1.1

2 Let $p \geq 5$ be a prime number, $M = \{1, 2, \ldots, p-1\}$. Define

$$T = \{(n, x_n) : p|(nx_n - 1) \quad \text{in which} \quad n, x_n \in M\}.$$

Find the nonnegative least residue of $\sum\limits_{(n,x_n)\in T} n\left[\dfrac{nx_n}{p}\right]$ modulo p, where $[x]$ represents the greatest integer less than or equal to x.

Solution For each $n \in \{1, 2, \ldots, p-1\}, (n, p) = 1$, clearly there exists $x_n \in M$ such that

$$nx_n \equiv 1 \pmod{p}, \qquad ①$$

or $p|(nx_n - 1)$. For uniqueness, suppose $x_n' \in M$ satisfies $p|nx_n' - 1$, then $p|n(x_n - x_n') \Rightarrow p|x_n - x_n' \Rightarrow x_n = x_n'$.

Since $p|nx_n - 1$, we may assume $nx_n = 1 + m_n p (1 \leq n \leq p-1)$. From ①, it follows that

$$\sum_{n=1}^{p-1} n\left[\frac{nx_n}{p}\right] = \sum_{n=1}^{p-1} n.\frac{nx_n - 1}{p} = \sum_{n=1}^{p-1} nm_n.$$

Denote $s = \sum\limits_{n=1}^{p-1} nm_n = \sum\limits_{n=1}^{\frac{p-1}{2}} nm_n + \sum\limits_{n=1}^{\frac{p-1}{2}} (p-n)m_{p-n}$ as desired, and in ① replace n by $p-n$ to obtain $p|(p-n)x_{p-n} - 1$. Similarly, $(p-n)x_{p-n} = 1 + m_{p-n}p$, which gives

$$nx_{p-n} \equiv -1 \pmod{p}. \qquad ②$$

Adding ① and ② yields $n(x_{p-n} + x_n) \equiv 0 \pmod{p}$, and thus

$$x_{p-n} + x_n = p. \qquad ③$$

Multiply ① and ② respectively by n and $p-n$ to obtain

$$n^2x_n \equiv n + nm_np \pmod p;$$
$$(p-n)^2x_{p-n} \equiv p-n+(p-n)m_{p-n}p \pmod p.$$

Adding the above equations and taking the summations over n, we have

$$\sum_{n=1}^{\frac{p-1}{2}} n^2x_n + \sum_{n=1}^{\frac{p-1}{2}}(p-n)^2x_{p-n} = \frac{p(p-1)}{2} + \sum_{n=1}^{\frac{p-1}{2}}[nm_n + (p-n)m_{p-n}].$$

According to ② and ③, the above is equivalent to

$$\begin{aligned}\frac{p(p-1)}{2} + ps &= p^2\sum_{n=1}^{\frac{p-1}{2}} x_{p-n} - 2p\sum_{n=1}^{\frac{p-1}{2}} nx_{p-n} + \sum_{n=1}^{\frac{p-1}{2}} n^2(x_{p-n}+x_n)\\ &= p^2\sum_{n=1}^{\frac{p-1}{2}} x_{p-n} - 2p\sum_{n=1}^{\frac{p-1}{2}} nx_{p-n} + \sum_{n=1}^{\frac{p-1}{2}} n^2p.\end{aligned}$$

Finally,

$$\begin{aligned}\frac{p-1}{2} + s &= p\sum_{n=1}^{\frac{p-1}{2}} x_{p-n} - 2\sum_{n=1}^{\frac{p-1}{2}} nx_{p-n} + \sum_{n=1}^{\frac{p-1}{2}} n^2\\ &\equiv 0 - 2\sum_{n=1}^{\frac{p-1}{2}}(-1) + 0\\ &\equiv 2\cdot\frac{p-1}{2} \pmod p,\end{aligned}$$

that is, $s \equiv \dfrac{p-1}{2} \pmod p$. □

3 Given $a, b, c \geq 0, a^2+b^2+c^2 \leq 1$. Prove:

$$\frac{a}{a^2+bc+1} + \frac{b}{b^2+ca+1} + \frac{c}{c^2+ab+1} + 3abc < \sqrt{3}.$$

Solution We use $\sum$ to represent the cyclic sum. First, notice that

$$\sum a - \sum\frac{a}{a^2+bc+1} = \sum\frac{a^3+abc}{a^2+bc+1} \geq \frac{\left(\sum(a^3+abc)\right)^2}{\sum(a^3+abc)(a^2+bc+1)},$$

where the last inequality comes from Cauchy-Schwarz inequality. Meanwhile,

$$\begin{aligned}
&\sum(a^3+abc)(a^2+bc+1)\\
&=\sum(a^3+abc)+\sum a(a^2+bc)(a^2+bc)\\
&=\sum a^3+3abc+\sum a^5+\sum abc(2a^2+bc)\\
&\le \sum a^3+3abc+\sum a^5+3\sum abc.a^2\\
&\le \sum a^3+3abc+\sum a^5+3abc\\
&\le 2\left(\sum a^3+3abc\right).
\end{aligned}$$

In the above relation, the equality cannot be attained as the requirements $a=b=c=1$ and $a^2+b^2+c^2=1$ cannot be true simultaneously. It follows that

$$\begin{aligned}
\sum a-\sum\frac{a}{a^2+bc+1} &\ge \frac{\left(\sum(a^3+abc)\right)^2}{\sum(a^3+abc)(a^2+bc+1)} > \frac{\left(\sum a^3+3abc\right)^2}{2\left(\sum a^3+3abc\right)}\\
&=\frac{1}{2}\left(\sum a^3+3abc\right)\ge\frac{1}{2}(3abc+3abc)=3abc,
\end{aligned}$$

and hence

$$\sum\frac{a}{a^2+bc+1}<\sum a-3abc\le\sqrt{3\left(\sum a^2\right)}-3abc=\sqrt{3}-3abc,$$

that is,

$$\frac{a}{a^2+bc+1}+\frac{b}{b^2+ca+1}+\frac{c}{c^2+ab+1}+3abc<\sqrt{3}. \qquad \square$$

4 For a positive integer k, if it is possible to remove a number from $M_k=\{1,2,\ldots,k\}$, such that the sum of the remaining $k-1$ numbers in M_k is a perfect square, then k is called a "Taurus number"; for instance, 7 is a Taurus number, as removing 3 from $\{1,2,3,4,5,6,7\}$ results in the sum $1+2+4+5+6+7=5^2$.

(1) Determine with reasoning whether 2021 is a Taurus number;
(2) Find $f(n)$ (in terms of n), the number of Taurus numbers among $1,2,\ldots,n$.

Solution (1) 2021 is a Taurus number, since removing 1190 from $M_{2021} = \{1, 2, \ldots, 2021\}$ results in the sum of the remaining 2020 numbers equal to $\sum_{k=1}^{2021} k - 1190 = 1429^2$.

(2) The answer is

$$f(n) = \begin{cases} \left[\sqrt{\dfrac{n(n-1)}{2}}\right], & \text{if } \dfrac{n(n+1)}{2} \text{ is a perfect square,} \\ \left[\sqrt{\dfrac{n(n+1)}{2}}\right], & \text{if } \dfrac{n(n+1)}{2} \text{ is not a perfect square.} \end{cases} \qquad ①$$

Alternatively,

$$f(n) = \left[\sqrt{\frac{n^2+n-2}{2}}\right]. \qquad ②$$

We give two methods as follows.

Method 1 To begin, notice that the largest perfect square not exceeding the sum $\sum_{k=1}^{n} k = \dfrac{n(n+1)}{2}$ is $\left[\sqrt{\dfrac{n(n+1)}{2}}\right]^2$. Compare it with $f(n)$ for the first few n values, as illustrated in the table: (By Chen Haoran)

n	1	2	3	4	5	6	7	8	9	10	11	12	13
$n(n+1)/2$	1	3	6	10	15	21	28	36	45	55	66	78	91
$[\sqrt{n(n+1)/2}]^2$	1	1	4	9	9	16	25	36	36	49	64	64	81
$f(n)$	0	1	2	3	3	4	5	5	6	7	8	8	9

we conjecture ① is the answer. Let

$$\varphi(n) = \frac{n(n+1)}{2} - \left[\sqrt{\frac{n(n+1)}{2}}\right]^2.$$

It is clear that n is a Taurus number if and only if $\varphi(n) \in M_n$, namely, $1 \le \varphi(n) \le n$. For a non-Taurus number n, either $\varphi(n) = 0$ (call it type A), or $\varphi(n) \ge n+1$ (call it type B).

Lemma 1 *Define* $d = \sqrt{\dfrac{n(n+1)}{2}} - \sqrt{\dfrac{n(n-1)}{2}}$. *For* $n > 1, 0 < d < 1$.

Proof of lemma 1 Notice that

$$d=\sqrt{\frac{n(n+1)}{2}}-\sqrt{\frac{n(n-1)}{2}}=\frac{\frac{n(n+1)}{2}-\frac{n(n-1)}{2}}{\sqrt{\frac{n(n+1)}{2}}+\sqrt{\frac{n(n-1)}{2}}}$$

$$=\frac{n}{\sqrt{\frac{n(n+1)}{2}}+\sqrt{\frac{n(n-1)}{2}}}=\frac{\sqrt{2}}{\sqrt{1+\frac{1}{n}}+\sqrt{1-\frac{1}{n}}}<1,$$

$$\left(\text{since } d^2=\frac{2}{2+2\sqrt{1-\frac{1}{n^2}}}=\frac{1}{1+\sqrt{1-\frac{1}{n^2}}}<1\right)$$

and hence $0<d<1$.

From lemma 1, the integer parts of $\sqrt{\frac{n(n-1)}{2}}$ and $\sqrt{\frac{n(n+1)}{2}}$ differ by at most 1. Hence

$$\left[\sqrt{\frac{n(n+1)}{2}}\right]-\left[\sqrt{\frac{n(n-1)}{2}}\right]=0 \text{ or } 1.$$

Lemma 2 *For integer $n>1$, if $\frac{n(n-1)}{2}$ is not a perfect square, then*

$$\frac{n(n-1)}{2}\geq\left[\sqrt{\frac{n(n-1)}{2}}\right]^2+1.$$

Proof of lemma 2 Given that $\frac{n(n-1)}{2}$ is not a perfect square, we have $\left[\sqrt{\frac{n(n-1)}{2}}\right]<\sqrt{\frac{n(n-1)}{2}}$, taking squares to yield $\left[\sqrt{\frac{n(n-1)}{2}}\right]^2<\left(\sqrt{\frac{n(n-1)}{2}}\right)^2=\frac{n(n-1)}{2}$ which is an integer. Therefore, $\frac{n(n-1)}{2}\geq\left[\sqrt{\frac{n(n-1)}{2}}\right]^2+1$.

Now we verify the conjecture ① by induction on n. Obviously it is true when $n=1$. Assume it is true for every positive integer up to $n-1$, and consider ① for n. There are two situations.

(1) If $\varphi(n) = \frac{n(n+1)}{2} - \left[\sqrt{\frac{n(n+1)}{2}}\right]^2 > 0$, define

$$d_0 = \left[\sqrt{\frac{n(n+1)}{2}}\right] - \left[\sqrt{\frac{n(n-1)}{2}}\right].$$

According to lemma 1, $d_0 = 0$ or 1. If $d_0 = 0$, $\left[\sqrt{\frac{n(n+1)}{2}}\right]^2 = \left[\sqrt{\frac{n(n-1)}{2}}\right]^2$, and then

$$\begin{aligned}\varphi(n) &= \frac{n(n+1)}{2} - \left[\sqrt{\frac{n(n+1)}{2}}\right]^2 \\ &= \frac{n(n+1)}{2} - \left[\sqrt{\frac{n(n-1)}{2}}\right]^2 \\ &= n + \frac{n(n-1)}{2} - \left[\sqrt{\frac{n(n-1)}{2}}\right]^2 \\ &\geq n+1.\end{aligned}$$

Clearly, n is not a Taurus number, and all the Taurus numbers in M_n belong to M_{n-1} as well. Hence,

$$f(n) = f(n-1) = \left[\sqrt{\frac{n(n-1)}{2}}\right] = \left[\sqrt{\frac{n(n+1)}{2}}\right].$$

If $d_0 = 1$, $\left[\sqrt{\frac{n(n+1)}{2}}\right] - \left[\sqrt{\frac{n(n-1)}{2}}\right] = 1$, and then

$$\begin{aligned}\varphi(n) &= \frac{n(n+1)}{2} - \left[\sqrt{\frac{n(n+1)}{2}}\right]^2 \\ &= \frac{n(n+1)}{2} - \left(\left[\sqrt{\frac{n(n-1)}{2}}\right] + 1\right)^2.\end{aligned}$$

It suffices to show

$$\frac{n(n+1)}{2} - \left[\sqrt{\frac{n(n+1)}{2}}\right]^2 \leq n, \tag{3}$$

equivalently, $\dfrac{n(n+1)}{2}-\left(\left[\sqrt{\dfrac{n(n-1)}{2}}\right]+1\right)^2 \le n$, or

$$\frac{n(n-1)}{2} \le \left(\left[\sqrt{\frac{n(n-1)}{2}}\right]+1\right)^2. \qquad ④$$

Since the left-hand side is an integer, it follows that $\dfrac{n(n-1)}{2} = \left(\sqrt{\dfrac{n(n-1)}{2}}\right)^2 < \left(\left[\sqrt{\dfrac{n(n-1)}{2}}\right]+1\right)^2$, and ④ holds, which implies ③, so n is a Taurus number.

There is one more Taurus number in $M_n = \{1, 2, \ldots, n\}$ than in $M_{n-1} = \{1, 2, \ldots, n-1\}$. Thus,

$$f(n) = f(n-1) + 1 = \left[\sqrt{\frac{n(n-1)}{2}}\right] + 1 = \left[\sqrt{\frac{n(n+1)}{2}}\right].$$

(2) If $\varphi(n) = 0$, that is, $S(n) = 1 + 2 + \cdots + n = \dfrac{n(n+1)}{2}$ is a perfect square. By lemma 1, $d = \sqrt{\dfrac{n(n+1)}{2}} - \sqrt{\dfrac{n(n-1)}{2}} < 1$, and

$$\left[\sqrt{\frac{n(n+1)}{2}}\right] - \left[\sqrt{\frac{n(n-1)}{2}}\right] = 1.$$

By assumption, $\sqrt{\dfrac{n(n+1)}{2}}$ is an integer, and the above equation becomes $\sqrt{\dfrac{n(n+1)}{2}} - 1 = \left[\sqrt{\dfrac{n(n-1)}{2}}\right]$, as the preceding perfect square of $\dfrac{n(n+1)}{2} = \left(\sqrt{\dfrac{n(n+1)}{2}}\right)^2$ is $\left(\sqrt{\dfrac{n(n+1)}{2}} - 1\right)^2$. Note when $n > 1$, $2n(n+1) > (n+1)^2$, hence

$$\left(\sqrt{\frac{n(n+1)}{2}}\right)^2 - \left(\sqrt{\frac{n(n+1)}{2}} - 1\right)^2 = 2\sqrt{\frac{n(n+1)}{2}} - 1 > n.$$

This is to say, by removing any number from $M_n = \{1, 2, \ldots, n\}$, the sum of the remaining $n-1$ numbers is always less than $\frac{n(n+1)}{2}$ $\left(\text{which is } \left[\sqrt{\frac{n(n+1)}{2}}\right]^2\right)$ and greater than $\left(\sqrt{\frac{n(n+1)}{2}}-1\right)^2$, thereby not a perfect square; n is not a Taurus number. As a result, the number of Taurus numbers in $M_n = \{1, 2, \ldots, n\}$ is the same as that in $M_{n-1} = \{1, 2, \ldots, n-1\}$, which gives $f(n) = \left[\sqrt{\frac{n(n-1)}{2}}\right]$.

The induction is completed and ① is verified.

Method 2 If n_0 is a Taurus number, there exists $k \in \{1, 2, \ldots, n_0\}$, such that $\frac{n_0(n_0+1)}{2} - k = m^2$, where m is a positive integer; m must be the largest integer satisfying $\frac{n_0(n_0-1)}{2} \leq m^2 < \frac{n_0(n_0+1)}{2}$.

From $\frac{n_0(n_0-1)}{2} \leq m^2$, it follows that $(2n_0-1)^2 \leq 1+8m^2$, or $n_0 \leq \frac{1+\sqrt{1+8m^2}}{2}$. For the largest positive integer n_0 satisfying this inequality, $n_0 = \left[\frac{1+\sqrt{1+8m^2}}{2}\right]$. Denote $\varphi(m) = \left[\frac{1+\sqrt{1+8m^2}}{2}\right]$, $\varphi(m)$ is strictly increasing for positive integers m, since

$$\begin{aligned}
\frac{1+\sqrt{1+8(m+1)^2}}{2} - \frac{1+\sqrt{1+8m^2}}{2} &= \frac{\sqrt{1+8(m+1)^2}-\sqrt{1+8m^2}}{2} \\
&= \frac{8(m+1)^2-8m^2}{2(\sqrt{1+8(m+1)^2}+\sqrt{1+8m^2})} \\
&= \frac{4(2m+1)}{\sqrt{1+8(m+1)^2}+\sqrt{1+8m^2}} \\
&> \frac{4(2m+1)}{3(m+1)+3m} = \frac{8m+4}{6m+3} > 1.
\end{aligned}$$

Therefore, $\varphi(m+1)-\varphi(m) = \left[\frac{1+\sqrt{1+8(m+1)^2}}{2}\right] - \left[\frac{1+\sqrt{1+8m^2}}{2}\right] \geq 1$. Furthermore, it implies that all the Taurus numbers have a 1-1 correspondence with all $\varphi(m)(m = 1, 2, \ldots)$ values. Consequently, the number of Taurus numbers not exceeding n is equal to the number of positive integers m satisfying $\varphi(m) \leq n$, that is, the largest m satisfying $\dfrac{1+\sqrt{1+8m^2}}{2} < n+1$, or $m^2 < \dfrac{n^2+n}{2}$: as the right-hand side is an integer,

$$m^2 \leq \frac{n^2+n}{2} - 1 = \frac{n^2+n-2}{2},$$

and $m = \left[\sqrt{\dfrac{n^2+n-2}{2}}\right] = f(n)$. This justifies ②. □

Remark The formulas ① and ② are equivalent. Indeed, when $\dfrac{n(n+1)}{2}$ is not a perfect square,

$$\left[\sqrt{\frac{n(n+1)}{2}}\right] = \left[\sqrt{\frac{n(n+1)}{2} - 1}\right] = \left[\sqrt{\frac{n^2+n-2}{2}}\right];$$

when $\dfrac{n(n+1)}{2} = m^2$ is a perfect square,

$$\frac{n^2+n-2}{2} = \frac{n(n+1)}{2} - 1 = m^2 - 1,$$

$$\left[\sqrt{\frac{n^2+n-2}{2}}\right] = [\sqrt{m^2-1}] = m - 1,$$

Since $\dfrac{n(n+1)}{2} = m^2$, we have $\dfrac{n(n-1)}{2} = m^2 - n \geq (m-1)^2$, $\dfrac{n(n-1)}{2} = m^2 - n < m^2$, and thus $\left[\sqrt{\dfrac{n(n-1)}{2}}\right] = m - 1$. So, when $\dfrac{n(n+1)}{2}$ is a perfect square, it is also true that

$$\left[\sqrt{\frac{n(n-1)}{2}}\right] = \left[\sqrt{\frac{n^2+n-2}{2}}\right].$$

The inequality $m^2 - n \geq (m-1)^2$ is equivalent to

$$2m - 1 \geq n = \frac{\sqrt{1+8m^2}-1}{2},$$

which holds for all positive integers m.

Second Day
(8 am – 12 pm; July 29, 2021)

5 Let $A = \{a_1, a_2, \ldots, a_n, b_1, b_2, \ldots, b_n\}$ be a $2n$-set, $B_i \subseteq A (i = 1, 2, \ldots, m)$. If $\cup_{i=1}^m B_i = A$, the ordered m-tuple $(B_1, B_2, \ldots, B_m)$ is called an "ordered m-covering" of A; for the ordered m-covering $(B_1, B_2, \ldots, B_m)$, if every $B_i (i = 1, 2, \ldots, m)$ does not contain both a_j and $b_j (j = 1, 2, \ldots, n)$, then $(B_1, B_2, \ldots, B_m)$ is called a "no-matching ordered m-covering" of A. Let $a(n, m)$ represent the number of ordered m-coverings of A; $b(n, m)$ represent the number of no-matching ordered m-coverings of A.

(1) Find $a(n, m)$ and $b(n, m)$;

(2) Suppose for some integers $m \geq 2$ and $n \geq 1$, $\dfrac{a(n, m)}{b(n, m)} \leq 2021$. Find the largest possible value of m.

Solution (1) (Modified by Chen Haoran) First, we find $a(n, m)$. Use $2n$ steps to determine an ordered m-covering $(B_1, B_2, \ldots, B_m)$ of A: for $i = 1, 2, \ldots, n$, in the ith step, determine which of the sets $B_1, B_2, \ldots, B_m$ contain a_i, and which do not contain a_i; in the $(i + n)$th step, determine which of the sets $B_1, B_2, \ldots, B_m$ contain b_i, and which do not contain b_i. Notice that at least one of $B_1, B_2, \ldots, B_m$ contains a_i or b_i, so there are $\sum\limits_{j=1}^{m} \binom{m}{j} = 2^m - 1$ ways to finish each step. By the multiplication principle, the number of ordered m-coverings of A is

$$a(n, m) = (2^m - 1)^{2n}.$$

Next, we find $b(n, m)$. By definition of no-matching ordered m-coverings, $m \geq 2$ is required. Similar to (1), use n steps to determine a no-matching ordered m-covering $(B_1, B_2, \ldots, B_m)$ of A: for $i = 1, 2, \ldots, n$, in the ith step, determine which of the sets $\mathrm{B}_1, B_2, \ldots, B_m$ contain a_i, which contain b_i, and which contain neither a_i nor b_i (none can contain both a_i and b_i). There are 3^m ways in total, 2^m of which do not cover a_i, 2^m of which do not cover b_i, and one of which covers neither. Hence, there are $3^m - 2^{m+1} + 1$ ways to finish each step. By the multiplication principle, the number of no-matching ordered m-coverings of A is

$$b(n, m) = (3^m - 2^{m+1} + 1)^n.$$

(2) From (1), we have

$$\frac{a(n,m)}{b(n,m)}=\frac{(2^m-1)^{2n}}{(3^m-2^{m+1}+1)^n}\le 2021,\quad \text{or}\quad \frac{(4^m-2^{m+1}+1)^n}{(3^m-2^{m+1}+1)^n}\le 2021.$$

Clearly, $\frac{(4^m-2^{m+1}+1)^n}{(3^m-2^{m+1}+1)^n}$ is increasing with respect to n. To find the largest possible value of m, set $n=1$, that is, $\frac{4^m-2^{m+1}+1}{3^m-2^{m+1}+1}\le 2021$. Since

$$\frac{4^m}{3^m}\le\frac{4^m-2^{m+1}+1}{3^m-2^{m+1}+1}\le\frac{4^m-2^{m+1}}{3^m-2^{m+1}}=\frac{2^m-2}{1.5^m-2},$$

it is easy to verify $\frac{4^m-2^{m+1}+1}{3^m-2^{m+1}+1}$ is increasing in m when $m\ge 4$. If $\frac{4^m}{3^m}>2021, m>\log_{4/3}2021$, then

$$m\ge[\log_{4/3}2021]+1=27.$$

So it must require $m\le[\log_{4/3}2021]=26$. On the other hand, when $m\ge 6$, $\frac{2^m-2}{1.5^m-2}\le\frac{2^m}{1.5^m}+1\le 2021$, and this gives $m\le\log_{4/3}2020$, $m\le[\log_{4/3}2020]=26$.

In conclusion, for $\frac{a(n,m)}{b(n,m)}\le 2021$ to be true, the largest possible value of m is 26, attained when $n=1$. □

6 As shown in Fig. 6.1, in the cyclic quadrilateral $ABCD$, the bisector of $\angle BAD$ meets side BC at E, and M is the midpoint of AE. The exterior bisector of $\angle BCD$ crosses the extension of AD at F; the line MF meets side AB at G. If $AB=2AD$, prove: $MF=2MG$.

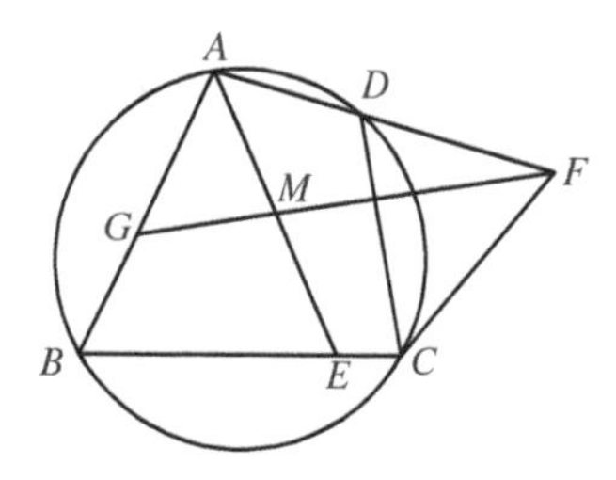

Fig. 6.1

Solution It is known that $\angle EAF = \frac{1}{2}\angle BAD, \angle ECF = 90^\circ + \frac{1}{2}\angle BCD$, and hence

$$\angle EAF + \angle ECF = 90^\circ + \frac{1}{2}(\angle BAD + \angle BCD) = 180^\circ,$$

which implies that A, E, C, F all lie on a circle, say ω.

Let the extension of CD beyond D cross circle ω at point K. As shown in Fig. 6.2, drop perpendiculars from F to the lines AE and AK, with feet X and Y, respectively.

Note that CF is the exterior bisector of $\angle ECK$, thus F is the midpoint of $\overset{\frown}{ECK}$ of ω, and AF bisects $\angle CAK$.

Since A, B, C, D are concyclic, A, E, C, K are also concyclic, we infer that $\angle ADK = \angle ABE, \angle AKD = \angle AEB$, hence $\triangle ADK \backsim \triangle ABE$, $\frac{AK}{AE} = \frac{AD}{AB} = \frac{1}{2}$, that is, $AK = \frac{1}{2}AE$.

On the other hand, as AF bisects $\angle CAK$, $\triangle FKY \cong \triangle FEX, KY = EX = \frac{1}{2}(AE - AK) = \frac{1}{4}AE$. Therefore, X is the midpoint of EM. From $AK = \frac{1}{2}AE = AM$, it follows that $\triangle AKF \cong \triangle AMF$, and

$$\angle AFG = \angle AFM = \angle AFK = \angle AEK. \quad ①$$

Moreover, by properties of cyclic quadrilaterals,

$$\angle AEK = \angle ACK = \angle ACD = \angle ABD. \quad ②$$

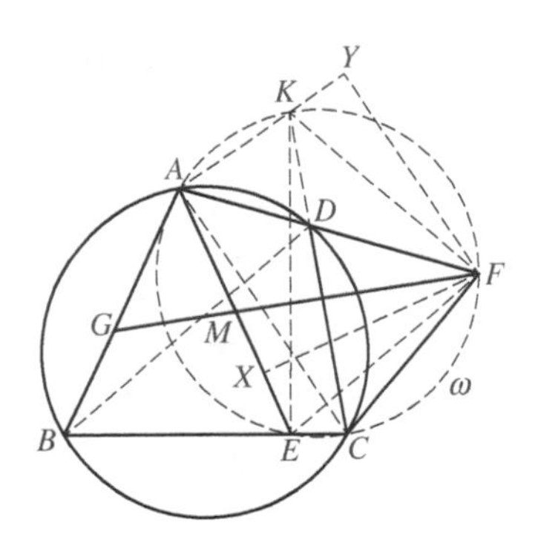

Fig. 6.2

From ① and ②, $\angle AFG = \angle ABD$. Thus, $\triangle AFG \backsim \triangle ABD$. Finally, by the angle bisector theorem, $\frac{MF}{MG} = \frac{AF}{AG} = \frac{AB}{AD} = 2$, namely, $MF = 2MG$. □

7 Find all odd integer pairs $\{a, b\}, a, b > 1$, such that

$$7\varphi^2(a) - \varphi(ab) + 11\varphi^2(b) = 2(a^2 + b^2), \qquad ①$$

where $\varphi(n)$ represents the number of positive integers less than n and relatively prime to n.

Solution We denote by $p^\alpha \| b$ that the highest power of p that divides b is p^α, that is, $p^\alpha | b$, but $p^{\alpha+1} \nmid b$.

To begin, notice that if $\{a, b\}$ is a solution of ①, p is a prime and $p^2|a, p^2|b$, then $\left\{\frac{a}{p}, \frac{b}{p}\right\}$ is another solution. Henceforth, we assume for each prime p, at least one of a, b is coprime with p or divisible exactly by p.

Since $a, b > 2, 2|\varphi(a), 2|\varphi(b)$. If $2\|\varphi(a), 2\|\varphi(b)$, it is easy to see that

$$7\varphi^2(a) \geq 7\left(\frac{2}{3}\right)^2 a^2 > 2.5a^2,$$

$$11\varphi^2(b) \geq 11\left(\frac{2}{3}\right)^2 b^2 > 2.5b^2.$$

Notice that $\varphi(ab) < ab \leq 0.5a^2 + 0.5b^2$, the left-hand side of ① is larger hence equality cannot hold. This indicates either $4|\varphi(a)$ or $4|\varphi(b)$. First, assume $4|\varphi(a)$. If $16|\varphi(ab)$, take ① modulo 16 to get $2(a^2 + b^2) \equiv 4 \equiv 11\varphi^2(b) \pmod{16}$, so $2\|\varphi(b)$, yet $11\varphi^2(b) = 12 \pmod{16}$, a contradiction. We infer that 16 does not divide $\varphi(ab)$. The above argument also applies when $4|\varphi(b)$.

As a result, there are two situations to consider: either $4\|\varphi(ab)$, or $8\|\varphi(ab)$. In any case, ab has at most three distinct prime factors; moreover, if indeed three factors, then they are all of the form $4k + 3$.

Case 1, ab is a prime power, say the prime factor is p. According to $4|\varphi(ab)$, p must be of the form $4k + 1$, and $p|a, p|b$. Then

$$7\varphi^2(a) > 2.5a^2, \quad 11\varphi^2(b) > 2.5b^2. \qquad ②$$

As stated before, ① cannot hold in this circumstance.

Case 2, ab contains two distinct prime factors, say p, q. We have the following observations. First, notice that p or q must be 3, otherwise ② holds, a contradiction. Second, a and b cannot be relatively prime, as otherwise a and b each has only one prime factor, and ② holds again. Third, if $pq|a$ and $pq|b$, assuming $q > p$, then $p = 3$. This implies $16|\varphi(a), 16|\varphi(b)$, and $4\|\varphi(ab)$, hence p, q are both primes of the form $4k+3$. If $q > 7$, ② holds again. Therefore, $q = 7, \{p, q\} = \{3, 7\}$.

If $q^2|a$ or $q^2|b$, assume it is a, then $q^2|\varphi^2(b), q^2|\varphi(ab)$, and so $q^2|11\varphi^2(b), q|\varphi(b)$. Since $q > p$, we have $q \nmid p-1, q^2|b$, and q^2 divides both a and b, but this violates the assumption at the beginning of the proof. Therefore, $q\|a, q\|b$. Let $a = 3^\alpha \cdot 7, b = 3^\beta \cdot 7$. (i) if $\alpha = \beta$, then ① implies $17\varphi^2(a) = 4a^2, 17|a$, a contradiction. (ii) $\alpha > \beta$, then $\beta = 1$, and the left-hand side of ① equals

$$\begin{aligned}
&28 \cdot 6^2 \cdot 3^{2\alpha-2} - 12 \cdot 7 \cdot 3^{\alpha+\beta-1} + 44 \cdot 6^2 \cdot 3^{2\beta-2} \\
&> (28 \cdot 36 - 12 \cdot 7) \cdot 3^{2\alpha-2} + 44 \cdot 6^2 \cdot 3^{2\beta-2} \\
&> 49(18 \cdot 3^{2\alpha-2} + 18 \cdot 3^{2\beta-2}) \\
&= 2(a^2 + b^2),
\end{aligned}$$

again, contradiction. (iii) $\alpha < \beta$, no solutions, either. In all, there are no solutions when $pq|a$ and $pq|b$.

Now we handle the last situation: a and b are not relatively prime, pq does not divide both a and b. Suppose a and b have the common factor p. If $pq|b$, then a has only one prime factor, causing

$$11\varphi^2(b) \geq 11\left(\frac{2}{3} \cdot \frac{4}{5}\right)^2 > 2.5, 7\varphi^2(a) \geq 7\left(\frac{2}{3}\right)^2 > 2.5,$$

and ② holds, contradiction. Thus, $pq|a, p|b$, and $(q, a) = 1$. Similar to the above argument, it must be $p = 3$. If $q > 7$, ② holds again. Hence, $q = 5$ or 7.

(i) $q = 5$, then $3 \nmid (q-1)$. If $3^2|a$ or $3^2|b$, assume $3^2|a$. Take ① modulo 3^2 to obtain $3|11$, impossible. The situation is similar when $3^2|b$. So it must be $3\|a, 3\|b$, and $d = 3$. If $5^2|a$, take ① modulo 5 to get $11\varphi^2(b) = 2b^2 \pmod 5$. This implies that 2 is a quadratic residue modulo 5, which is absurd. Hence, $5\|a, a = 15$. After verification, $(a, b) = (15, 3)$ is a solution.

(ii) $q = 7$, if $3^2|a$, then $3\|b, b = 3, 3|11\varphi(b)$, contradiction. Hence $3\|a$. Let $a = 3 \cdot 7^\beta, b = 3^\alpha$. If $\alpha \geq 3$, take ① modulo 27 to obtain

$$7 \cdot 2^2 \cdot 6^2 \cdot 7^{2\beta-2} \equiv 2 \cdot 7^{2\alpha} \cdot 9 \pmod{27},$$

which can be simplified to $16 \equiv 14 \pmod 3$, absurd. Thus, b =3 or 9, indicating

$$\begin{aligned} 7\varphi^2(a) - \varphi(ab) &= 1008 \cdot 7^{2\alpha-2} - 18 \cdot 6 \cdot 7^{\alpha-1} \\ &\geq 900 \cdot 7^{2\alpha-2} > 882 \cdot 7^{2\alpha-2} = 2c^2. \end{aligned}$$

Meanwhile, $11\varphi^2(b) > 2b^2$. Now ① becomes unbalanced, and no solutions exist.

Case 3, ab contains three distinct prime factors, each of which has the form $4k + 3$. In this circumstance, $8|\varphi(ab)$, which gives $4\|\varphi(a)$ or $4\|\varphi(b)$, that is, a or b has only one prime factor. If it is a, it can be easily checked that

$$7\varphi^2(a) \geq 7\left(\frac{2}{3}\right)^2 a^2 > 2.5a^2,$$

$$11\varphi^2(b) \geq 11\left(\frac{2}{3} \cdot \frac{6}{7} \cdot \frac{10}{11}\right)^2 b^2 > 2.5b^2,$$

and ② holds. So, b has only one prime factor, say p, and the other two prime factors of cd are q and r. First, $3|a$, or else ① holds. If a and b are coprime, then $3 \nmid b, p \geq 7$, and

$$11\varphi^2(b) \geq 11 \cdot \frac{36}{49} \cdot b^2 > 8b^2.$$

Meanwhile, a has only two prime factors, hence $7\varphi^2(a) \geq 7\left(\frac{2}{3} \cdot \frac{36}{49}\right)^2 a^2 > 2.25$, yielding

$$\begin{aligned} 7\varphi^2(a) + 11\varphi^2(b) &> 2.25a^2 + 8b^2 > 2a^2 + 2b^2 + (0.25a^2 + 4b^2) \\ &> 2a^2 + 2b^2 + \varphi(ab), \end{aligned}$$

a contradiction. Hence, $(a, b) \neq 1, p|a$. If $p^2|a$, by the method of descent, $p = b$. Based on the previous argument, $p^2\|11\varphi(b)$, but this is impossible, hence $p\|a$. Take ① modulo p twice, to derive $p|7\varphi^2(b)$. So $p = 7$ or p divides one of $q - 1$ and $r - 1$, say $q - 1$.

If $p \geq 11$, then $q \geq 19$. A similar argument reveals that the left-hand side of ① is larger and equality cannot hold. Therefore, $p = 3$ or $p = 7$.

If $p = 7$, then $7|a, 7|b, 7^2|2(a^2 + b^2)$. If $7^2|b$, then $7||a$, and we must have $7^2|\varphi(ab), 7^2|\varphi^2(b)$, indicating $7^2|7\varphi(b^2), 7|(r-1)$. The smallest prime r of the form $4k+3$ satisfying $7|(r-1)$ is $r = 43$, which gives

$$7\varphi^2(a) \geq \left(\frac{2}{3} \cdot \frac{6}{7} \cdot \frac{42}{43}\right)^2 > 2.18.$$

In the meantime, $11\varphi^2(b) \geq 11\left(\frac{6}{7}\right)^2 > 8$, thus

$$\begin{aligned} 7\varphi^2(a) + 11\varphi^2(b) &\geq 2(a^2+b^2) + (0.18a^2 + 6b^2) \\ &\geq 2(a^2+b^2) + \varphi(ab), \end{aligned}$$

a contradiction.

If $p = 3$, let $b = p^\alpha$. Notice that

$$\varphi^2(a) = \left(\frac{2}{3} \cdot \frac{q-1}{q} \cdot \frac{r-1}{r}\right)^2 a^2, \quad \varphi^2(b) = \frac{4}{9}b^2,$$

$$\varphi(ab) = \frac{2}{3} \cdot \frac{q-1}{q} \cdot \frac{r-1}{r} ab.$$

By the AM-GM inequality, we have $\frac{1}{2}\varphi^2(a) + 2b^2 \geq 2\varphi(ab) > \varphi(ab)$. Since $11\varphi^2(b) = \frac{44}{9}b^2 > 2b^2 + 2b^2$, in case $6.5\varphi^2(a) > 2a^2$, then ① becomes unbalanced, contradiction.

If $q, r \neq 7$, we know

$$6.5\left(\frac{2}{3} \cdot \frac{q-1}{q} \cdot \frac{r-1}{r}\right)^2 \geq \frac{4}{9} \cdot \frac{100}{121} \cdot \frac{324}{361} > 2,$$

and again the left-hand side of ① is larger. Hence, q or r must be 7. Assume $q = 7, r \geq 11$. Let $c = 3.7^\beta r^\gamma$, and $3|(p-1)$. If $\alpha \geq 3$, then $27|\varphi(ab), 27|\varphi^2(b)$. Take ① modulo 27 to find

$$7 \cdot 36 \cdot 4 \cdot (r-1)^2 \cdot 7^{2\beta-2} \cdot r^{2\gamma-2} \equiv 2 \cdot 3^2 \cdot 7^{2\beta} \cdot r^{2\gamma} \pmod{27},$$

which can be simplified to

$$8(r-1)^2 = 7r^2 \pmod 3.$$

Since $3 \nmid r$, it follows that $3 \nmid (r-1)$, and 2 is a quadratic residue modulo 3, a contradiction.

Therefore, $\alpha = 1$ or 2, and $d = 3$ or 9. Since $3|\varphi(a), \varphi(ab)$, taking ① modulo 3 yields $3|11\varphi^2(b)$. Hence $3^2|b$, and $b = 9$. Let $e = 7^\alpha r^\gamma$. Plug $c = 3 \cdot 7^\alpha \cdot r^\gamma$ and $b = 9$ into ① to obtain

$$14\varphi^2(e) - 9\varphi(e) + 117 = 9e^2. \qquad ③$$

If $\gamma \geq 2$, then $7|\varphi(e), 7|117$, which is not true.

We deduce that $\beta \geq 2$ is impossible, $\beta = 1$. Let $f = r^\gamma$ and simplify ③ to get

$$56\varphi^2(f) - 6\varphi(f) + 13 = 49f^2.$$

Similarly, if $\gamma > 1$, then $r|13, r = 13$. But 13 is a prime of the form $4k+1$, contradiction.

If $\gamma = 1$, then $\varphi(f) = r - 1$, and ③ reduces to

$$7r^2 - 118r + 85 = 0,$$

which does not have integer solutions.

The final conclusion: with the assumption that for each prime p, at least one of a, b is relatively prime to p or divisible exactly by p, there is only one solution

$$\{a, b\} = \{15, 3\}.$$

By the method of descent, all solutions of ① are

$$\{a, b\} = \{15 \cdot 3^\beta, 3 \cdot 3^\beta\},$$

where β is any nonnegative integer. □

8 The integer sequence $\{z_i\}$ satisfies: for every $i = 1, 2, \ldots, z_i \in \{0, 1, \ldots, 9\}, z_i \equiv i - 1 \pmod{10}$. Suppose there are 2021 nonnegative real numbers $x_1, x_2, \ldots, x_{2021}$, satisfying

$$\sum_{i=1}^{k} x_i \geq \sum_{i=1}^{k} z_i, \ \sum_{i=1}^{k} x_i \leq \sum_{i=1}^{k} z_i + \sum_{i=1}^{k} \frac{10-i}{50} z_{k+i},$$

for $k = 1, 2, \ldots, 2021$. Find the least possible value of $\sum_{i=1}^{2021} x_i^2$.

Solution To begin, let $k = 10l + j + 1, l \in \mathbb{N}, j \in \{0, 1, \ldots, 9\}$, and denote $C_k = \frac{1}{50}\sum\limits_{i=1}^{10}(10 - i)z_{k+i}$. By definition of $\{z_i\}$,

$$C_k = \frac{1}{50}(45(j+1) + 120 - 5j(j+1)).$$

Denote $D_k = \sum\limits_{i=1}^{k} x_i, S_k = \sum\limits_{i=1}^{k} z_i$. Clearly,

$$S_k = 45l + \frac{j(j+1)}{2}. \qquad ①$$

For $k = 1, 2, \ldots, 2021$, it is required that

$$D_k \le S_k + C_k = 45l + A_j, \qquad ②$$

where $A_j = 0.9j + 3.3 + 0.8\frac{j(j+1)}{2}$.
On the other hand, for k =1, 2, $\ldots$, 2021, it is required that

$$45l + \frac{j(j+1)}{2} \le D_k.$$

Suppose $x_1^*, \ldots, x_{2021}^*$ satisfy the inequalities and $\sum\limits_{i=1}^{2021} x_i^2$ attains the least possible value. Let $k_0 = 0$ and $1 \le k_1 < \cdots < k_N < 2021$ be all ordinals k such that $S_k = D_k, k \in \{1, 2, \ldots, 2020\}$. We call $k_1, \ldots, k_l$ reset positions. The objective function $G := \sum\limits_{i=1}^{2021} x_i^2$ is convex in each of the variables $x_1, \ldots, x_{2021}$.

Lemma *For every $k, 1 \le k < 2021$: if k is not a reset position, then $x_k^* \le x_{k+1}^*$, and in case $x_k^* < x_{k+1}^*, D_k = S_k + C_k$ must hold; if k is a reset position, then $x_k^* \ge x_{k+1}^*$.*

Proof of lemma If k is not a reset position, then $D_k > S_k$. Suppose $x_k^* > x_{k+1}^*$, and we make the adjustment $\underset{\wedge}{x}_k^* = x_k^* - \varepsilon, \underset{\wedge}{x}_{k+1}^* = x_k^* + \varepsilon, \varepsilon > 0$. It can be verified that when $\varepsilon > 0$ is sufficiently small, the value of G decreases due to the convexity, while all the restrictions are met. This violates the optimal assumption of $x_1^*, \ldots, x_{2021}^*$.

Similarly, if $x_k^* < x_{k+1}^*$, make the adjustment $\underset{\wedge}{x}_k^* = x_k^* + \varepsilon, \underset{\wedge}{x}_{k+1}^* = x_k^* - \varepsilon$, and the value of G decreases. Yet the only condition that may be broken is $D_k \le S_k + C_k$. Hence, $D_k = S_k + C_k$.

On the other hand, if k is a reset position, then $D_k = S_k < S_k + C_k$. Suppose $x_k^* < x_{k+1}^*$, and we make the adjustment $\underset{\wedge}{x}_k^* = x_k^* + \varepsilon, \underset{\wedge}{x}_{k+1}^* = x_{k+1}^* - \varepsilon, \varepsilon > 0$. In the same manner, it can be verified when $\varepsilon > 0$ is sufficiently small, G decreases while all the restrictions are met. This completes the proof of the lemma.

The lemma reveals the following fact: between two reset positions, x_k^* is increasing; at the reset position $k, x_k^* \geq x_{k+1}^*$ is decreasing. We claim that for $k = 1, 2, \ldots, 2020$, if $k \equiv 0 \pmod{10}$, then k is a reset position.

When $k \equiv 0 \pmod{10}$, k can be written as $10l + 9 + 1$. From ①, it follows that $S_k = S_{k+1} = 45l + 45$; from ②, $A_k = A_9 = 47.4$. Also, $A_{k+1} = A_0 = 3.3$. In fact, $A_k = A_{k \pmod{10}}$, and the values of $A_0, \ldots, A_9$ are

$$3.3, 5.0, 7.5, 10.8, 14.9, 19.8, 25.5, 32.0, 39.3, 47.4,$$

respectively, which form a convex increasing sequence. Evidently, we have

$$x_{k+1}^* = D_{k+1} - D_k \leq 45(l+1) + A_0 - 45l - \frac{9 \cdot 10}{2} = 3.3.$$

Assume the preceding reset position is $k' = 10l' + j' + 1 < k$. By the lemma, $x_{k'+1}^* \leq x_{k'+2}^* \leq \cdots \leq x_k^*$, and

$$x_{k'+1}^* + x_{k'+2}^* + \cdots + x_k^* = S_k - S_{k'} = 45(l - l') + 45 - \frac{j'(j'+1)}{2}.$$

Since $45 - \dfrac{j'(j'+1)}{2} \geq 4.5(9 - j')$ is true for every $j' = 0, 1, 2, \ldots, 9$, and among $x_{k'+1}^*, \ldots, x_k^*$ there are exactly $10(l - l') + (9 - j')$ numbers. It follows that the average of $x_{k'+1}^* + x_{k'+2}^* + \cdots + x_k^*$ is greater than or equal to 4.5. As $x_{k'+1}^* \leq x_{k'+2}^* \leq \cdots \leq x_k^*$ is increasing, $x_k^* \geq 4.5 > 3.3 \geq x_{k+1}^*$; according to the lemma, k must be a reset position.

Now, it is clear that every 10th ordinal is a reset position, namely $10l \in \{k_1, \ldots, k_N\}$. As the restrictions are periodically reset, it suffices to consider the optimization problem between consecutive reset positions. Furthermore, $S_{2020} = D_{2020}$, and we can just take $x_{2021}^* = z_1 = 0$ to meet the restriction, $x_{2021}^* = 0$. The optimization problem to deal with is essentially the following:

Find $G' = \min\limits_{x_1, \ldots, x_{10}} \sum\limits_{i=1}^{10} x_i^2$, such that $\sum\limits_{i=1}^{k} x_i \geq S_k = \sum\limits_{i=1}^{k} z_i$, and $\sum\limits_{i=1}^{k} x_i \leq A_{k-1}$ hold for $k = 1, 2, \ldots, 10$; moreover, $\sum\limits_{i=1}^{10} x_i = S_{10} = 45$. (This corresponds to the least possible value $G = 202G' + (x_{2021}^*)^2 = 202G'$.) Here, $S_1, \ldots, S_{10}$ are, respectively,

$$0, 1, 3, 6, 10, 15, 21, 28, 36, 45.$$

Consider a relaxed problem first: let $x_1, \ldots, x_{10} \geq 0, \sum_{i=1}^{10} x_i = 45$, and $S_k = \sum_{i=1}^{k} x_i \leq A_{k-1}$ for $k = 1, 2, \ldots, 9$. Find the minimum of $G'' = \sum_{i=1}^{10} x_i^2$. If the optimal solution also satisfies $D_k \geq S_k, k = 1, 2, \ldots, 9$, then $G'' = G'$. Define $A'_i = A_i (i = 0, 1, \ldots, 8), A'_9 = 45$.

First, as the lemma indicates, for G'' to be minimal, the sequence must satisfy $x_1 \leq x_2 \leq \cdots \leq x_{10}$; in addition, if $x_k < x_{k+1}$, then $S_k = A'_k$. We call this k a leap position. Let $k_0 = 0$ and $1 \leq k_1 < k_2 < \cdots < k_M < 10$ be the leap positions. Denote $k_{M+1} = 10$ and $A'_{-1} = 0$. For each $j \in \{1, 2, \ldots, M\}$, the numbers between the $(j-1)$th and the jth leap positions, that is, all $x_t (k_{j-1} + 1 \leq t \leq k_j)$ are equal, say to x_j. Since $x_j < x_{k_j+1}, x_k > x_j$ for all $k \geq k_j + 1$. Define

$$f_j(k) = \frac{A'_{k-1} - A'_{k_{j-1}-1}}{k - k_{j-1}}, \quad k = k_{j-1} + 1, \ldots, 10. \qquad ③$$

By definition, we have $D_{k_{j-1}} = A'_{k_{j-1}-1}$; as x_i is increasing,

$$x_j \leq \frac{1}{k - k_{j-1}} \sum_{i=k_{j-1}}^{k} x_i \leq \frac{1}{k - k_{j-1}} (A'_{k-1} - A'_{k_{j-1}-1}) = f_j(k),$$

where inequalities are strict for all $k \geq k_j + 1$. Hence $f_j(k), k = k_{j-1} + 1, \ldots, 10$, attains the minimum at $k = k_j$; in other words, k_j makes $f_j(k)$, $k = k_{j-1} + 1, \ldots,$ 10 minimal. If there are a few such k values, then k_j must be the largest one. Based on this observation, we can find k_j in the order $j = 1, 2, \ldots$. The sequence $A'_j, j = 0, 1, \ldots, 9$, is

$$3.3, 5.0, 7.5, 10.8, 14.9, 19.8, 25.5, 32.0, 39.3, 45.$$

Then, from ③, the first k should be $k_1 = 3$, and $x_1 = x_2 = x_3 = \frac{A'_2}{3} = 2.5$; the second one is $k_2 = 4$, and $x_4 = A'_3 - A'_2 = 3.3$, and so on, $k_3 = 5, x_5 = A'_4 - A'_3 = 4.1, k_4 = 6, x_6 = A'_5 - A'_4 = 4.9, k_5 = 7, x_7 = A'_6 - A'_5 = 5.7, k_6 = 10, x_8 = x_9 = x_{10} = 6.5$. So, an optimal solution for G'' is

$$(x'_1, \ldots, x'_{10}) = (2.5, 2.5, 2.5, 3.3, 4.1, 4.9, 5.7, 6.5, 6.5, 6.5).$$

For $k = 1, 2, \ldots, 7, D_k = A'_k > S_k$; for $k = 8, 9, 10$, it is

$$D_8 = 32 > S_8 = 28, \quad D_9 = 38.5 > S_9 = 36, \quad D_{10} = 45 = S_{10}.$$

The solution satisfies $D_k \geq S_k$ for $k = 1, 2, \ldots, 10$. Hence, it optimizes G' as well. The least possible value of G' is given by

$$\sum_{i=1}^{10} (x_i')^2 = 229.7.$$

Correspondingly, the least possible value of G is $202 \times 229.7 = 46399.4$.

□

Northern Star of Hope Mathematical Summer Camp

2021 (Tongliao, Inner Mongolia)

The 4th Northern Star of Hope Mathematical Summer Camp (NSHMSC) was held on July 25–29, 2021 at Tongliao No. 5 Senior High School, Tongliao, Inner Mongolia Autonomous Region. During this 2-day competition, each participant worked on a set of four problems (15 marks for each problem, 120 marks in total) on each day. In the end, there were 24 first prizes, 55 second prizes, and 28 third prizes. Tang Jinqi from the Beijing United Team won the first place in the individual score; teams from Jinan Licheng No. 2 High School, The High School Attached to Northeast Normal University, and Tsinghua University High School won the top three in the group scores.

NSHMSC, established under the initiative of Mr. Qiu Zonghu, former vice chairman of the Chinese Mathematical Olympiad Committee, has become a mathematics competition and exchange activity that is widely participated by top high schools in Northern China. The purpose of NSHMSC is to strengthen the exchange and cooperation of mathematics competitions in northern China, and to promote the high school mathematics competition level in this area. A total of 16 teams from Beijing, Hebei, Jilin, Liaoning, Inner Mongolia, Shandong, Shaanxi and Tianjin participated in the competition; most contestants are first year high school students.

The 4th NSHMSC committee chair: Liu Ruochuan (Peking University); committee members: Leng Fusheng (Chinese Academy of Sciences), Zou Jin (Gaosi Education), Jin Chunlai (Shenzhen Middle School), Tang Xiaoxu (The High School Affiliated to Renmin University of China), Wang Chao (Northeast Yucai School), Zhang Hongtao (Tongliao No. 5 Senior High School), Zhang Lu (Gaosi Education), Su Qizhou (Gaosi Education).

First Day
(8 – 12 pm; July 27, 2021)

1 Let E, F be two points outside of the parallelogram $ABCD$, such that $\triangle ABE \backsim \triangle CFB$; the lines DA and FB meet at G, the lines DC and EB meet at H, as shown in Fig. 1.1.
Prove that D, E, F, G, and H are concyclic.

(Contributed by Wang Chao)

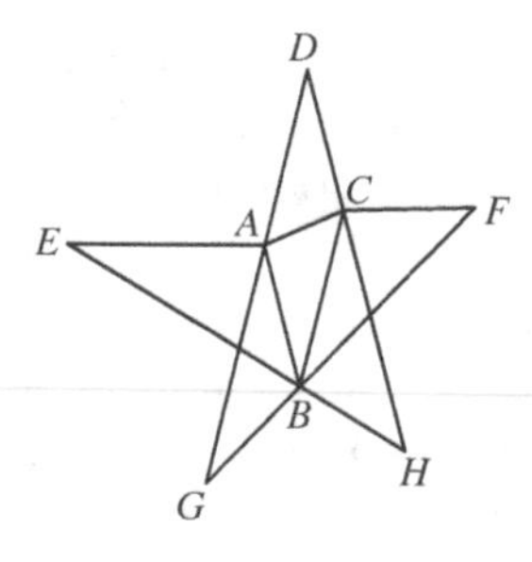

Fig. 1.1

Solution Connect EG and FH, as shown in Fig. 1.2.

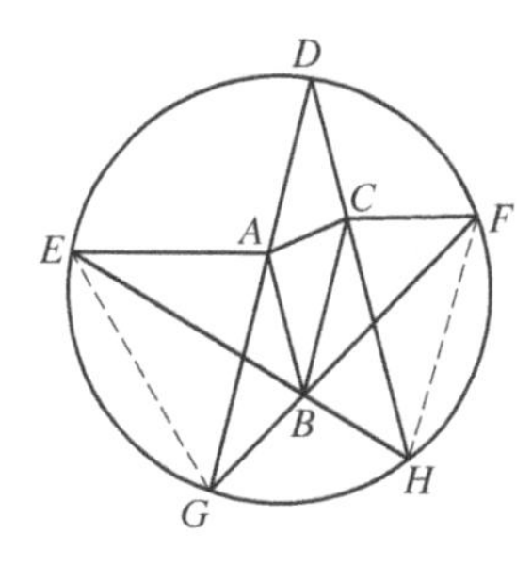

Fig. 1.2

Since $\triangle ABE \backsim \triangle CFB$, we have $\angle ABE = \angle CFB$, $\angle AEB = \angle CBF$; since $ABCD$ is a parallelogram, $AB // CD$, $BC // AD$. It follows that $\angle AEB = \angle CBF = \angle AGB$, and A, B, G, E are concyclic; $\angle CFB = \angle ABE = \angle CHD$, and B, C, F, H are concyclic. Consequently, $\angle GEB = \angle GAB = \angle GDH = \angle BCH = \angle BFH$, that is, $\angle GEH = \angle GDH = \angle GFH$, indicating that D, E, F, G, and H all lie on a circle. □

2 A primary school has several classes, each consisting of an equal number of boys and girls. The size of each class does not exceed $2n$, where n is a positive integer. The headmaster plans to establish a few activity groups, such that: (a) there are no more than $2n$ groups, and each student belongs to exactly one group; (b) any two students in a class belong to different groups; (c) for each group, the numbers of boys and girls differ by at most 1. Prove that the plan is always feasible.

(Contributed by Su Qizhou)

We give two solutions as follows.

Solution 1 Suppose there are k classes. Apply induction on n. When $n = 1$, each class consists of one boy and one girl. Let A, B be two activity groups: $\left\lceil \frac{k}{2} \right\rceil$ boys join A, and their classmates (girls) join B; the rest $\left\lfloor \frac{k}{2} \right\rfloor$ boys join B, and their classmates (girls) join A. Since $\left| \left\lceil \frac{k}{2} \right\rceil - \left\lfloor \frac{k}{2} \right\rfloor \right| \leq 1$, it meets the requirements.

Suppose when $n = t$ $(t \in \mathbb{Z}_+)$, the plan is feasible. For $n = t + 1$, each class has no more than $2(t+1)$ students. Choose one boy and one girl from each class, and let them join groups A and B as the case $n = 1$. Now each class has no more than $2t$ students who are to join no more than $2t$ groups. By the induction hypothesis, this is always doable. Clearly, the requirements are all met. This completes the induction and the proof. □

Solution 2 First, establish $2n$ activity groups with blank lists. We register one class at a time, until all the classes are registered, and show that the conditions are all satisfied at each time.

For the first class, let the $2n' \leq 2n$ students register in distinct groups. Clearly, the conditions are satisfied. Moreover, there are $n' \leq n$ groups with one more boy on the list, and $n' \leq n$ groups with one more girl on the list. Suppose a certain number of classes have registered, such that there are $t \leq n$ groups with one more boy and $t \leq n$ groups with one more girl (since there are equal numbers of boys and girls), and the other groups have equal numbers of boys and girls.

Now, for the next class, say with $2m$ students, let t boys register in the groups with one more girl, t girls register in the groups with one more boy, and the other students (if $m > t$) register in the groups with equal numbers of boys and girls. Evidently, conditions (a) and (b) are met, and there are $|t-m| \leq n$ groups with one more boy, the same number of groups with one more girl, all other groups with equal numbers of boys and girls, hence (c) is met, too.

Register every class in the above way, and the goal is achieved. □

3 For n real numbers $a_1, a_2, \ldots, a_n$, let

$$b_k = \frac{a_k + a_{k+1}}{2}, \quad c_k = \frac{a_{k-1} + a_k + a_{k+1}}{3} \quad (k = 1, 2, \ldots, n),$$

where $a_0 = a_n$, $a_{n+1} = a_1$. Find the maximum value of λ, $\lambda > 0$, such that the inequality

$$\sum_{k=1}^{n} (a_k - b_k)^2 \geq \lambda \cdot \sum_{k=1}^{n} (a_k - c_k)^2$$

holds for any $n \geq 3$ and $a_1, a_2, \ldots, a_n$.

(Contributed by Zhang Lu)

Solution 1 The maximum value of λ is $\frac{9}{16}$.

When $n = 4$, let $a_1 = 0$, $a_2 = 1$, $a_3 = 0$, $a_4 = 1$. We have

$$b_1 = b_2 = b_3 = b_4 = \frac{1}{2}, \quad c_1 = c_3 = \frac{2}{3}, \quad c_2 = c_4 = \frac{1}{3}, \quad \text{and} \quad 1 \geq \frac{16}{9}\lambda.$$

Hence, $\lambda \leq \frac{9}{16}$.

It suffices to prove: $\sum_{k=1}^{n} (a_k - b_k)^2 \geq \frac{9}{16} \sum_{k=1}^{n} (a_k - c_k)^2$.

In fact,

$$(a_k - c_k)^2 = \frac{1}{9}(a_{k-1} + a_{k+1} - 2a_k)^2 < \frac{2}{9}((a_{k-1} - a_k)^2 + (a_{k+1} - a_k)^2),$$

as $(a_k - b_k)^2 = \frac{1}{4}(a_k - a_{k+1})^2$, it follows that

$$\frac{9}{16}\sum_{k=1}^{n}(a_k - c_k)^2 \le \frac{1}{8}\sum_{k=1}^{n}((a_{k-1} - a_k)^2 + (a_{k+1} - a_k)^2)$$

$$= \frac{1}{4}\sum_{k=1}^{n}(a_k - a_{k+1})^2 = \sum_{k=1}^{n}(a_k - b_k)^2.$$

Therefore, the maximum value of λ is $\frac{9}{16}$. □

Solution 2 The proof for $\lambda \le \frac{9}{16}$ is the same as in the first solution. Another proof for $\sum_{k=1}^{n}(a_k - b_k)^2 \ge \frac{9}{16}\sum_{k=1}^{n}(a_k - c_k)^2$ is given as follows.

$$\begin{aligned}
&\sum_{k=1}^{n}(a_k - b_k)^2 - \frac{9}{16}\sum_{k=1}^{n}(a_k - c_k)^2 \\
&= \frac{1}{4}\sum_{k=1}^{n}(a_k - a_{k+1})^2 - \frac{1}{16}\sum_{k=1}^{n}(a_{k-1} + a_{k+1} - 2a_k)^2 \\
&= \frac{1}{4}\sum_{k=1}^{n}(a_k^2 + a_{k+1}^2 - 2a_k a_{k+1}) \\
&\quad - \frac{1}{16}\sum_{k=1}^{n}(a_{k-1}^2 + a_{k+1}^2 + 4a_k^2 - 4a_{k-1}a_k - 4a_k a_{k+1} + 2a_{k-1}a_{k+1}) \\
&= \frac{1}{8}\sum_{k=1}^{n}a_k^2 - \frac{1}{8}\sum_{k=1}^{n}a_{k-1}a_{k+1} \\
&= \frac{1}{16}\sum_{k=1}^{n}(a_{k-1} - a_{k+1})^2 \ge 0.
\end{aligned}$$

□

4 We call $S \subseteq \mathbb{N}_+$ a "nice set", if there exists $f\colon S \to \mathbb{N}_+$, such that for any a, b, $c \in S$, $\{a, b, c\}$ is a geometric sequence if and only if $\{f(a), f(b), f(c)\}$ is an arithmetic sequence.

(1) Prove that $\mathbb{N}_+$ is not a nice set;

(2) Is $T = \{x \in \mathbb{N}_+ \,|\, \text{for any prime } p,\ p^{2021} \text{ does not divide } x\}$ a nice set? Give your reasons.

(Contributed by Zhang Lu)

Solution

(1) Assume that $\mathbb{N}_+$ is a nice set and $f\colon \mathbb{N}_+ \to \mathbb{N}_+$ has the desired property. If x_1, x_2 satisfy $f(x_1) = f(x_2)$, then $\{f(x_1), f(x_1), f(x_2)\}$ is an arithmetic sequence, which implies that $\{x_1, x_1, x_2\}$ is a geometric sequence, and $x_1 = x_2$. So, f is one-to-one.

Notice that $\{1, 2, 4, \ldots, 2^k, \ldots\}$ is an infinite geometric sequence of positive integers. As the problem indicates, $\{f(1), f(2), f(4), \ldots, f(2^k), \ldots\}$ must be an infinite arithmetic sequence of positive integers. Let $f(2^k) = f(1) + kd_2$, where the common difference d_2 is a positive integer.

Likewise, we can let $f(3^k) = f(1) + kd_3$, where d_3 is a positive integer. Then

$$f(2^{d_3}) = f(1) + d_3 \cdot d_2 = f(3^{d_2}).$$

Since f is one-to-one, $2^{d_3} = 3^{d_2}$, but this is impossible.

Hence, $\mathbb{N}_+$ is not a nice set.

(2) T is a nice set.

Let all prime numbers be $p_1, p_2, p_3, \ldots$. For any positive integer $n \in T$, let $n = \prod_{i=1}^{\infty} p_i^{x_i}$ be the prime factorization, in which $x_i \leq 2020$ is a nonnegative integer for each i, and only finitely many x_i are not 0. Define

$$f(n) = 1 + \sum_{i=1}^{\infty} x_i \cdot 4041^i.$$

We show that f satisfies the problem condition.

Clearly, $f\colon T \to \mathbb{N}_+$, and for any three integers in T,

$$a = \prod_{i=1}^{\infty} p_i^{\alpha_i}, \quad b = \prod_{i=1}^{\infty} p_i^{\beta_i}, \quad c = \prod_{i=1}^{\infty} p_i^{\gamma_i},$$

$\{a, b, c\}$ is geometric if and only if $\alpha_i + \gamma_i = 2\beta_i$ holds for every positive integer i. On the other hand, $\{f(a), f(b), f(c)\}$ is arithmetic if and only if

$$\sum_{i=1}^{\infty} \alpha_i \cdot 4041^i + \sum_{i=1}^{\infty} \gamma_i \cdot 4041^i = 2\sum_{i=1}^{\infty} \beta_i \cdot 4041^i,$$

that is,

$$\sum_{i=1}^{\infty}(\alpha_i + \gamma_i) \cdot 4041^i = \sum_{i=1}^{\infty} 2\beta_i \cdot 4041^i. \qquad ①$$

Notice that both $\alpha_i + \gamma_i$ and $2\beta_i$ are nonnegative integers less than or equal to 4040. We can see that ① gives two representations of the same number in base 4041, which must agree. Hence, ① holds if and only if $\alpha_i + \gamma_i = 2\beta_i$ is true for every positive integer i. It turns out that $\{a, b, c\}$ is geometric if and only if $\{f(a), f(b), f(c)\}$ is arithmetic. Thus, f satisfied the problem condition and T is a nice set. □

Second Day
(8 – 12 pm; July 28, 2021)

5 Find the largest positive integer n, such that there exist n positive integers $x_1 < x_2 < \cdots < x_n$ satisfying

$$x_1 + x_1x_2 + \cdots + x_1x_2\cdots x_n = 2021. \qquad (*)$$

(Contributed by Zhang Lu)

Solution The largest n is 4.

When $n = 4$, take $x_1 = 1$, $x_2 = 4$, $x_3 = 6$, $x_4 = 83$, and (*) holds.

Claim When $n \geq 5$, (*) has no positive integer solutions.

Proof of claim Suppose on the contrary that $n \geq 5$, $x_1 < x_2 < \cdots < x_n$ satisfy (*). From $2021 > x_1x_2\cdots x_n \geq x_1x_2x_3x_4x_5 > x_1^5$, it follows that $x_1 < 5$; also, $x_1 \mid 2021$. Hence, $x_1 = 1$. Now,

$$x_2 + x_2x_3 + \cdots + x_2x_3\cdots x_n = 2020.$$

From $2020 > x_2x_3\cdots x_n \geq x_2x_3x_4x_5 > x_2^4$, we have $x_2 < 10$; also, $x_2 \mid 2020$. Hence, $x_2 = 2$, 4 or 5.

If $x_2 = 2$, $x_3 + x_3x_4 + \cdots + x_3x_4\cdots x_n = 1009$, $x_3 > 2$.

From $1009 > x_3x_4\cdots x_n \geq x_3x_4x_5 > x_3^3$, we have $x_3 < 11$, yet $x_3 \mid 1009$ has no integer solutions.

If $x_2 = 5$, $x_3 + x_3x_4 + \cdots + x_3x_4\cdots x_n = 403$, $x_3 > 5$.

From $403 > x_3x_4\cdots x_n \geq x_3x_4x_5 > x_3^3$, we have $x_3 < 8$, yet $x_3 \mid 403$ has no integer solutions.

If $x_2 = 4$, $x_3 + x_3x_4 + \cdots + x_3x_4\cdots x_n = 504$, $x_3 > 4$.

From $504 > x_3x_4\cdots x_n \geq x_3x_4x_5 \geq x_3(x_3+1)(x_3+2)$, we have $x_3 < 7$; also, $x_3 \mid 504$. Thus, $x_3 = 6$, $x_4 + x_4x_5 + \cdots + x_4x_5\cdots x_n = 83$. However, 83 is a prime number, $x_4 < x_5$, indicating that $x_4 = 1$, a contradiction.

The Claim is verified, and the largest n is 4. □

6 For real numbers $x, y, z \in [0,1]$, let M be the minimum of the three numbers $x - xy - xz + yz$, $y - yx - yz + xz$, and $z - zx - zy + xy$. Find the maximum possible value of M.

(Contributed by Zhang Hongtao)

Solution The maximum possible value of M is $\frac{1}{4}$.

First, when $x = y = z = \frac{1}{2}$, $M = \min\left\{\frac{1}{4}, \frac{1}{4}, \frac{1}{4}\right\} = \frac{1}{4}$. In the following, we show $M \leq \frac{1}{4}$.

Note that among the three numbers $x - \frac{1}{2}$, $y - \frac{1}{2}$, $z - \frac{1}{2}$, two are nonnegative or two are non-positive, say they are $x - \frac{1}{2}$ and $y - \frac{1}{2}$, $\left(x - \frac{1}{2}\right)\left(y - \frac{1}{2}\right) \geq 0$. This implies that

$$\begin{aligned} 2M &\leq (x - xy - xz + yz) + (y - yx - yz + xz) = x + y - 2xy \\ &= -2\left(x - \frac{1}{2}\right)\left(y - \frac{1}{2}\right) + \frac{1}{2} \leq \frac{1}{2}, \end{aligned}$$

and thus $M \leq \frac{1}{4}$. □

7 As shown in Fig. 7.1, A, B, C, D, and E all lie on the circle ω. It is known that $AB // CE$, and AD bisects the segment BE. The tangent of circle ω through B meets the line CA at F. Prove: $FC = FD$.

(Contributed by Jin Chunlai)

We give three solutions as follows.

Solution 1 Let the line DF and circle ω meet at D and G. Connect AE, DE, BC, BD, BG, and CG, as shown in Fig. 7.2.

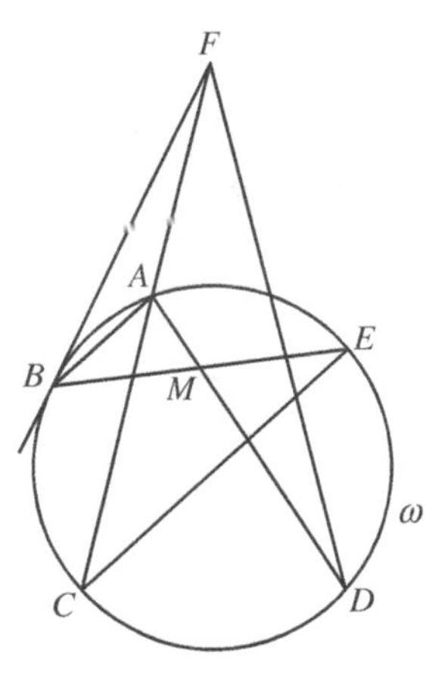

Fig. 7.1

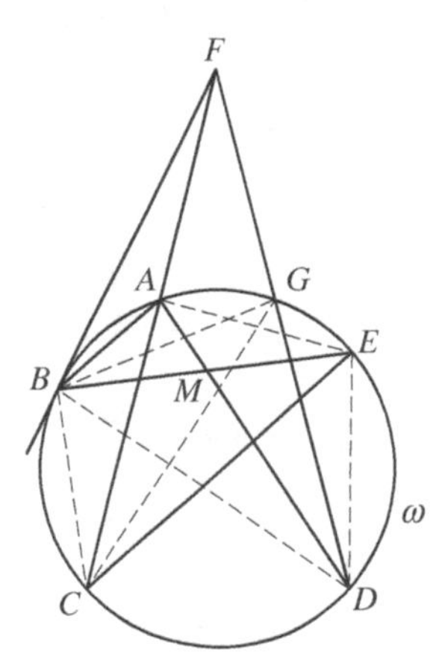

Fig. 7.2

From the given conditions, it is straightforward to find $\triangle FBG \backsim \triangle FDB$, $\triangle FAD \backsim \triangle FGC$, $\triangle FAB \backsim \triangle FBC$, which further imply that

$$\frac{BG}{BD} = \frac{BF}{DF}, \quad \frac{AD}{CG} = \frac{DF}{CF}, \quad \frac{BC}{AB} = \frac{CF}{BF}.$$

Multiply the three equations to get

$$\frac{BG}{BD} \cdot \frac{AD}{CG} \cdot \frac{BC}{AB} = \frac{BF}{DF} \cdot \frac{DF}{CF} \cdot \frac{CF}{BF} = 1,$$

or

$$\frac{BG}{CG} = \frac{BD}{AD} \cdot \frac{AB}{BC}. \qquad ①$$

Let AD and BE intersect at M, $BM = EM$.

Note that $\triangle ABM \backsim \triangle EDM$, $\triangle AEM \backsim \triangle BDM$. Hence, $\frac{AB}{DE} = \frac{BM}{DM}$, $\frac{AE}{BD} = \frac{EM}{DM}$. So we have $\frac{AB}{DE} = \frac{AE}{BD}$, $DE = BD \cdot \frac{AB}{AE}$, and

$$\frac{DE}{AD} = \frac{BD}{AD} \cdot \frac{AB}{AE}. \qquad ②$$

In the meantime, from $AB // CE$, we find $BC = AE$, together with ①, ②,

$$\frac{BG}{CG} = \frac{DE}{AD}.$$

That $BC = AE$ also leads to $\angle BGC = \angle EDA$, $\triangle BGC \backsim \triangle EDA$, $\angle CBG = \angle AED$, and $CG = AD$. Finally, from $\triangle FAD \backsim FGC$, we deduce $\frac{CF}{DF} = \frac{CG}{AD} = 1$, namely, $CF = DF$. □

Solution 2 As shown in Fig. 7.3, let O be the centre of circle ω, and M be the intersection of AD and BE. It is known that M is the midpoint of BE.

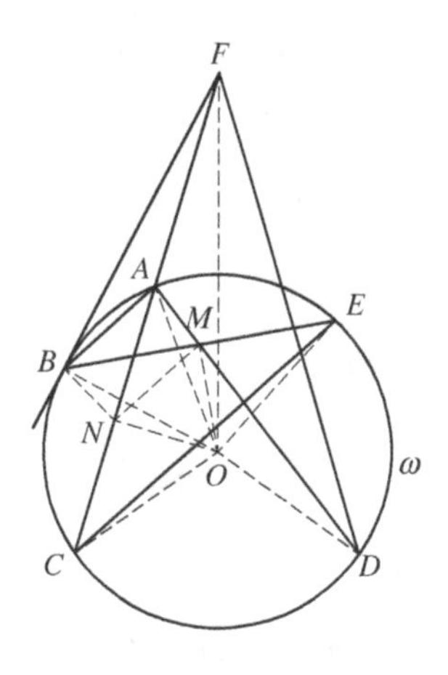

Fig. 7.3

Let N be the midpoint of AC. Connect OA, OB, OC, OD, OE, OF, OM, ON, BN, and MN. Since $AB // CE$, $AC = BE$; also, $OA = OB = OC = OE$, thus $\triangle OAC \cong \triangle OBE$.

Since M, N are the midpoints of BE, AC, respectively, we have $\triangle OAN \cong \triangle OBM$. This indicates that $OM = ON$, $\angle AON = \angle BOM$, $\angle BON = \angle AOM$. Thus, $\triangle OBN \cong \triangle OAM$, which gives

$$\angle OBN = \angle OAM. \quad ①$$

On the other hand, BF touches circle ω at B, $\angle OBF = 90°$; N is the midpoint of AC, $\angle ONF = 90°$. It turns out that $\angle OBF = \angle ONF$, and O, N, B, F are concyclic, which implies

$$\angle OBN = \angle OFN. \quad ②$$

Combine ① and ② to obtain $\angle OFN = \angle OAM$; also, $\angle OAM = \angle ODA$, hence

$$\angle OFN = \angle ODA,$$

and O, A, F, D are concyclic. Then

$$\angle OCF = \angle OCA = \angle OAC = \angle ODF; \quad ③$$

in addition, $OA = OD$, which implies $\angle OFA = \angle OFD$, that is,

$$\angle OFC = \angle OFD. \quad ④$$

By ③, ④, and $OC = OD$, it follows $\triangle OCF \cong \triangle ODF$. Hence, $CF = DF$.

Solution 3 As shown in Fig. 7.4, take O as the centre of circle ω. Connect OA, OC, OB, OD, and OE.

Let AD and BE meet at M. It is given that M is the midpoint of BE.

Let the tangent of circle ω through E intersect BF at S. Clearly, S, M, O are collinear; O, B, S, E are concyclic. Since A, B, D, E are concyclic, we find

$$OM \cdot SM = BM \cdot EM = AM \cdot DM,$$

which implies that O, A, S, D are concyclic.

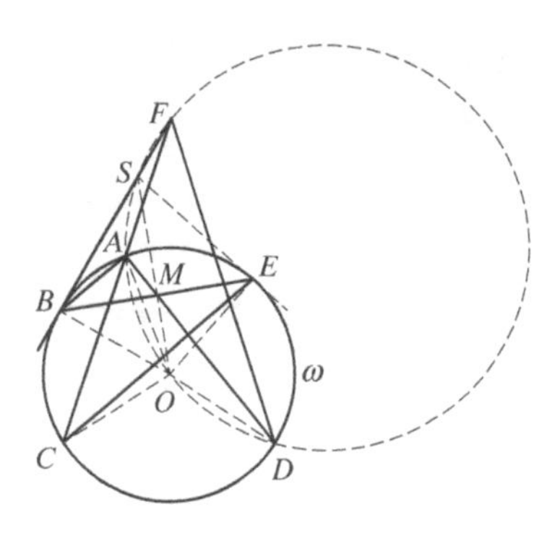

Fig. 7.4

Furthermore, as O, B, S, E are concyclic, $OB = OE$, and $AC = BE$, it follows

$$\angle OAC = \angle OEB = \angle OSB, \angle OAF = \angle OSF,$$

and O, A, S, F are concyclic. Therefore, O, A, S, F, D are all concyclic. The rest is the same as in the second proof. □

8 Given 2021 distinct prime numbers $p_1, p_2, \ldots, p_{2021}$, let $S = \left\{\frac{p_i}{p_j} \mid i, j \in \{1, 2, \ldots, 2021\},\ i \neq j\right\}$. Alice and Bob play the following game: they take turns to write a number in S on the blackboard, the number must be different from any number already written, and the game ends when someone writes a number such that the product of several (distinct) numbers on the blackboard equals 1, with that person being the loser. If Alice writes first, who has a winning strategy? Explain why.

(Contributed by Leng Fusheng)

Solution Bob has a winning strategy. We give two solutions as follows.

Solution 1 Consider $\frac{a}{b}$ and $\frac{b}{a}$ as a pair. There are $\binom{2021}{2}$ pairs in S.

If there are more than $\binom{2021}{2}$ numbers on the blackboard, by the pigeonhole principle, two numbers in a certain pair are both written, and the game must end instantly.

Claim If there are fewer than $\binom{2021}{2}$ numbers on the blackboard, then a player can always write another number from an unwritten pair without losing the game.

Proof of claim We use the language of graph theory. Let $p_1, p_2, \ldots, p_{2021}$ be the vertices of a directed graph G. If $\frac{p_i}{p_j}$ is written, then add an edge $p_i \to p_j$ to G. As the written numbers are distinct, there is at most one directed edge from p_i to p_j. The game ends when G has a directed cycle: the product of the numbers corresponding to the edges in the cycle is equal to 1, as each involved prime number appears exactly once in the numerator and once in the denominator. Conversely, if the product of several numbers written on the blackboard is 1, consider all the prime numbers appearing in the product and the corresponding vertices in G: each vertex has the equal indegree and outdegree. Clearly, G has a directed cycle.

If the number of edges in G is less than $\binom{2021}{2}$, some two vertices say p_i and p_j are not adjacent by an edge (the pair $\frac{p_i}{p_j}$ and $\frac{p_j}{p_i}$ is unwritten). Then directed paths from p_i to p_j and from p_j to p_i cannot exist at the same time. Indeed, if this is possible, then a directed path is formed from p_i to p_j then to p_i. Start from any vertex say p_{k_1} and move along the path until it reaches a vertex already visited, say $p_{k_j} \to p_{k_i}$ $(j > i)$, then $p_{k_i} \to p_{k_{i+1}} \to \cdots \to p_{k_j} \to p_{k_i}$ is a directed cycle (no vertex is repeated). This is contradictory. Suppose G does not contain a directed path from p_i to p_j. A player can safely add the edge $p_j \to p_i$, or write $\frac{p_j}{p_i}$ on the blackboard without losing the game. The claim is verified.

Note that $\binom{2021}{2} = 2021 \cdot 1010$ is an even number. When $\binom{2021}{2}$ numbers are written, all $\binom{2021}{2}$ pairs of vertices of G are adjacent. It is Alice's turn and she must add another edge and lose the game. $\square$

Solution 2 Here is another proof of the claim. Replace 2021 by n and it is generally true that: if a directed, cycle-free graph G has fewer than $\binom{n}{2}$ edges, then one edge can be added to G such that the resulting graph is still cycle-free.

We use induction on n. When $n = 2$, the claim is obviously true. Assume it is true when $n = k$. Consider a directed, cycle-free graph G with $n = k + 1$ vertices. If every vertex has positive indegree and outdegree, then we choose any vertex p_{k_1} and move along an outward edge to p_{k_2}, and so on, until $p_{k_j} \to p_{k_i} (j > i)$ where p_{k_i} has already been visited. Then $p_{k_i} \to p_{k_{i+1}} \to \cdots \to p_{k_j} \to p_{k_i}$ is a directed cycle, a contradiction. Thus, G must have a vertex v whose indegree or outdegree is 0, say the latter.

If there exists a vertex u not adjacent to v, then adding the edge from u to v will suffice. Otherwise, there is a directed edge from each vertex to v, for a total of $n - 1 = k$ edges. Remove v and the edges incident to it. There are k vertices and fewer than $\binom{k+1}{2} - k = \binom{k}{2}$ edges left. By the induction hypothesis, adding an edge to them will not generate a cycle within them, or involving v. This completes the induction. $\square$

International Mathematical Olympiad

2021 (Saint Petersburg, Russia)

The 62nd International Mathematical Olympiad (IMO) was held from July 14th to July 24th, 2021 in Saint Petersburg, Russia. A total of 619 contestants from 107 countries or regions participated the competition. Due to the pandemic, the competition was conducted online. In order to ensure the confidentiality and fairness of the competition, each participating country or region set up an independent test centre according to the requirements of the organizing committee. In addition to remote video proctoring, neutral IMO commissioners conducted on-site proctoring. At three hours before the test, the test problems were sent to the team leader and observer A for translation; at 30 minutes before the test, the confirmed test problems were sent to the IMO commissioner for printing and distributing to the contestants. The organizing committee required that each participating team must select the starting time between 7:30 am and 12 noon GMT. The Chinese team chose to start at 7:30 am GMT, or 15:30 Beijing time.

The opening ceremony of 62nd IMO was held online on July 18th, 2021. After two tests on July 19th and 20th, the Chinese team won the group first place with a total of 208 marks. In the second day test, all six members received full scores. Wang Yichuan was the only contestant in this IMO who received full score (42 marks). All the Chinese team members won gold medals and entered the top 20.

The Chinese team:

Leader: Xiao Liang (Peking University)

Deputy leader: Qu Zhenhua (East China Normal University)

Observers: Xiong Bin (East China Normal University), Fu Yunhao (Southern University of Science and Technology), Wang Bin (Institute of Mathematics and Systems Science, Chinese Academy of Sciences)

Team members:

Wang Yichuan (No. 2 High School of East China Normal University), 12th grade, 42 marks, gold medalist

Peng Yebo (Shenzhen Middle School), 11th grade, 36 marks, gold medalist

Wei Chen (Beijing National Day School), 12th grade, 36 marks, gold medalist

Xia Yuxing (No. 1 Middle School affiliated to Central China Normal University), 11th grade, 32 marks, gold medalist

Chen Ruitao (The High School Affiliated to Renmin University of China), 11th grade, 31 marks, gold medalist

Feng Chenxu (Shenzhen Middle School), 10th grade, 31 marks, gold medalist

The top 10 teams in the total scores:

1.	China	208 marks
2.	Russia	183 marks
3.	South Korea	172 marks
4.	United States	165 marks
5.	Canada	151 marks
6.	Ukraine	149 marks
7.	Israel	139 marks
7.	Italy	139 marks
9.	Chinese Taipei	131 marks
9.	UK	131 marks

Gold medal: score $\geq$24 marks;

Silver medal: score $\geq$19 marks;

Bronze medal: score $\geq$12 marks.

From July 13th to July 17th, the Chinese team was trained at School of Mathematical Sciences and Beijing International Centre for Mathematical Research, Peking University. Academician Tian Gang, Chairman of the Chinese Mathematical Society; Professor Gong Fuzhou, Secretary-General of the Chinese Mathematical Society; Professor Peng Liangang, Director of the Mathematical Competition Committee of the Chinese Mathematical Society; Professor Chen Dayue, Dean of School of Mathematical Sciences, Peking University; Professor Sun Zhaojun, Associate Dean; Professor Wu Jianping, Professor Liu Bin, had guided and discussed with the students.

First Day
(15:30 – 20:00; July 19, 2021)

Problem 1. Let $n > 100$ be an integer. Ivan writes the numbers $n, n + 1, \ldots, 2n$ each on different cards. He then shuffles these $n + 1$ cards, and divides them into two piles. Prove that at least one of the piles contains two cards such that the sum of their numbers is a perfect square.

(Contributed by Australia)

Solution It suffices to find three integers $a, b, c \in [n, 2n]$, such that for some integer k,

$$a + b = (2k - 1)^2, \quad a + c = (2k)^2, \quad b + c = (2k + 1)^2.$$

In this way, two of a, b, c must be put into one pile, and their sum is a perfect square. Solving the equations yields $a = 2k^2 - 4k$, $b = 2k^2 + 1$, $c = 2k^2 + 4k$. We must require that $n \leq 2k^2 - 4k$ and $2k^2 + 4k \leq 2n$. Let $2m^2 - 2 < n \leq 2(m+1)^2 - 2$, where m is an integer. From the inequalities, we find that k must satisfy $m + 2 \leq k \leq \sqrt{n+1} - 1$. If $m = 7$, $n \geq 99$ is needed; if $m \geq 8$, $\sqrt{n+1} - 1 \geq \sqrt{2}m - 1 > m + 2$ always hold. Hence, the conclusion is true for $n \geq 99$. □

Remark The statement is untrue when $n = 98$. For example, Ivan can put all even numbers between 98 and 126, all odd numbers between 129 and 161, and all even numbers between 162 and 196, into the first pile; he puts the other numbers into the second pile. No two numbers in a pile add up to a perfect square.

Problem 2. Show that the inequality

$$\sum_{i=1}^{n}\sum_{j=1}^{n}\sqrt{|x_i - x_j|} \le \sum_{i=1}^{n}\sum_{j=1}^{n}\sqrt{|x_i + x_j|} \qquad ①$$

holds for all real numbers $x_1, \ldots, x_n$.

(Contributed by Canada)

Solution We apply induction on n. When $n = 0$ or $n = 1$, the inequality is obviously true; suppose ① holds for fewer than n variables. Now consider ① for n variables. If $x_i = 0$, then the terms on two sides that involve x_i are equal, and by removing x_i from $\{x_1, x_2, \ldots, x_n\}$ we can reduce it to $n - 1$ variables. Similarly, if $x_i = -x_j \neq 0$, then the terms involving x_i or x_j are

$$2\sqrt{2|x_i|} + \sum_{\substack{k=1,\ldots,n \\ k\neq i,j}} \left(\sqrt{|x_k - x_i|} + \sqrt{|x_k + x_i|}\right)$$

on both sides. By removing x_i and x_j from $\{x_1, x_2, \ldots, x_n\}$ we can reduce it to $n - 2$ variables. In either case, the induction hypothesis leads to the conclusion.

In the following, assume $x_i + x_j \neq 0$ for all $i, j \in \{1, 2, \ldots, n\}$ (possibly $i = j$). We wish to apply a translation to all variables $x_i \mapsto x_i + a$ such that some two variables add up to 0. Note that a translation does not change the value on the left-hand side of ①.

If there exists $a > 0$ such that $x_i + x_j + a = 0$ for some $i, j \in \{1, 2, \ldots, n\}$, denote the minimal of all such a as a_+; otherwise, denote $a_+ = \infty$. If there exists $a < 0$ such that $x_i + x_j + a = 0$ for some $i, j \in \{1, 2, \ldots, n\}$, denote the maximal of all such a as a_-; otherwise, denote $a_- = -\infty$. Since the functions $\sqrt{x}$ and $\sqrt{-x}$ are concave down in their natural domains, it follows that for every i, j, the function $\sqrt{x_i + x_j + a}$ is concave down in $[a_-, a_+]$, and likewise for the sum $\sum_{i,j=1}^{n} \sqrt{x_i + x_j + a}$. Hence, for $b = a_+$ or $b = a_-$ ($b \neq \pm\infty$),

$$\sum_{i,j=1}^{n} \sqrt{|x_i + x_j|} \ge \sum_{i,j=1}^{n} \sqrt{|x_i + x_j + b|}. \qquad ②$$

(Note: if $a_+ = \infty$ or $a_- = -\infty$, the sum is infinite and ② does not hold.)

Take $y_i = x_i + \frac{b}{2}$. Then for some i, j, $y_i + y_j = 0$. We have

$$\begin{aligned}\sum_{i=1}^{n}\sum_{j=1}^{n}\sqrt{|x_i - x_j|} &= \sum_{i=1}^{n}\sum_{j=1}^{n}\sqrt{|y_i - y_j|}\\ &\leq \sum_{i=1}^{n}\sum_{j=1}^{n}\sqrt{|y_i + y_j|}\\ &\leq \sum_{i=1}^{n}\sum_{j=1}^{n}\sqrt{|x_i + x_j|}\end{aligned}$$

which is ①. Here, the first inequality comes from the induction hypothesis (as $y_i + y_j = 0$). This completes the induction. □

Problem 3. Let D be an interior point of the acute triangle ABC with $AB > AC$ so that $\angle DAB = \angle CAD$. The point E on the segment AC satisfies $\angle ADE = \angle BCD$, the point F on the segment AB satisfies $\angle FDA = \angle DBC$, and the point X on the line AC satisfies $CX = BX$. Let O_1 and O_2 be the circumcentres of the triangles ADC and EXD, respectively. Prove that the lines BC, EF, and O_1O_2 are concurrent.

(Contributed by Ukraine)

Solution Let Q be the isogonal conjugate of point D in $\triangle ABC$. As $\angle BAD = \angle DAC$, Q is on AD, and $\angle QBA = \angle DBC = \angle FDA$. It follows that Q, D, F, B are concyclic; likewise, Q, D, E, C are concyclic. We have $AF \cdot AB = AD \cdot AQ = AE \cdot AC$, and thereby B, F, E, C concyclic, as illustrated in Fig. 3.1.

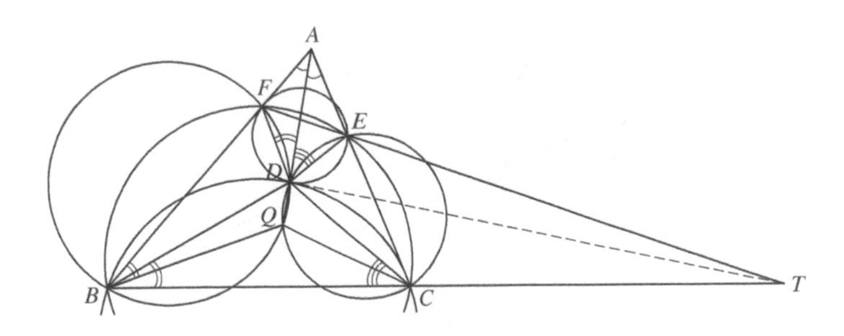

Fig. 3.1

Lemma *Let the extensions of BC and FE beyond C and E meet at T. Then $TD^2 = TB \cdot TC = TF \cdot TE$.*

Proof of lemma First, $\odot DEF$ and $\odot BDC$ are tangent to each other, since

$$\begin{aligned}\angle BDF &= \angle AFD - \angle ABD \\ &= (180^\circ - \angle FAD - \angle FDA) - (\angle ABC - \angle DBC) \\ &= 180^\circ - \angle FAD - \angle ABC \\ &= 180^\circ - \angle DAE - \angle FEA \\ &= \angle FED + \angle ADE \\ &= \angle FED + \angle DCB.\end{aligned}$$

Next, as B, C, E, F are concyclic, the powers of T with respect to $\odot BDC$ and $\odot EDF$ are equal. Therefore, the radical axis through T is the common tangent at D, $TD^2 = TB \cdot TC = TF \cdot TE$. The lemma is verified.

As illustrated in Fig. 3.2, let the line TA and $\odot ABC$ intersect at another point M. Notice that B, C, E, F are concyclic, A, M, C, B are concyclic, and by the lemma, we find

$$TM \cdot TA = TF \cdot TE = TB \cdot TC = TD^2,$$

implying that A, M, E, F are concyclic.

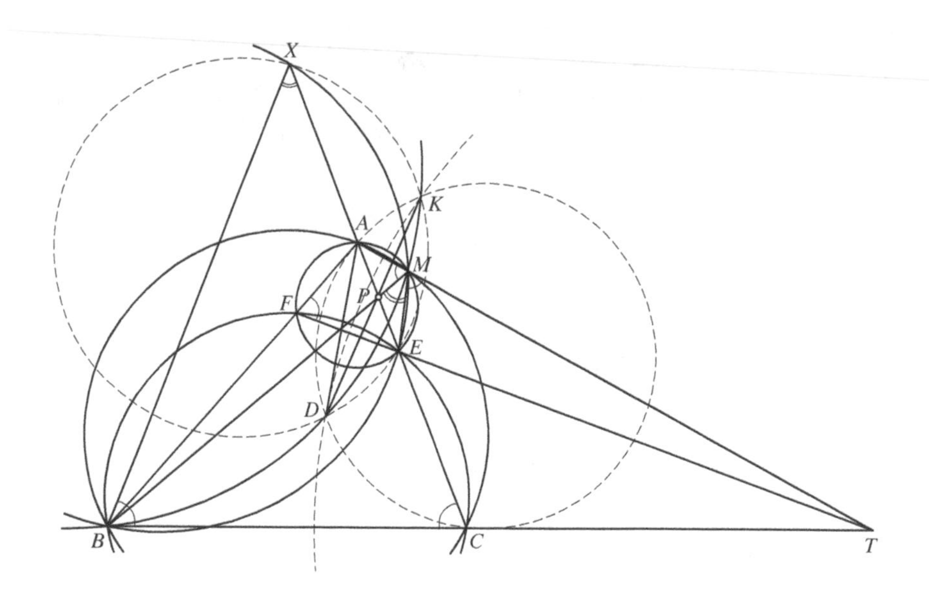

Fig. 3.2

Consider the inversion transformation centred at T of radius TD: M is mapped to A, B is mapped to C, and $\odot MBD$ is mapped to $\odot ACD$. As their common point D lies on the inversion circle, so does the other common point K, $TK = TD$. It follows that T and the centres of $\odot KDE$ and $\odot ADC$ all lie on the perpendicular bisector of KD.

As O_1 is the circumcentre of $\triangle ADC$, it remains to prove D, K, E, X lie on a circle, whose centre is O_2.

Observe that BM, DK, and AC are the radical axes of pairs from $\odot ABCM$, $\odot ACDK$, and $\odot BMDK$, respectively; they are concurrent at P. In addition, M lies on $\odot AEF$. Thus,

$$\begin{aligned}\measuredangle(EX, XB) &= \measuredangle(CX, XB) = \measuredangle(XC, BC) + \measuredangle(BC, BX)\\ &= 2\measuredangle(AC, CB) = \measuredangle(AC, CB) + \measuredangle(EF, FA)\\ &= \measuredangle(AM, BM) + \measuredangle(EM, MA)\\ &= \measuredangle(EM, BM).\end{aligned}$$

This implies that M, E, X, B are concyclic, and hence

$$PE \cdot PX = PM \cdot PB = PK \cdot PD.$$

Therefore, E, K, D, and X are concyclic. □

Second Day
(15:30 – 20:00; July 20, 2021)

Problem 4. Let Γ be a circle with centre I, and $ABCD$ be a convex quadrilateral such that each of the segments AB, BC, CD and DA is tangent to Γ. Let Ω be the circumcircle of the triangle AIC. The extension of BA beyond A meets Ω at X, and the extension of BC beyond C meets Ω at Z. The extensions of AD and CD beyond D meet Ω at Y and T, respectively. Prove that

$$AD + DT + TX + XA = CD + DY + YZ + ZC.$$

(Contributed by Poland)

Solution To begin, notice that I is the intersection of the external angle bisector of $\angle TCZ$ and the circumcircle Ω of $\triangle TCZ$. Thus, I is the midpoint of the arc TCZ, $IT = IZ$. Likewise, I is the midpoint of the arc YAX, $IX = IY$. Let O be the centre of Ω. The pairs X and Y, T and Z are symmetric about the line IO, hence $XT = YZ$.

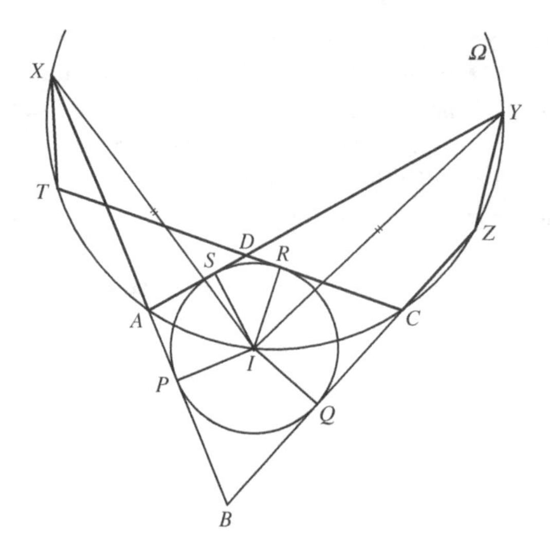

Fig. 4.1

Let the inscribed circle of $ABCD$ touch the four sides at P, Q, R, and S, respectively, as illustrated in Fig. 4.1. Since $IP = IS$, $IX = IY$, the right triangles IXP and IYS are congruent. Likewise, the right triangles IRT and IQZ are congruent. They imply $XP = YS$, $RT = QZ$.

Finally, from $AS = AP$, $CQ = RC$, $SD = DR$, we obtain

$$\begin{aligned} AD + DT + TX + XA &= TX + XP + RT = YZ + SY + QZ \\ &= CD + DY + YZ + ZC, \end{aligned}$$

as desired. □

Problem 5. Two squirrels, Bushy and Jumpy, have collected 2021 walnuts for the winter. Jumpy numbers the walnuts from 1 through 2021, and digs 2021 little holes in a circular pattern in the ground around their favorite tree. The next morning Jumpy notices that Bushy had placed one walnut into each hole, but had paid no attention to the numbering. Unhappy, Jumpy decides to reorder the walnuts by performing a sequence of 2021 moves. In the kth move, Jumpy swaps the positions of the two walnuts adjacent to walnut k. Prove that there exists a value of k such that, in the kth move, Jumpy swaps some walnuts a and b such that $a < k < b$.

(Contributed by Spain)

Solution 1 Assume that the statement is untrue. In the kth move, if Jumpy swaps a and b with $a, b < k$, then we say k is a "big" walnut; if $a, b > k$, then we say k is a "small" walnut. After the kth move, we call every walnut with number $a < k$ *operated* and every walnut with number $b > k$ *unoperated*.

At every moment, the operated walnuts are divided by the unoperated walnuts into several intervals (each interval consists of consecutive operated walnuts and the neighbours on the two ends are unoperated ones). We use induction to prove that, after k moves, each interval contains an odd number of walnuts. For $k = 1$, the assertion is obvious. In the kth move, there are two possibilities:

(i) If k is a small walnut, then its neighbours are both unoperated before and after the move, k itself is an interval of length 1.
(ii) If k is a big walnut, then its neighbours are both operated, say they belong to intervals of lengths p and q. After the move, they merge into one interval of length $p+q+1$. By the induction hypothesis, p, q are both odd, and the new interval's length $p+q+1$ is also odd.

Now the induction is complete and the assertion is verified. Notice that after 2020 moves, the 2020 operated walnuts form a single interval of length 2020 which is an even number. This contradiction indicates that during the process, Jumpy must have swapped walnuts a, b adjacent to k such that $a < k < b$. □

Solution 2 As in the first proof, we assume the statement is untrue and define the big and small walnuts in the same way. In the kth move, the walnut with number k is called *being operated*, and the two adjacent walnuts are called *being swapped*.

Claim 1 If k is a small walnut, then in the kth move, Jumpy swaps two big walnuts.

Claim 2 If k is a big walnut, then in the kth move, Jumpy swaps two small walnuts.

Proof of Claim 1 Assume that in the kth move, a small walnut a is swapped. Clearly $a > k$ since k is small. Before the ath move, a has been swapped (for example in the kth move). Assume the last time a is swapped before the ath move is in the k'th move, then $a > k' \geq k$ and k' is also a small walnut. Before and after the k'th move, a is adjacent to k'. After the k'th move, let walnut k' be in the hole P. Walnut a is not swapped after the k'th move and before the ath move. However in the ath move, the walnut in hole P is numbered bigger than a since a is small. Thus in some move after the k'th and before the ath, the walnut in the hole P is changed from an operated one to an unoperated one. Let it be the lth move, $k' < l < a$. In the lth move, the two swapped walnuts is an operated one (which is numbered smaller than l) and an unoperated one (which is numbered larger than l). Contrary to the assumption.

Proof of Claim 2 Claim 2 follows from claim 1 if we reverse the time and treat walnut i as walnut $2022 - i$. The moves are proceeded reversely starting from the final configuration and finishing at the initial configuration. However the big walnuts are the original small walnuts and the small walnuts are the original big walnuts. And claim 1 in the reversal moves implies claim 2 in the original moves.

From claim 1 and claim 2, we see that two swapped walnuts at each move are both big or both small. In particular, the size of walnuts in each hole remains unchanged during the moves. Thus we may define a hole to be a big hole if the walnut in it is big, or a small hole otherwise. We shall prove that the adjacent holes of a big hole is small, and the adjacent holes of a small hole is big. Then it follows that the big holes and small holes are arranged on the circle alternatively. However, 2021 is odd. Thus, this is impossible.

To prove that the two adjacent holes of a small hole are big, consider a small hole P which initially contains a small walnut a. Walnut a cannot be swapped before a th move, otherwise say a is swapped in kth move where $k < a$, then k is small and a is big by claim 1. Absurd. So a remains in P until the ath move when it is operated, then the two adjacent holes are big since a is small. The fact that the two adjacent holes of a big hole are small again follows from the reversal argument as in the proof of claim 2. This completes the proof. □

Problem 6. Let $m > 2$ be an integer, A be a finite set of (not necessarily positive) integers, and $B_1, B_2, B_3, \ldots, B_m$ be subsets of A. Assume that for each $k = 1, 2, \ldots, m$ the sum of the elements of B_k is m^k. Prove that A contains at least $m/2$ elements.

(Contributed by Austria)

Solution Let $A = \{a_1, \ldots, a_k\}$. Suppose the statement is not true: let $k = |A| < \frac{m}{2}$.

Consider $f(c_1, \ldots, c_m) := c_1 m + c_2 m^2 + \cdots + c_m m^m$, $c_j \in \{0, 1, \ldots, m - 1\}$ for $j = 1, \ldots, m$. Of all these m^m sums, each one is the representation of some number in base m, and they are all distinct. Since each m^j ($1 \leq j \leq m$) is the sum of all elements of B_j, we may rewrite $f(c_1, \ldots, c_m)$ as the sum

$$\alpha_1 a_1 + \cdots + \alpha_k a_k,$$

where each $\alpha_i \in \{0, 1, \ldots, m(m-1)\}$. The total number of sums as above is

$$(m(m-1)+1)^k < m^{2k} < m^m,$$

yet the sums $f(c_1, \ldots, c_m)$ are proven to be all distinct. This contradiction overturns our assumption and hence $k = |A| \geq \frac{m}{2}$. $\square$

International Mathematical Olympiad

2022 (Oslo, Norway)

The 63rd International Mathematical Olympiad (IMO) was held from July 6th to July 16th, 2022 in Oslo, Norway. Due to the pandemic, the Chinese team participated online. A total of 589 contestants from 104 countries or regions participated the competition. For the Chinese team, all six members won gold medals with full scores. This is the second time in the history of IMO that all six members of a team receive full scores (the first time was USA in 1994).

The Chinese team:

Leader: Xiao Liang (Peking University)

Deputy leader: Qu Zhenhua (East China Normal University)

Observers: Xiong Bin (East China Normal University), Fu Yunhao (Southern University of Science and Technology)

Team members:

Zhang Zhicheng (No. 1 Middle School affiliated to Central China Normal University), 11th grade, 42 marks, gold medalist

Zhang Yiran (Shanghai High School), 12th grade, 42 marks, gold medalist

Liao Yubo (The High School Affiliated to Renmin University of China), 11th grade, 42 marks, gold medalist

Jiang Cheng (Shanghai High School), 11th grade, 42 marks, gold medalist

Qu Xiaoyu (Chongqing Bashu Secondary School), 10th grade, 42 marks, gold medalist

Liu Jiayu (Changsha Yali High School), 11th grade, 42 marks, gold medalist

The top 10 teams in the total scores:

1.	China	252 marks
2.	South Korea	208 marks
3.	United States of America	207 marks
4.	Vietnam	196 marks
5.	Romania	194 marks
6.	Thailand	193 marks
7.	Germany	192 marks
8.	Iran	191 marks
8.	Japan	191 marks
10.	Israel	188 marks
10.	Italy	188 marks

Gold medal: score $\geq$34 marks;

Silver medal: score $\geq$29 marks;

Bronze medal: score $\geq$23 marks.

Between May 23rd and 29th, the Chinse team received online training by coaches Ye Zhonghao, Yang Yuan, Wu Hao, Fu Yunhao, Jin Chunlai and Wang Yichuan.

From June 19th, the Chinese team was trained at Peking University, Beijing. Professor Gong Fuzhou, Secretary-General of the Chinese Mathematical Society; Professor Chen Dayue, Dean of School of Mathematical Sciences, Peking University; Professor Sun Zhaojun, Associate Dean; Professor Liu Bin; and coaches Wang Bin, Qu Zhenhua, Zhou Xingjian, Wei Chen, Chen Ruitao, Xiao Liang, Fu Yunhao, gave reports or had discussions with the team members.

First Day
(2:30 pm – 7 pm; July 11, 2022)

Problem 1. The Bank of Oslo issues two types of coin: aluminium (denoted A) and bronze (denoted B). Marianne has n aluminium coins and n bronze coins, arranged in a row in some arbitrary initial order. A chain is any subsequence of consecutive coins of the same type. Given a fixed positive integer $k \le 2n$, Marianne repeatedly performs the following operation: she identifies the longest chain containing the kth coin from the left, and moves all coins in that chain to the left end of the row. For example, if $n = 4$ and $k = 4$, the process starting from the ordering $AABBBABA$ would be

$$AAB\underline{B}BABA \to BBB\underline{A}AABA \to AAA\underline{B}BBBA \to BBB\underline{B}AAAA$$
$$\to BBB\underline{B}AAAA \to \cdots.$$

Find all pairs (n, k) with $1 \le k \le 2n$ such that for every initial ordering, at some moment during the process, the leftmost n coins will all be of the same type.

Solution The desired are all pairs (n, k) that satisfy $n \le k \le \dfrac{3n+1}{2}$.

As defined in the problem, a chain is any subsequence of consecutive coins of the same type. We call it a "block", if it is a chain but not contained in a longer chain. For $M = A$ or B, let M^x represent a block of M coins of length x. We are interested in whether an initial ordering will turn into a 2-block sequence A^nB^n or B^nA^n after finitely many operations.

(1) Claim If $k < n$ or $k > \dfrac{3n+1}{2}$, then the sequence cannot always turn into A^nB^n or B^nA^n.

Proof of claim If $k < n$, the sequence $A^{n-1}B^nA^1$ does not change after an operation, and cannot turn into a 2-block sequence. If $k > \dfrac{3n+1}{2}$, let $m = \left\lfloor \dfrac{n}{2} \right\rfloor$, $l = \left\lceil \dfrac{n}{2} \right\rceil$, and the initial ordering be $A^mB^mA^lB^l$. As $k > \dfrac{3n+1}{2} \ge m + l + l \ge m + m + l$ the kth coin always belongs to the rightmost block, and the process is

$$A^mB^mA^lB^l \to B^lA^mB^mA^l \to A^lB^lA^mB^m \to B^mA^lB^lA^m$$
$$\to A^mB^mA^lB^l \to \cdots,$$

a 4-periodic cycle. It cannot turn into a 2-block sequence.

(2) Claim If $n \le k \le \dfrac{3n+1}{2}$, any initial ordering will turn into a 2-block sequence.

Proof of claim Notice that after an operation, the number of blocks does not increase. Eventually, it will be stabilized, say at s blocks. It suffices to prove $s = 2$. Suppose $s > 2$. As $k \geq n$ and $s > 2$, the kth coin cannot belong to the leftmost block, and for the current operation, the block being moved is not the leftmost one. If it is a middle one, then after the operation, the two blocks on its sides will merge into one block, reducing the number of blocks by 1, a contradiction. Therefore, when stabilized, at each operation the block being moved to the left is always the rightmost (sth) one. Since $k \leq \dfrac{3n+1}{2}$, the length of the sth block is at least $2n - k + 1 \geq \dfrac{n+1}{2}$; since s does not decrease after the operation, the first and the sth blocks are of different types, and hence s is even. Let $s = 2t$, $t \geq 2$, and suppose one of the stable orderings is $X_1X_2\cdots X_{2t}$, in which every X_i represents a block. The process is

$$X_1X_2\cdots X_{2t} \to X_{2t}X_1\cdots X_{2t-1} \to \cdots \to X_2X_3\cdots X_1$$
$$\to X_1X_2\cdots X_{2t} \to \cdots.$$

Based on the previous argument, at any moment the rightmost block has length $\geq \dfrac{n+1}{2}$. So, each block has length $\geq \dfrac{n+1}{2}$, and the total length of the sequence is at least $4 \cdot \dfrac{n+1}{2} > 2n$, a contradiction. Now, the claim and the conclusion follow. □

Problem 2. Let $\mathbb{R}_+$ denote the set of positive real numbers. Find all functions $f : \mathbb{R}_+ \to \mathbb{R}_+$ such that for each $x \in \mathbb{R}_+$, there is exactly one $y \in \mathbb{R}_+$ satisfying

$$xf(y) + yf(x) \leq 2.$$

Solution The desired function is unique: $f(x) = \dfrac{1}{x}$, $\forall x \in \mathbb{R}_+$.

Sufficiency. It is straightforward that for any $x, y > 0$,

$$xf(y) + yf(x) = \frac{x}{y} + \frac{y}{x} \geq 2,$$

and equality holds only when $x = y$. This implies that for fixed $x > 0$, $xf(y) + yf(x) \leq 2$ if and only if $y = x$.

Necessity. We give two solutions as follows.

Solution 1 Call (x, y) a "good pair" if $xf(y) + yf(x) \leq 2$. Notice that if (x, y) is a good pair, then (y, x) is also a good pair. The problem requires that for each $x > 0$, there exists a unique $y > 0$ such that (x, y) is a good pair. We assert for each good pair (x, y), $x = y$. Otherwise $x \neq y$, (x, x) and (y, y) are not good pairs, and $xf(x) > 1$, $yf(y) > 1$. By the AM-GM inequality,

$$xf(y) + yf(x) \geq 2\sqrt{xf(y) \cdot yf(x)} = 2\sqrt{xf(x) \cdot yf(y)} > 2,$$

opposing the assumption that (x, y) is a good pair.

From the above assertion, it follows that for every $x > 0$, the only good pair is (x, x), and $xf(x) \leq 1$. Hence, $f(x) \leq \frac{1}{x}$, $\forall x \in \mathbb{R}_+$. Let $y = \frac{1}{f(x)}$. We have

$$xf(y) + yf(x) \leq x \cdot \frac{1}{y} + \frac{1}{f(x)} \cdot f(x) = xf(x) + 1 \leq 2,$$

indicating that $\left(x, \frac{1}{f(x)}\right)$ is a good pair. Since (x, x) is the only good pair, $x = \frac{1}{f(x)}$, and the conclusion follows. $\square$

Solution 2 For $x > 0$, let $\sigma(x)$ be the unique positive real number satisfying $xf(y) + yf(x) \leq 2$. First, we prove two statements.

(1) The function f is strictly decreasing: for $0 < y_1 < y_2$, $f(y_1) > f(y_2)$. Otherwise, suppose $f(y_1) \leq f(y_2)$. Let $x = \sigma(y_2)$. We have

$$xf(y_1) + y_1 f(x) < xf(y_2) + y_2 f(x) \leq 2,$$

opposing the uniqueness of $\sigma(x)$.

(2) For any $x > 0$, let $y = \sigma(x)$, and $xf(y) + yf(x) = 2$. Otherwise, suppose $xf(y) + yf(x) < 2$. Let $y' = y + \varepsilon$, where $\varepsilon = \frac{2 - xf(y) - yf(x)}{f(x)} > 0$. By (1), we have

$$xf(y') + y'f(x) < xf(y) + yf(x) + \varepsilon f(x) = 2,$$

opposing the uniqueness of $\sigma(x)$.

From (2), it follows that for any $x, y > 0$, $xf(y) + yf(x) \geq 2$. Taking $y = x$, we obtain $f(x) \geq \frac{1}{x}$, and furthermore

$$2 = f(x)\sigma(x) + f(\sigma(x))x \geq \frac{\sigma(x)}{x} + \frac{x}{\sigma(x)},$$

yielding $\sigma(x) = x$. In (2), it exactly gives $f(x) = \frac{1}{x}$. $\square$

Problem 3. Let k be a positive integer and let S be a finite set of odd prime numbers. Prove that there is at most one way (up to rotation and reflection) to place the elements of S around a circle such that the product of any two neighbours is of the form x^2+x+k for some positive integer x.

Solution We will allow $x=0$ in x^2+x+k and prove the modified statement.

Say an unordered prime pair $\{p,q\}$ ($p \neq q$) is good if there exists a nonnegative integer x such that $pq = x^2+x+k$. We need the following propositions:

(a) for each prime $r \geq 3$, there are at most two odd primes smaller than r, each forming a good pair with r;
(b) if in (a) there are indeed p and q such that (p,r) and (q,r) are good pairs, then (p,q) is also a good pair.

Once (a) and (b) are verified, we induct on $|S|$: for $|S| \leq 3$, the statement is obvious; assume the statement is true for $|S| = n$. Consider $|S| = n+1$: let r be the largest prime in S, by (a) the neighbours of r around the circle are uniquely determined (up to reflection); by (b) after removing r the circle of length n is still valid under the problem condition. By the induction hypothesis, there is at most one valid circle of length n. Hence, there is at most one valid circle of length $n+1$, and the modified statement is justified.

Verification of (a). Consider

$$x^2+x+k \equiv 0 \pmod{r}. \qquad ①$$

According to Lagrange's theorem, ① has at most two integer solutions $0 \leq x < r$.

Verification of (b).

Method 1 Suppose there exist odd primes p and $q, p<q<r$, and nonnegative integers x, y such that

$$x^2+x+k = pr, \quad y^2+y+k = qr.$$

As $p<q<r$, it must be $0 \leq x < y \leq r-1$, and x, y are the two roots of ①. By Viète's formulas, $x+y \equiv -1 \pmod{r}$; from the range of x and y, we infer that $x+y=r-1$. Let $K=4k-1$, $X=2x+1$, $Y=2y+1$. Rewrite the above equations as

$$4pr = X^2 + K, \quad 4qr = Y^2 + K,$$

where $X+Y=2r$. Multiply the two equations to find

$$\begin{aligned}16pqr^2 &= (X^2+K)(Y^2+K) = (XY-K)^2 + K(X+Y)^2 \\ &= (XY-K)^2 + 4Kr^2.\end{aligned}$$

Hence,

$$4pq = \left(\frac{XY-K}{2r}\right)^2 + K.$$

Since $Z = \left|\frac{XY-K}{2r}\right|$ is a rational number and its square $Z^2 = 4pq - K$ is an integer, we deduce that Z itself is an integer. It is easy to see that Z is odd, say $Z = 2z+1$. Then

$$pq = z^2 + z + k,$$

implying that $\{p, q\}$ is a good pair.

Method 2 Similar to method 1, assume that

$$x^2 + x + k = pr, \quad y^2 + y + k = qr.$$

The subtraction of the two equations gives

$$(x+y+1)(x-y) = r(p-q).$$

Now, as in method 1, we have $x+y=r-1$. Hence $x-y=p-q$, and

$$x = \frac{1}{2}(r+p-q-1), \quad y = \frac{1}{2}(r+q-p-1).$$

It follows that

$$\begin{aligned}z &= pr - x^2 - x \\ &= \frac{1}{4}(4pr - (r+p-q-1)^2 - 2(r+p-q-1)) \\ &= \frac{1}{4}(4pr - (r+p-q)^2 + 1) \\ &= \frac{1}{4}(2pq + 2pr + 2qr - p^2 - q^2 - r^2 + 1)\end{aligned}$$

is symmetric in p, q, r, indicating that

$$pq = z^2 + z + k,$$

where $z = \frac{1}{2}(p + q - r - 1)$ Hence, $\{p, q\}$ is also a good pair. □

Remark The construction of a valid cycle appears nontrivial, at least for some k values. For $k = 41$, the following 385 odd primes form a valid cycle:

53, 4357, 104173, 65921, 36383, 99527, 193789, 2089123, 1010357, 2465263, 319169, 15559, 3449, 2647, 1951, 152297, 542189, 119773, 91151, 66431, 222137, 1336799, 469069, 45613, 1047941, 656291, 355867, 146669, 874879, 2213327, 305119, 3336209, 1623467, 520963, 794201, 1124833, 28697, 15683, 42557, 6571, 39607, 1238833, 835421, 2653681, 5494387, 9357539, 511223, 1515317, 8868173, 114079681, 59334071, 22324807, 3051889, 5120939, 7722467, 266239, 693809, 3931783, 1322317, 100469, 13913, 74419, 23977, 1361, 62983, 935021, 512657, 1394849, 216259, 45827, 31393, 100787, 1193989, 600979, 209543, 357661, 545141, 19681, 10691, 28867, 165089, 2118023, 6271891, 12626693, 21182429, 1100467, 413089, 772867, 1244423, 1827757, 55889, 1558873, 5110711, 1024427, 601759, 290869, 91757, 951109, 452033, 136471, 190031, 4423, 9239, 15809, 24133, 115811, 275911, 34211, 877, 6653, 88001, 46261, 317741, 121523, 232439, 379009, 17827, 2699, 15937, 497729, 335539, 205223, 106781, 1394413, 4140947, 8346383, 43984757, 14010721, 21133961, 729451, 4997297, 1908223, 278051, 529747, 40213, 768107, 456821, 1325351, 225961, 1501921, 562763, 75527, 5519, 9337, 14153, 499, 1399, 2753, 14401, 94583, 245107, 35171, 397093, 195907, 2505623, 34680911, 18542791, 7415917, 144797293, 455529251, 86675291, 252704911, 43385123, 109207907, 204884269, 330414209, 14926789, 1300289, 486769, 2723989, 907757, 1458871, 65063, 4561, 124427, 81343, 252887, 2980139, 1496779, 3779057, 519193, 47381, 135283, 268267, 446333, 669481, 22541, 54167, 99439, 158357, 6823, 32497, 1390709, 998029, 670343, 5180017, 13936673, 2123491, 4391941, 407651, 209953, 77249, 867653, 427117, 141079, 9539, 227, 1439, 18679, 9749, 25453, 3697, 42139, 122327, 712303, 244261, 20873, 52051, 589997, 4310569, 1711069, 291563, 3731527, 11045429, 129098443, 64620427, 162661963, 22233269, 37295047, 1936969, 5033449, 725537, 1353973, 6964457, 2176871, 97231, 7001, 11351, 55673, 16747, 169003, 1218571, 479957, 2779783, 949609, 4975787, 1577959, 2365007, 3310753, 79349, 23189, 107209, 688907, 252583, 30677, 523, 941, 25981, 205103, 85087, 1011233, 509659, 178259, 950479, 6262847, 2333693, 305497, 3199319, 9148267, 1527563, 466801, 17033, 9967, 323003, 4724099, 14278309, 2576557, 1075021, 6462593,

2266021, 63922471, 209814503, 42117791, 131659867, 270892249, 24845153, 12104557, 3896003, 219491, 135913, 406397, 72269, 191689, 2197697, 1001273, 2727311, 368227, 1911601, 601883, 892057, 28559, 4783, 60497, 31259, 80909, 457697, 153733, 11587, 1481, 26161, 15193, 7187, 2143, 21517, 10079, 207643, 1604381, 657661, 126227, 372313, 2176331, 748337, 64969, 844867, 2507291, 29317943, 14677801, 36952793, 69332267, 111816223, 5052241, 8479717, 441263, 3020431, 1152751, 13179611, 38280013, 6536771, 16319657, 91442699, 30501409, 49082027, 72061511, 2199433, 167597, 317963, 23869, 2927, 3833, 17327, 110879, 285517, 40543, 4861, 21683, 50527, 565319, 277829, 687917, 3846023, 25542677, 174261149, 66370753, 9565711, 1280791, 91393, 6011, 7283, 31859, 8677, 10193, 43987, 11831, 13591, 127843, 358229, 58067, 15473, 65839, 17477, 74099, 19603, 82847, 21851, 61.

Second Day
(2:30 pm – 7 pm; July 12, 2022)

Problem 4. Let $ABCDE$ be a convex pentagon such that $BC = DE$. Assume that there is a point T inside $ABCDE$ with $TB = TD$, $TC = TE$ and $\angle ABT = \angle TEA$. Let line AB intersect lines CD and CT at points P and Q, respectively. Assume that the points P, B, A, Q occur on their line in that order. Let line AE intersect lines CD and DT at points R and S, respectively. Assume that the points R, E, A, S occur on their line in that order. Prove that the points P, S, Q, R lie on a circle.

Solution 1 From the given conditions, it follows $BC = DE$, $CT = ET$, and $TB = TD$. Hence, $\triangle TBC$ and $\triangle TDE$ are congruent; in particular, $\angle BTC = \angle DTE$. In $\triangle TBQ$ and $\triangle TES$, we have $\angle TBQ = \angle SET$ and $\angle QTB = 180° - \angle BTC = 180° - \angle DTE = \angle ETS$, and hence they are similar triangles, implying $\angle TSE = \angle BQT$ and

$$\frac{TD}{TQ} = \frac{TB}{TQ} = \frac{TE}{TS} = \frac{TC}{TS}.$$

We see that $TD \cdot TS = TC \cdot TQ$, and C, D, Q, S concyclic. (An alternative approach is to use similar triangles $\triangle TCS$ and $\triangle TDQ$ to derive $\angle CQD = \angle CSD$.) Now, $\angle DCQ = \angle DSQ$,

$$\angle RPQ = \angle RCQ - \angle PQC = \angle DSQ - \angle DSR = \angle RSQ,$$

and thus P, Q, R, S are concyclic. □

Remark 1 (Another proof for C, D, Q, S concyclic from $\triangle TBQ \backsim \triangle TES$) Notice that $\triangle TBD \backsim \triangle TCE \backsim \triangle TEC$: together with $\triangle TBQ \backsim \triangle TES$, we find similar (concave) quadrilaterals $\triangle TBDQ \backsim \triangle TECS$. Then $\triangle TDQ \backsim \triangle TCS$ and C, D, Q, S lie on a circle.

Second proof. As in the first proof, we find that $\triangle TBC$ and $\triangle TDE$ are congruent. Let DT and BA, CT and EA meet at V, W, respectively, as shown in Fig. 4.1:

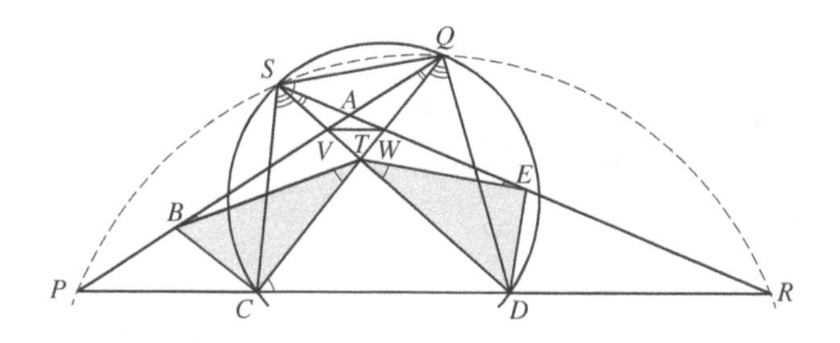

Fig. 4.1

In $\triangle BCQ$ and $\triangle DES$, we find

$$\begin{aligned}\angle VSW = \angle DSE &= 180^\circ - \angle SED - \angle EDS \\ &= 180^\circ - \angle AET - \angle TED - \angle EDT \\ &= 180^\circ - \angle TBA - \angle TCB - \angle CBT \\ &= 180^\circ - \angle QCB - \angle CBQ = \angle BQC = \angle VQW,\end{aligned}$$

and hence V, S, Q, W are concyclic. In particular, $\angle WVQ = \angle WSQ$. Since

$$\angle VTB = 180^\circ - \angle BTC - \angle CTD = 180^\circ - \angle CTD - \angle DTE = \angle ETW,$$

and given that $\angle TBV = \angle WET$, we arrive at $\triangle VTB \backsim \triangle WTE$. Hence,

$$\frac{VT}{WT} = \frac{BT}{ET} = \frac{DT}{CT},$$

and $CD \,//\, VW$. Finally, the conclusion follows from $\angle RPQ = \angle WVQ = \angle WSQ = \angle RSQ$.

Remark 2 (Another approach for $CD \,//\, VW$ in the second proof) As in the first proof, we find that V, S, Q, W are concyclic, $\angle BVD = \angle QVS = \angle QWS = \angle CWE$. Moreover, $\triangle TBC$ and $\triangle TDE$ are congruent, and hence $\triangle TBD \backsim \triangle TCE$: as they are isosceles triangles, $\angle TBD = \angle TEC$, $\angle TDB = \angle TCE$. It follows that $\triangle VBD \backsim \triangle WEC$, and BT, ET divide $\angle VBD$, $\angle WEC$, respectively at the same ratio. Consequently, $\dfrac{VT}{TD} = \dfrac{WT}{TC}$, and $CD\,//\,VW$.

Remark 3 (Still another approach for $CD\,//\,VW$ in the second proof) As V, W, Q, S are concyclic; C, D, Q, S are concyclic, the conclusion follows from Reims' theorem. One can also use $\angle SDC = \angle SQC = \angle DVW$.

Problem 5. Find all triples (a, b, p) of positive integers with p prime and

$$a^p = b! + p.$$

Solution There are only two such triples: $(2, 2, 2)$ and $(3, 4, 3)$. It is straightforward to check them. We shall prove that no other triples exist.

Clearly, $a > 1$. Consider three situations as follows.

(1) $a < p$. If $a \leq b$, then $a \mid (a^p - b!) = p$, contradicting the assumption $1 < a < p$; if $a > b$, then $b! \leq a! < a^p - p$, the second inequality only requiring $p > a > 1$.

(2) $a > p$. Then $b! = a^p - p > p^p - p \geq p!$, $b > p$, and $a^p = b! + p$ is a multiple of p. As $b! = a^p - p$, we have $p \parallel b$, and $b < 2p$. If $a < p^2$, then $\dfrac{a}{p}$ divides a^p and $b!$, hence divides p as well, contradicting $1 < \dfrac{a}{p} < p$; if $a \geq p^2$, then it is contradicting $a^p \geq (p^2)^p > (2p-1)! + p \geq b! + p$.

(3) $a = p$. Then $b! = p^p - p$. Try $p = 2, 3, 5$ to get the two triples $(2, 2, 2)$ and $(3, 4, 3)$. Now assume $p \geq 7$. From $b! = p^p - p > p!$, it follows $b \geq p + 1$, and further

$$\begin{aligned} v_2((p+1)!) &\leq v_2(b!) \\ &= v_2(p^{p-1} - 1) = 2v_2(p-1) + v_2(p+1) - 1 \\ &= v_2\left(\frac{p-1}{2} \cdot (p-1) \cdot (p+1)\right). \end{aligned}$$

Since $\dfrac{p-1}{2}$, $p - 1$, $p + 1$ are distinct factors of $(p+1)!$ and $p + 1 \geq 8$, there are four or more even numbers among $1, 2, \ldots, p + 1$, which is impossible.

Second approach for $a = p \geq 5$: according to Zsigmondy's theorem, there exists a prime q that divides $p^{p-1} - 1$ but not $p^k - 1$ for any $k < p - 1$.

Thus, $p \neq q$, and $q \equiv 1 \pmod{p-1}$. We must have $b \geq 2p-1$, yet

$$b! \geq (2p-1)! > (2p-1)\cdot(2p-2)\cdot\cdots\cdot(p+1)\cdot p > p^p > p^p - p$$

leads to a contradiction.

Third approach for $a = p \geq 5$: as $b > p \geq 5$, the required equation modulo $(p+1)^2$ gives

$$\begin{aligned} p^p - p &= (p+1-1)^p - p \\ &\equiv p\cdot(p+1)(-1)^{p-1} + (-1)^p - p \\ &= p^2 - 1 \not\equiv 0 \pmod{(p+1)^2}. \end{aligned}$$

However, as $p \geq 5$, 2, $\dfrac{p+1}{2} < p$ are distinct, and $(p+1) \mid p!$. It follows that

$$(p+1)^2 \mid (p+1)!$$

which is contrary to $(p+1)^2 \nmid b!$. □

Problem 6. Let n be a positive integer. A *Nordic* square is an $n \times n$ board containing all the integers from 1 to n^2 so that each cell contains exactly one number. Two different cells are considered adjacent if they share a common side. Every cell that is adjacent only to cells containing larger numbers is called a *valley*. An *uphill* path is a sequence of one or more cells such that:

(i) the first cell in the sequence is a valley;
(ii) each subsequent cell in the sequence is adjacent to the previous cell;
(iii) the numbers written in the cells in the sequence are in increasing order.

Find, as a function of n, the smallest possible total number of uphill paths in a Nordic square.

Solution The smallest possible number of uphill paths is $2n^2 - 2n + 1$.

Let table A be the original $n \times n$ board. Define table B as another $n \times n$ board, and in each cell of B we write the number of uphill paths ending at that cell. The total number of uphill paths in A is equal to the sum of all entries of B: call this number S.

There is at least one valley in A, namely the cell with number 1. Each valley in A corresponds to 1 in B, and vice versa. Let T be the sum of all non-valley entries of B. Consider a pair of neighbouring cells whose numbers are x and $y, x < y$: they contribute at least 1 to T, as there is always an uphill path through x and ending at y. Therefore, $T \geq 2n^2 - 2n$, the number of pairs of neighbouring cells, and

$$S = T + \text{number of valleys} \geq 2n^2 - 2n + 1.$$

It remains to prove the lower bound $2n^2 - 2n + 1$ is attainable. To this end, we label some cells of B with dots (they will correspond to valleys in A): the dotted cells form a tree in B while the unlabeled cells are all isolated. Then, construct the table A as follows:

First, choose any dotted cell and write 1; then, consecutively write $2, 3, \ldots$ in the dotted cells such that a new number is always written next to an already written number; finally, write the remaining large numbers arbitrarily in the unlabeled cells.

Clearly, there is only one cell with number 1, and for each pair of neighbouring cells, the one with the larger number is the end of exactly one uphill path. So, the number of uphill paths is indeed $2n^2 - 2n + 1$.

We label the cells in detail as follows: for 1×1 table, label the unique cell; for 2×2 table, label any three cells; for $n \times n$ $(n \geq 3)$ tables, define

$$s = 2 \quad \text{if } n \equiv 0, \quad 2 \pmod 3; \qquad s = 1 \quad \text{if } n \equiv 1 \pmod 3.$$

In the first column, label $(1, i)$ as long as $i \neq 6k + s$ (here, $(1, 1)$ is the lower left cell and $(1, n)$ is the upper left cell). In the second column, label $(2, j)$ if $j = 6k + s - 1$, $6k + s$ or $6k + s + 1$. Evidently, the dotted cells in the first two columns are connected. Now we expand the dotted cells $(2, 6k + s)$ and $(1, 6k + s + 3)$ to the right: for each dotted (i, j), label all $(i + l, j)$ and $(i + 2l, j \pm 1)$ where l is any positive integer (we only label cells within the table). It is easy to see that the dotted cells form a tree and the unlabeled cells $(1, 6k+s)$, $(2+2l+1, 6k+s\pm 1)$, $(2+2l, 6k+s+3\pm 1)$ are all isolated (since, if they are in the same column, the distance is at least 2; if in adjacent columns, then they are in different rows). □

The following figures illustrate the labeling for $n = 3, 4, 5, 6, 7$ and give tables A and B for $n = 5$ (there are other tables that take the same minimum value).

$n = 3$

•	•	
	•	•
•	•	

$n = 4$

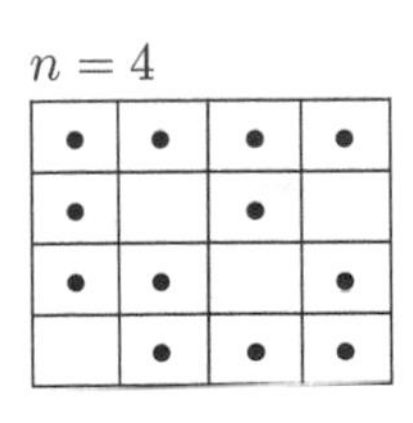

$n = 5$

•	•	•	•	•
•		•		•
•	•		•	
	•	•	•	•
•	•		•	

$n = 6$

•		•		•	
•	•	•	•	•	•
•		•		•	
•	•		•		•
	•	•	•	•	•
•	•		•		•

$n = 7$

	•	•	•	•	•	•
•	•		•		•	
•		•		•		•
•	•	•	•	•	•	•
•		•		•		•
•	•		•		•	
	•	•	•	•	•	•

Table A for $n = 5$.

12	13	14	15	16
11	24	17	25	18
10	9	22	8	23
21	3	4	5	6
1	2	19	7	20

Table B for $n = 5$.

1	1	1	1	1
1	4	1	4	1
1	1	4	1	3
3	1	1	1	1
1	1	3	1	2